D1677527

Edited by
Francis S. Markland, Stephen Swenson, and Radu Minea

Tumor Angiogenesis

Related Titles

Allgayer, H., Rehder, H., Fulda, S. (eds.)

Hereditary Tumors

From Genes to Clinical Consequences

2009

ISBN: 978-3-527-32028-8

Meyers, R. A. (ed.)

Cancer

From Mechanisms to Therapeutic Approaches

2007

ISBN: 978-3-527-31768-4

Staton, C., Lewis, C., Bicknell, R. (eds.)

Angiogenesis Assays

A Critical Appraisal of Current Techniques

2007

ISBN: 978-0-470-01600-8

zur Hausen, H.

Infections Causing Human Cancer

2006

ISBN: 978-3-527-31056-2

Edited by Francis S. Markland, Stephen Swenson, and Radu Minea

Tumor Angiogenesis

From Molecular Mechanisms to Targeted Therapy

WILEY-BLACKWELL

The Editors

Francis S. Markland
University of Southern California
Keck School of Medicine
1303 North Mission Road
Los Angeles, CA 90033
USA

Stephen Swenson
University of Southern California
Keck School of Medicine
1303 North Mission Road
Los Angeles, CA 90033
USA

Radu Minea
University of Southern California
Keck School of Medicine
1303 North Mission Road
Los Angeles, CA 90033
USA

Cover
Growth factor loops in the tumor microenvironment (with kind permission by Andrew Dudley and Lena Claesson-Welsh).

Limit of Liability/Disclaimer of Warranty: While the publisher and authors have used their best efforts in preparing this book, they make no representations or warranties with respect to the accuracy or completeness of the contents of this book and specifically disclaim any implied warranties of merchantability or fitness for a particular purpose. No warranty can be created or extended by sales representatives or written sales materials. The Advice and strategies contained herein may not be suitable for your situation. You should consult with a professional where appropriate. Neither the publisher nor authors shall be liable for any loss of profit or any other commercial damages, including but not limited to special, incidental, consequential, or other damages.

Library of Congress Card No.: applied for

British Library Cataloguing-in-Publication Data
A catalogue record for this book is available from the British Library.

Bibliographic information published by the Deutsche Nationalbibliothek
The Deutsche Nationalbibliothek lists this publication in the Deutsche Nationalbibliografie; detailed bibliographic data are available on the Internet at http://dnb.d-nb.de.

© 2010 WILEY-VCH Verlag GmbH & Co. KGaA, Weinheim

Wiley-Blackwell is an imprint of John Wiley & Sons, formed by the merger of Wiley's global Scientific, Technical, and Medical business with Blackwell Publishing.

All rights reserved (including those of translation into other languages). No part of this book may be reproduced in any form – by photoprinting, microfilm, or any other means – nor transmitted or translated into a machine language without written permission from the publishers. Registered names, trademarks, etc. used in this book, even when not specifically marked as such, are not to be considered unprotected by law.

Cover Design Adam Design, Weinheim
Typesetting Laserwords Private Limited, Chennai, India
Printing and Binding Strauss GmbH, Mörlenbach

Printed in the Federal Republic of Germany
Printed on acid-free paper

ISBN: 978-3-527-32091-2

Contents

Tumor Angiogenesis – From Molecular Mechanisms to Targeted Therapy. Edited by Francis S. Markland, Stephen Swenson, and Radu Minea
Copyright © 2010 WILEY-VCH Verlag GmbH & Co. KGaA, Weinheim
ISBN: 978-3-527-32091-2

Preface

This book is dedicated to the memory of Dr. Judah Folkman who pioneered the field of angiogenesis in the early 1970s with the publication of his first paper, in 1971, on the role of blood vessels in the growth of tumors; he continued doing seminal work in the field until his untimely death on January 14, 2008.

In this volume, various aspects of angiogenesis and the application of antiangiogenic therapy for the treatment of cancer are discussed. We have organized the book into four sections following an introductory chapter by Drs. Donati and Lorenzet, "Overview on Tumor Angiogenesis". These sections are (1) Mechanisms of angiogenesis and lymph-angiogenesis, (2) Signal transduction and angiogenesis, (3) Therapeutic approaches and angiogenesis, and (4) Imaging and biomarkers in angiogenesis. These sections and the chapters contained therein offer an updated view of the field that has grown exponentially since the early days when Dr. Folkman was doing his pioneering investigations.

In the first section, Drs. Dudley and Claesson-Welsh discuss the molecular mechanisms of tumor angiogenesis. Their chapter describes the mechanisms of normal and tumor vessel formation and the important factors operating in these processes. Further, different mechanisms to explain how tumors may escape anti-angiogenic therapy are discussed. One of the long-term challenges will be to better characterize the biology of tumor endothelial cells (EC) at different stages of tumor progression. Rigorous characterization of tumor EC may be necessary in order to effectively design antiangiogenic strategies. In the second chapter of this section, Dr. Ribatti summarizes knowledge on regulators of angiogenesis and their role in tumor progression. Several classic factors including vascular endothelial growth factor and fibroblast growth factor-2 are discussed. In addition, numerous endogenous peptides, "nonclassic" factors, including erythropoietin, angiotensin-II, endothelins, adrenomedullin, leptin, adiponectin, neuropeptide-Y, vasoactive intestinal peptide, and substance P, play a regulatory role in angiogenesis. In the third chapter, Dr. Takakura discusses the role of accessory cells in angiogenesis. Many reports suggest that cells of the hematopoietic lineage are mobilized and then entrapped in tumor tissues where they function as accessory cells, promoting the sprouting of resident ECs by releasing angiogenic signals. Potential accessory cells for tumor angiogenesis include tumor-associated fibroblasts, pericytes, mesenchymal stem cells, and hematopoietic cells. In the next chapter, Dr. Bikfalvi compares

Tumor Angiogenesis – From Molecular Mechanisms to Targeted Therapy. Edited by Francis S. Markland, Stephen Swenson, and Radu Minea
Copyright © 2010 WILEY-VCH Verlag GmbH & Co. KGaA, Weinheim
ISBN: 978-3-527-32091-2

tumor angiogenesis to developmental angiogenesis. He discusses angiogenesis as a fundamental process that occurs during development and in pathology. Both processes share a number of common features, but tumor angiogenesis is viewed as imperfect developmental angiogenesis that has acquired some specific additional mechanisms. In the final chapter of this section, Drs. Ramu and Hong discuss lymphatic vascular biology. This field has emerged over the last decade with the realization that tumors may actively induce lymphangiogenesis in addition to angiogenesis. There appears to be extensive cross talk between the blood and lymphatic vascular systems during tumor development. Thus, tumor-associated lymphatics represents an important target for the development of new therapeutic agents aimed at tumor metastasis.

In the second section of the book, the importance and role of integrins in tumor angiogenesis is discussed in a chapter written by Drs. Akalu, Contois, and Brooks. Integrins are a major family of cell surface molecules, which facilitate bidirectional communication between cells and the extracellular matrix. Dr. Brooks and colleagues have focused their discussion on the multiple roles by which integrins may regulate angiogenesis. In the final chapter of this section, Drs. Abrahams, Daly, Eichten, Li, Noguera-Troise, and Thurston review several of the major non-integrin signaling pathways that show promise as targets for tumor angiogenesis including the VEGF pathway, the Delta-Notch and angiopoietin-Tie pathways, and the TGF-β and ephrin/Eph pathways. For each pathway, multiple ligands and receptors are involved, and much work remains to be done to understand how these pathways interact in the process of tumor angiogenesis.

In the third section, Drs. Swenson, Minea, and Markland discuss the antiangiogenic role of disintegrins, peptides originally derived from snake venom, but now available in recombinant form. Disintegrins are Arg-Gly-Asp (or other tripeptide motif) containing peptides that target several tumor or endothelial cell integrins. The authors discuss how disintegrins may prove useful as antiangiogenic agents for cancer therapy. In the next chapter, Drs. Vemuri and Libutti discuss the discovery of vascular endothelial growth factor (VEGF) and the VEGF-based antiangiogenic therapeutics, which are approved and in clinical use, including monoclonal antibodies, small molecule tyrosine kinase inhibitors, and inhibitors of mTOR (the mammalian target of Rapamycin). The hope expressed in this chapter is to find better targets and more powerful targeted therapeutics. In the third chapter in this section, Drs. Longo, Torino, and Gasparini discuss the combination of antiangiogenic agents with various other anticancer agents including antiangiogenic therapy and immunomodulation, anti-VEGF and anti-EGFR/HER-2 combinations, and antiangiogenic therapy with radiation or chemoradiation therapy. The authors discuss preclinical studies and clinical trials, and also examine the challenges of study design for combined antiangiogenic therapy. In the final chapter in this section, Drs. Rixe, Kelly, and Giaccone discuss tumor specificity of antiangiogenic agents. In view of the lack of efficacy of antiangiogenic agents alone, which merely arrest tumor growth, the combination of antiangiogenics with other cytotoxic therapeutics is being explored, but this frequently leads to a loss of specificity. The challenge will be to combine antiangiogenics with other specific targeted therapies to produce a relevant clinical benefit in cancer patients.

In the fourth section of the book, Drs. Turkbey, Ravizzini and Choyke discuss *in vivo* tumor imaging techniques including ultrasound, computed tomography, magnetic resonance imaging, positron emission tomography, and targeted imaging approaches. Targeted imaging is aimed at VEGF and VEGFR-2, integrins, matrix metalloproteinases, and the extracellular matrix. The authors propose that imaging approaches utilizing targeted angiogenesis-specific molecules will enable noninvasive surveillance of angiogenesis that may aid in the development of targeted anticancer agents. In the final chapter of the book, Dr. Platero discusses the identification of biomarkers that can be used to establish drug efficacy. From early preclinical use both in cell lines and in animals, to their use as pharmacodynamic biomarkers in early clinical trials and to their application in everyday use in the clinic, these examples point to the usefulness of biomarkers and the impact they are having on the drug development process.

We feel this book will be a most useful and up-to-date addition to the bookshelves of investigators involved in the field of angiogenesis and in its control in cancer.

Los Angeles, February 2010

Francis S. Markland
Stephen Swenson
Radu Minea

List of Contributors

Cristina Abrahams
Regeneron Pharmaceuticals
777 Old Saw Mill River Road
Tarrytown
New York 10591
USA

Abebe Akalu
Maine Medical Center Research Institute
Center for Molecular Medicine
81 Research Drive
Scarborough
Maine 04074
USA

Maria Benedetta Donati
Catholic University
RE ARTU Research Laboratories
"John Paul II" Center for High Technology Research and Education in Biomedical Sciences
Largo A. Gemelli, 1
86100 Campobasso
Italy

Andreas Bikfalvi
University Bordeaux I
INSERM U920
Avenue des Facultés
33405 Talence
France

Peter C. Brooks
Maine Medical Center Research Institute
Center for Molecular Medicine
81 Research Drive
Scarborough
Maine 04074
USA

Lena Claesson-Welsh
Uppsala University
Department of Genetics and Pathology
Rudbeck Laboratory
Dag Hammarskjöldsv. 20
751 85 Uppsala
Sweden

Peter L. Choyke
National Institutes of Health
National Cancer Institute
Molecular Imaging Program
9000 Rockville Pike
Bethesda
Maryland 20892-1088
USA

Tumor Angiogenesis – From Molecular Mechanisms to Targeted Therapy. Edited by Francis S. Markland, Stephen Swenson, and Radu Minea
Copyright © 2010 WILEY-VCH Verlag GmbH & Co. KGaA, Weinheim
ISBN: 978-3-527-32091-2

Liangru Contois
Maine Medical Center Research Institute
Center for Molecular Medicine
81 Research Drive
Scarborough
Maine 04074
USA

Christopher Daly
Regeneron Pharmaceuticals
777 Old Saw Mill River Road
Tarrytown
New York 10591
USA

Andrew C. Dudley
Harvard Medical School &
Children's Hospital Boston
Boston MA 02115
USA

Alexandra Eichten
Regeneron Pharmaceuticals
777 Old Saw Mill River Road
Tarrytown
New York 10591
USA

Giampietro Gasparini
San Filippo Neri Hospital
Division of Medical Oncology
via Martinotti, 20
Rome 00135
Italy

Giuseppe Giaccone
National Institutes of Health
National Cancer Institute
Center for Cancer Research
Medical Oncology Branch
10 Center Drive
Bethesda
Maryland 20892
USA

Young-Kwon Hong
University of Southern California
Departments of Surgery and Biochemistry and Molecular Biology
Norris Comprehensive Cancer Center
1441 Eastlake Ave.
Los Angeles
California 90033
USA

Ronan J. Kelly
National Institutes of Health
National Cancer Institute
Center for Cancer Research
Medical Oncology Branch
10 Center Drive
Bethesda
Maryland 20892
USA

Zhe Li
Regeneron Pharmaceuticals
777 Old Saw Mill River Road
Tarrytown
New York 10591
USA

Steven K. Libutti
Montefiore Medical Center
Department of Surgery
Medical Arts Pavilion
3400 Brainbridge Avenue
Bronx
New York 10467
USA

Raffaele Longo
San Filippo Neri Hospital
Division of Medical Oncology
via Martinotti, 20
Rome 00135
Italy

Roberto Lorenzet
Catholic University
RE ARTU Research Laboratories
"John Paul II" Center for High
Technology Research and
Education in Biomedical Sciences
Largo A. Gemelli, 1
86100 Campobasso
Italy

Francis S. Markland
University of Southern California
Department of Biochemistry and
Molecular Biology
Keck School of Medicine
1303 N. Mission Rd CRL106
Los Angeles
California 90033
USA

Radu Minea
University of Southern California
Department of Biochemistry and
Molecular Biology
Keck School of Medicine
1303 N. Mission Rd CRL106
Los Angeles
California 90033
USA

Irene Noguera-Troise
Regeneron Pharmaceuticals
777 Old Saw Mill River Road
Tarrytown
New York 10591
USA

J. Suso Platero
Centocor R&D
145 King of Prussia
Radnor
Pennsylvania
19087
USA

Swapnika Ramu
University of Southern California
Departments of Surgery and
Biochemistry and Molecular
Biology
Norris Comprehensive Cancer
Center
1441 Eastlake Ave.
Los Angeles
California 90033
USA

Gregory Ravizzini
National Institutes of Health
National Cancer Institute
Molecular Imaging Program
9000 Rockville Pike
Bethesda
Maryland 20892-1088
USA

Domenico Ribatti
University of Bari
Department of Human Anatomy
and Histology
School of Medicine
Piazza Giulio Cesare 11
Bari
I-70124
Italy

Olivier Rixe
National Institutes of Health
National Cancer Institute
Center for Cancer Research
Medical Oncology Branch
10 Center Drive
Bethesda
Maryland 20892
USA

Stephen Swenson
University of Southern California
Department of Biochemistry and
Molecular Biology
Keck School of Medicine
1303 N. Mission Rd CRL106
Los Angeles
California 90033
USA

Nobuyuki Takakura
Osaka University
Department of Signal
Transduction
Research Institute for Microbial
Diseases
3-1 Yamadaoka
Suita-shi
Osaka 565-0871
Japan

Gavin Thurston
Regeneron Pharmaceuticals
777 Old Saw Mill River Road
Tarrytown
New York 10591
USA

Francesco Torino
San Filippo Neri Hospital
Division of Medical Oncology
via Martinotti, 20
Rome 00135
Italy

Baris Turkbey
National Institutes of Health
National Cancer Institute
Molecular Imaging Program
9000 Rockville Pike
Bethesda
Maryland 20892-1088
USA

Chandu Vemuri
University of Michigan
Department of Surgery
2917 N Knightsbridge Circle
Ann Arbor
Michigan 48105
USA

Part I
Introduction

Tumor Angiogenesis – From Molecular Mechanisms to Targeted Therapy. Edited by Francis S. Markland, Stephen Swenson, and Radu Minea
Copyright © 2010 WILEY-VCH Verlag GmbH & Co. KGaA, Weinheim
ISBN: 978-3-527-32091-2

1
Overview on Tumor Angiogenesis

Maria Benedetta Donati and Roberto Lorenzet

1.1
Once upon a Time . . .

The concept that tumor growth can be accompanied by increased vascular proliferation was introduced over a 100 years ago (see [1]). During the first half of the last century, several investigators examined different tumors and the pattern of their vascular tree; however, this did not generate particular interest.

The general background of these studies was that cancer cells might produce growth factors promoting vascular development, which precedes rapid tumor growth [1]. But the available methods for further investigations on this subject were still poor and *in vitro* conditions for growing cancer cells were considered highly artifactual and a potential source of bias as compared with those *in vivo*. As it will be observed several times during the story of Folkman's discoveries, and more generally of scientific progress, the development of new methods may be instrumental to the elucidation of new physiopathological mechanisms.

Indeed, Folkman, at that time a young surgeon at Harvard Medical School in Boston, was already interested in the field of tumorigenesis and began by developing a technique for the *in vitro* growth of tumors in solid form, in a system where they could grow and metastasize as if in the living host. In the 1960s, he and his colleagues presented a method of cancer cell implantation in isolated perfused organs, where the "cohesiveness, shedding rate, invasiveness and behavior of tumor cells in circulation" could be explored [2]. Their assumption was that it could not only provide a method for estimating the metastatic potential of different tumors, but also offer a possibility to test chemotherapeutic agents and their therapeutic potential. Subsequently, Folkman and a group of fellow workers reported the isolation of a soluble factor, from human and animal tumors, able to stimulate new vessel formation in a rat dorsal air sac model [3].

The hypothesis that solid tumors are angiogenesis-dependent, and, acting through a tumor angiogenesis factor (TAF), can attract endothelial cells to form new vessels, was proposed. This was exactly in contrast to the existing dogma that tumors grow on preexisting vessels and invade them, and that the redness,

Tumor Angiogenesis – From Molecular Mechanisms to Targeted Therapy. Edited by Francis S. Markland, Stephen Swenson, and Radu Minea
Copyright © 2010 WILEY-VCH Verlag GmbH & Co. KGaA, Weinheim
ISBN: 978-3-527-32091-2

morphologically detectable in tumors, is due to unspecific inflammation, and not due to tumor vascular tissue.

In the same year as the "isolation" paper (1971), a fundamental report, published in the New England Journal of Medicine (NEJM), allowed Folkman and his coworkers to put forward the hypothesis, which would direct and inspire the efforts of the subsequent 37 years (1971–2008), that is the possibility that, by developing angiogenesis inhibitors, they could propose and clinically validate a new type of anticancer treatment [4].

1.2
A Challenge Across Two Centuries: Inhibition of Angiogenesis to Stop Tumor Growth

Within the decade immediately following Folkman's NEJM paper, the field of angiogenesis grew exponentially (165 papers as opposed to 6 papers between 1950 and 1971), and Folkman was the author of 32 (20%) of those papers. From a Pubmed search, the following logarithmic growth of publications on angiogenesis records 996 papers from 1981 to 1990, 8364 from 1991 to 2000 and 24 573 from 2001 until the week of Judah Folkman's demise.

If one looks at the sequence of observations (see [5]) which, during the above-mentioned period (37 years) led from the concept of a TAF till the present clinical applications, one of the basic discoveries is that there exists a possibility to induce "dormancy" (i.e., a condition associated with halt or blockade of growth) in solid tumors by preventing neovascularization in rabbit models of anterior chamber cancer implants [6]. Under these conditions, tumors, unable to trigger the formation of new vasculature, would survive in a very small size, reached only by simple diffusion. In this "avascular" phase of growth, tumors, devoid of their malignancy, could remain in the host until death without showing any symptoms [6]. From that moment the extensive search for factors inhibiting or inducing angiogenesis began; it was a very rational sequence of cellular and animal experiments that led to the present clinical applications.

In 1975, using the basic knowledge that capillaries are not present in *cartilage*, Folkman and his group showed that in the presence of neonatal scapular cartilage, capillary proliferation induced by tumors is inhibited [7]. This led to the conclusion that the human body in physiologic conditions is able to produce angiogenesis inhibitors, which in animal models may also inhibit pathologic vascularization. Since angiogenesis is a process typically taking place in the microvasculature, the newly developed umbilical cord endothelial cells model [8, 9] was not found to be perfectly suited. Therefore, an original method was developed to obtain long-term culture of *capillary endothelial cells* [10]. This opened new possibilities to investigate basic processes in proliferation and motility of capillary endothelial cells, which were already considered to play a pivotal role not only in tumor growth and metastasis, but also in diseases like diabetic retinopathy, psoriasis, or arthritis. Again, as mentioned above, the development of new methods opened the way to new discoveries. Many scientists from that moment on devoted their efforts to look for angiogenic factors. In the years that followed, an idea, again from Folkman's group, made the researchers' efforts easier. In 1984 they purified the basic fibroblast growth factor (bFGF) – one of the proangiogenic factors – after discovering its high *affinity to heparin* [11]. The latter was already known to have a huge structural similarity to heparan sulfate, a major glycosaminoglycan on the surface of endothelial cells – crucial cells for the angiogenic process. Heparin binding of growth factors also led to identification and purification of the vascular endothelial growth factor (VEGF) [1]. Since the discovery of bFGF and then VEGF, several other proangiogenic factors have been purified and described, which will be extensively dealt with in different chapters of this book. However VEGF, as the first vascular endothelium-specific growth factor [11], was shown to have a crucial role in tumor angiogenesis: this resulted in preparation of the first inhibitors against this growth factor, to achieve inhibition of tumor vasculogenesis [12]. The search for inhibitors of angiogenesis naturally existing in the human body led to the discovery and purification of two prototype factors in the 1990s, namely, *angiostatin* and *endostatin* [13, 14]. During animal experimentation, endostatin was especially found to have a number of advantages: lack of side effects, active on different types of tumors, and not inducing drug resistance [15].

1.3 Clinical Implications of Angiogenesis Inhibition

Endostatin entered clinical trials in a small number of patients in the United States, first as a single agent, and subsequently in combination with chemotherapy in patients with advanced non–small cell lung cancer [16]. On the other hand, the recognition of the fundamental role of VEGF as an endogenous trigger of angiogenesis has led to the development of specific inhibitors of this growth factor, such as bevacizumab, *a monoclonal antibody against VEGF* (see a report, published again in the NEJM, [17]).

Over 30 years after Folkman had first proposed the concept of the antiangiogenic treatment of cancers, the idea finally became clinical practice, as extensively described in dedicated chapters of this book.

In the mean time, looking for the mechanisms leading to angiogenesis, Folkman had proposed the concept of the so-called *angiogenic switch*, which could be defined as the conversion from a dormant to an activated state; as long as there is a balance in the human body between the activators and the inhibitors of angiogenesis, cancer cells present in different tissues remain in a dormant state, not growing and not presenting symptoms [18, 19]. When this balance is disturbed and proangiogenic factors are in excess, rapid vascularization and growth of tumors may occur [19, 20]. It still remains an unsolved issue as to which factors initiate the angiogenic switch, but more and more hypoxic conditions are considered a candidate trigger [21, 22]; in adults, the formation of new capillaries from postcapillary venules is induced by tissue hypoxia-inducible factor (HIF-1), whose promoter region encodes for the transcription of several genes, VEGF among others [23].

The concept of treating tumors by inhibition of angiogenesis, underlines from a physiopathological standpoint, the high relevance of host's cells in tumor development, since the target may not only be, as in chemotherapy, the tumor cell itself, but also the vascular cells of the host, abnormally proliferating within the tumor tissue.

As in many other biological fields, the search for new angiogenesis inhibitors has led recently to the recognition that drugs, already used for other indications, might also be found to possess antiangiogenic function [24]. After years of investigation, several agents used as antibiotics, nonsteroidal anti-inflammatory drugs, bone-targeted bisphosphonates, and many others were shown to inhibit angiogenesis when administered in long-term therapy in modified doses [24, 25]. Similar inhibitory influences on endothelial cell functions were linked to cytotoxic agents [26]. On the basis of this idea, a novel approach to the use of chemotherapy for tumor treatment, termed *antiangiogenic chemotherapy*, was introduced [27]. Chemotherapeutic agents, when administered at lower doses but more frequently, acted as antiangiogenics without generating side-effects typical of high-dose intermittent chemotherapy [26, 27]. By influencing genetically stable endothelial cells of the tumor bed, derived from the host endothelium, on the one hand, antiangiogenic chemotherapeutic and antiangiogenic agents reduce drug resistance, typical for conventional chemotherapy scheduling [26, 28]. On the other hand, antiangiogenic drugs administered alone or in combination with conventional chemotherapy and acting directly on endothelial function may increase the risk of thromboembolic events during therapy [29–31]. While the beneficial effects of anticoagulant treatment alone are not always obvious in cancer patients, inhibition of procoagulant activity may become a real challenge for clinicians when combined antiangiogenic and chemotherapy treatment will become of common use in oncology.

1.4 Angiogenesis: A Novel Bridge between Malignancy and Thrombosis

As mentioned above, the milestone in the understanding of tumor angiogenesis was achieved with the purification of *VEGF*, a potent endothelial mitogen from bovine pituitary follicular cell conditioned media [32], previously isolated as vascular permeability factor (VPF) by Senger and colleagues [33].

VEGF stimulates endothelial cell proliferation, migration, and survival, and mediates vascular permeability; in doing so, it facilitates the formation of a fibrin network for the new proliferating endothelial cells [34, 35]. This may be due to many cellular activities, which have been ascribed to VEGF, but particularly to its complex interactions with the main actor in the clotting cascade, that is, *tissue factor* (*TF*). In tumor angiogenesis, both VEGF and TF are expressed by tumor cells and by tumor-infiltrating cells of the host, such as fibroblasts, monocyte/macrophages, and endothelial cells [36].

TF is a 47-kDa integral membrane glycoprotein tightly associated with phospholipids, which is structured in three domains: an intracellular region, a single transmembrane domain, and an extracellular domain. The latter functions as a high-affinity receptor and cofactor for coagulation factor VII and its active form VIIa [37, 38].

Once TF is available for binding to factor VIIa, the resulting complex acquires a powerful catalytic activity toward its natural substrates coagulation factors IX and X, which are rapidly cleaved to their active derivatives IXa and Xa, respectively, thus leading to thrombin generation, which in turn cleaves fibrinogen to fibrin [39, 40]. Since the process is remarkably fast, TF segregation from cells in direct contact with the blood stream has to be continuously ensured. Accordingly, immunohistochemical studies have localized TF in extravascular sites, such as the adventitia of arteries and veins, and, more generally, in capsules surrounding organs, in cells of epithelial surfaces, that is, at tissue barriers between the body and the environment [41]. This pattern is consistent with the hypothesis that TF can be considered as a "hemostatic envelope," whose duty is to minimize blood loss when the integrity of the vasculature is compromised.

Although a primary role for TF in *in vivo* blood coagulation has been well established, in recent years a significant body of evidence linking TF to other biological processes has accumulated. Indeed, TF has also been implicated in embryonic blood vessel development [42–44]; intracellular signal transmission [45]; wound healing [46]; promotion of cell transmigration and adhesion [47, 48]; and the regulation of tumor growth, angiogenesis, and metastasis formation (reviewed in [36]).

1.4.1
The Association between Human Cancer and the Clotting System

This double-way connection has been recognized for more than a century. The occurrence of venous thromboembolism as a common complication in cancer patients was first underlined by Trousseau in 1865 [49], and the possibility of a relationship between the clotting mechanism and the development of metastases was later postulated by Billroth who described cancer cells within a thrombus and interpreted this finding as evidence of the spread of tumor cells by thromboembolism [50]. Indeed, significant hemostatic abnormalities and thrombotic and hemorrhagic complications have been observed in cancer patients, complications which represent a common cause of death in cancer patients [51].

Tumor cells express intrinsic procoagulant properties without the need of exposure to any inducing agent, which promote the local activation of the coagulation system (for review [51, 52]). Following the first observation that cancer tissues shorten the clotting time of normal recalcified plasma [53], it has then been shown that the main procoagulant activity of a wide variety of tumor cells has to be ascribed to TF associated with cell membranes [54]. The constitutive expression of TF by malignant cells has been shown to correlate with cell metastatic potential [55–58].

Mechanisms of *TF participation in tumor progression* are complex and still require full elucidation. In this respect, both proteolytic activity of the TF-VIIa complex [59] and TF-induced intracellular signaling, which may involve the engagement of the protease activated receptors (PARs) [60, 61] have been reported.

In addition to tumor cells, TF may be also expressed in malignancy by tumor-infiltrating and circulating monocytes. We have shown that both circulating monocytes and tumor-associated macrophages from the V2 rabbit carcinoma express higher TF activity than those from control animals [62]. Moreover, both peripheral blood monocytes cultured from cancer patients and tumor-associated macrophages harvested from some experimental tumors express markedly increased amounts of TF [63–65]. In addition, in vascular cells around human breast cancer cells, the *in situ* detection of TF correlated with the malignant phenotype, suggesting TF as a marker for the initiation of angiogenesis in cancer [66].

Thus, TF on tumor or host cell surface represents a major mediator of clotting activation at the tumor–host interface.

1.4.2
The Role of TF in Physiologic and Tumor Angiogenesis

In 1996, three different groups reported a markedly defective vessel formation of yolk sac vessels in TF knockout mice embryos followed by massive hemorrhaging of embryonic blood [42–44]. Later it was shown that TF null mouse embryos could be rescued by human TF lacking the cytoplasmic domain, suggesting that embryogenesis requires the extracellular protease activity of the TF:VIIa complex [67].

In tumor angiogenesis the role of TF is exerted at several levels. In a clotting-dependent mechanism, TF generates thrombin that is responsible for fibrin formation and both these molecules modulate the angiogenic properties of the tumor [68–71]. Thrombin, once formed, activates platelets that are an important reservoir of VEGF-A [72]. Conversely, in a clotting-independent manner, TF acquires a signaling function, which results in modulation of the expression of proangiogenic molecules, among which is VEGF [55]. In this respect, several pieces of evidence link TF and VEGF. Colocalization of TF and VEGF mRNA and protein has been demonstrated in several tumors [36]. In addition, Meth-A sarcoma cells transfected to overexpress TF grew more rapidly and formed larger and more vascularized tumors than control cells *in vivo* [73]. These cells released more mitogenic activity for endothelial cells in parallel with enhanced transcription of VEGF, and diminished transcription of thrombospondin 2, a well-known antiangiogenic molecule, indicating that the capability of the tumor to induce a supporting neovasculature is dependent upon TF expression by tumor cells. A positive correlation for angiogenesis and levels of TF expression was found in human prostate carcinoma [74] and in non–small cell lung carcinoma [75]. In human melanoma cells, TF overexpression induced upregulation of VEGF and a concomitant enhancement of angiogenic potential in SCID mice [76].

Of particular importance is the finding that VEGF induces TF expression in vascular cells. Induction of monocytes and endothelial cells TF by a Meth-A-derived

VEGF was observed [77]. In endothelial cells the TF upregulation was proposed to be mediated by EGR-1 [78]. Notably, it should be mentioned that, while VEGF induces TF expression, the latter can upregulate VEGF synthesis in cells of tumor stroma [79–81]. This procoagulant/angiogenic loop should not be surprising considering the overlapping transcription pathways responsible for the synthesis of both TF and VEGF [82]. An additional link between TF and VEGF is that hypoxia, which is a common feature of the neovascularized tumors, upregulates both molecules [23, 83]. The same procoagulant/angiogenic loop has recently been the subject of studies by our group, showing that a chemotherapeutic agent such as paclitaxel, at antiangiogenic concentrations, was able to inhibit the TF expression in both human mammary tumor cells and in host cells such as human monocytes and endothelial cells [84].

In the last decade, particular attention in tumor development was devoted to the PARs. PARs are G-protein-coupled seven-transmembrane receptors, which have the peculiarity to carry their own tethered ligands [85]. In addition to PAR-mediated angiogenesis in animal models *in vivo*, a correlation has been shown between PAR expression and malignancy in several types of human cancer [86, 87]. PAR-induced endothelial cell proliferation is inhibited by neutralizing anti-VEGF antibodies, while oncogenic transformation by genes belonging to the PAR signaling machinery increases VEGF expression, suggesting that PAR activation is sufficient to induce VEGF and consequently angiogenesis [88]. Although thrombin is considered the main ligand of PARs, in recent years it has been shown that the TF-factor VIIa–factor Xa complex cleaves PAR-1 and/or PAR-2 promoting tumor progression [89, 90].

From these different pieces of evidence, it can be concluded that the effects of TF expression on angiogenetic processes may further contribute to the complex TF modulation of tumor growth and place angiogenesis as a new bridge between thrombosis and malignancy.

1.5 Conclusions

Four decades after Judah Folkman put forward his pioneering hypothesis and started to fight with strong determination and endure endless scientific debate to demonstrate its validity, it would now be simply impossible to deal with solid or some hematological tumors without considering angiogenesis and antiangiogenic therapy. The development of knowledge in this field represents a paradigmatic example of an integrative approach from fundamental science to clinical medicine, as this book will extensively show. For those investigators, like our group, who have long been interested in the association between thrombosis and malignancy, angiogenesis represents a crucial mechanism linking both processes, and offering new targets for combined therapies.

Acknowledgments

The authors wish to honor the memory of Professor Judah Folkman, and are thankful to him for inspiration and fruitful discussions, particularly on the occasion of the 2nd "John Paul II" lecture (Campobasso, March 2006). The helpful support of Dr. Joanna Gozdzikiewicz (University of Bialystok, Poland) is gratefully acknowledged. The authors' work mentioned in this review was partially supported by the Italian Ministry of Health (Progetto Oncologico Ordinario RF CCM 2006-424501 Conv. 29/07).

References

1. Ferrara, N. (2002) VEGF and the quest for tumour angiogenesis factors. *Nat. Rev. Cancer*, **2**, 795–803.
2. Folkman, J., Cole, P., and Zimmerman, S. (1966) Tumor behavior in isolated perfused organs: in vitro growth and metastases of biopsy material in rabbit thyroid and canine intestinal segment. *Ann. Surg.*, **164**, 491–502.
3. Folkman, J., Merler, E., Abernathy, C. *et al.* (1971) Isolation of a tumor factor responsible for angiogenesis. *J. Exp. Med.*, **133**, 275–288.
4. Folkman, J. (1971) Tumor angiogenesis: therapeutic implications. *N. Engl. J. Med.*, **285**, 1182–1186.
5. Donati, M.B. and Gozdzikiewicz, J. (2008) Angiogenesis and the progress of vascular and tumor biology. A tribute to Judah Folkman. *Thromb. Haemost.*, **99**, 647–650.
6. Gimbrone M.A. Jr., Leapman S.B., Cotran R.S. *et al.* (1972) Tumor dormancy in vivo by prevention of neovascularization. *J. Exp. Med.*, **136**, 261–276.
7. Brem, H. and Folkman, J. (1975) Inhibition of tumor angiogenesis mediated by cartilage. *J. Exp. Med.*, **41**, 427–439.
8. Jaffe, E.A., Nachman, R.L., Becker, C.G. *et al.* (1973) Culture of human endothelial cells derived from umbilical veins: identification by morphologic and immunologic criteria. *J. Clin. Invest.*, **52**, 2745–2756.
9. Gimbrone, M.A. Jr., Cotran, R.S., and Folkman, J. (1974) Human vascular endothelial cells in culture. Growth and DNA synthesis. *J. Cell. Biol.*, **60**, 673–684.
10. Folkman, J., Haudenschild, C.C., and Zetter, B.R. (1979) Longterm culture of capillary endothelial cells. *Proc. Natl. Acad. Sci. U.S.A.*, **10**, 5217–5221.
11. Shing, Y., Folkman, J., Sullivan, R. *et al.* (1984) Heparin affinity: purification of a tumor-derived capillary endothelial cell growth factor. *Science*, **223**, 1296–1299.
12. Yancopoulos, G.D., Davis, S., Gale, N.W. *et al.* (2000) Vascular-specific growth factors and blood vessel formation. *Nature*, **407**, 242–248.
13. O'Reilly, M.S., Holmgren, L., Shing, Y. *et al.* (1994) Angiostatin: a novel angiogenesis inhibitor that mediates the suppression of metastases by a Lewis lung carcinoma. *Cell*, **79**, 315–328.
14. O'Reilly, M.S., Boehm, T., Shing, Y. *et al.* (1997) Endostatin: an endogenous inhibitor of angiogenesis and tumor growth. *Cell*, **88**, 277–285.
15. Folkman, J. (2006) Antiangiogenesis in cancer therapy – endostatin and its mechanisms of action. *Exp. Cell Res.*, **312**, 594–607.
16. Sun, Y., Wang, J., Liu, Y. *et al.* (2005) Results of phase III trial of EndostarTM (rh-endostatin, YH-16) in advanced non-small cell lung cancer (NSCLC) patients.ASCO Annual Meeting Proceedings 2005, p. 23, Abstract 16S.
17. Hurwitz, H., Fehrenbach, L., Novotny, W. *et al.* (2004) Bevacizumab plus irinotecan, fluorouracil, and leucovorin for metastatic colorectal cancer. *N. Engl. J. Med.*, **350**, 2335–2342.

18. Hanahan, D. and Folkman, J. (1996) Patterns and emerging mechanisms of the angiogenic switch during tumorigenesis. *Cell*, **86**, 353–364.
19. Udagawa, T., Fernandez, A., Achilles, E.G. *et al.* (2002) Persistence of microscopic human cancers in mice: alterations in the angiogenic balance accompanies loss of tumor dormancy. *FASEB J.*, **16**, 1361–1370.
20. Naumov, G.N., Bender, E., Zurakowski, D. *et al.* (2006) A model of human tumor dormancy: an angiogenic switch from the nonangiogenic phenotype. *J. Natl. Cancer Inst.*, **98**, 316–325.
21. Carmeliet, P. and Jain, R.K. (2000) Angiogenesis in cancer and other diseases. *Nature*, **407**, 249–257.
22. Simons, M. (2005) Angiogenesis: where do we stand now? *Circulation*, **111**, 1556–1566.
23. Pugh, C.W. and Ratcliffe, P.J. (2003) Regulation of angiogenesis by hypoxia: role of the HIF system. *Nat. Med.*, **6**, 677–684.
24. Folkman, J., Browder, T., and Palmblad, J. (2001) Angiogenesis research: guidelines for translation to clinical application. *Thromb. Haemost.*, **86**, 23–33.
25. Folkman, J. (2003) Angiogenesis inhibitors. *Cancer Biol. Ther.*, **2** (Suppl 1), S127–S133.
26. Kerbel, R.S. (1991) Inhibition of tumor angiogenesis as a strategy to circumvent acquired resistance to anticancer therapeutic agents. *Bioessays*, **13**, 31–36.
27. Browder, T., Butterfield, C.E., Kraling, B.M. *et al.* (2000) Antiangiogenic scheduling of chemotherapy improves efficacy against experimental drug-resistant cancer. *Cancer Res.*, **60**, 1878–1886.
28. Kerbel, R. and Folkman, J. (2002) Clinical translation of angiogenesis inhibitors. *Nat. Rev. Cancer*, **2**, 727–739.
29. Kuenen, B.C. (2003) Analysis of prothrombotic mechanisms and endothelial perturbation during treatment with angiogenesis inhibitors. *Pathophysiol. Haemost. Thromb.*, **33** (Suppl 1), 13–14.
30. Zangari, M., Anaissie, E., Barlogie, B. *et al.* (2001) Increased risk of deep-vein thrombosis in patients with multiple myeloma receiving thalidomide and chemotherapy. *Blood*, **98**, 1614–1615.
31. Caine, G.J. and Lip, G.Y. (2003) Thromboembolism associated with new anti-cancer treatment strategies in combination with conventional chemotherapy: new drugs, old risks? *Thromb. Haemost.*, **90**, 567–569.
32. Ferrara, N. and Henzel, W.J. (1989) Pituitary follicular cells secrete a novel heparin-binding growth factor specific for vascular endothelial cells. *Biochem. Biophys. Res. Commun.*, **161**, 851–858.
33. Senger, D.R., Galli, S.J., Dvorak, A.M., Perruzzi, C.A., Harvey, V.S., and Dvorak, H.F. (1983) Tumor cells secrete a vascular permeability factor that promotes accumulation of ascites fluid. *Science*, **219**, 983–985.
34. Ferrara, N., Gerber, H.P., and LeCouter, J. (2003) The biology of VEGF and its receptors. *Nat. Med.*, **6**, 669–676.
35. Ho, Q.T. and Kuo, C.J. (2007) Vascular endothelial growth factor: biology and therapeutic applications. *Int. J. Biochem. Cell. Biol.*, **39**, 1349–1357.
36. Donati, M.B. and Lorenzet, R. (2004) The role of blood clotting in tumor metastasis: a case for tissue factor, in *Thrombosis and Cancer* (eds G. Lugassy, A. Falanga, A.K. Kakkar, and F.R. Rickles), Martin Dunitz Ltd., London, pp. 47–58.
37. Nemerson, Y. (1988) Tissue factor and hemostasis. *Blood*, **71**, 1–8.
38. Edgington, T.S., Mackman, N., Brand, K., and Ruf, W. (1991) The structural biology of expression and function of tissue factor. *Thromb. Haemost.*, **66**, 67–79.
39. Lawson, J.H., Butenas, S., and Mann, K.G. (1992) The evaluation of complex-dependent alterations in human factor VIIa. *J. Biol. Chem.*, **267**, 4834–4843.
40. østerud, B. and Rapaport, S.I. (1977) Activation of factor IX by the reaction product of tissue factor and factor VII: additional pathway for initiating blood coagulation. *Proc. Natl. Acad. Sci. U.S.A.*, **74**, 5260–5264.
41. Drake, T.A., Morrissey, J.H., and Edgington, T.S. (1989) Selective cellular expression of tissue factor in human

tissues. Implications for disorders of hemostasis and thrombosis. *Am. J. Pathol.*, **134**, 1087–1097.
42. Bugge, T.H., Xiao, Q., Kombrinck, K.W., Flick, M.J., Holmback, K., Danton, M.J.S., Colbert, M.C., Witte, D.P., Fujikawa, K., Davie, E.W., and Degen, J.L. (1996) Fatal embryonic bleeding events in mice lacking tissue factor, the cell-associated initiator of blood coagulation. *Proc. Natl. Acad. Sci. U.S.A.*, **93**, 6258–6263.
43. Carmeliet, P., Mackman, N., Moons, L., Luther, T., Gressens, P., Van Vlaenderen, I., Demunck, H., Kasper, M., Breier, G., Evrard, P., Muller, M., Risau, W., Edgington, T., and Collen, D. (1996) Role of tissue factor in embryonic blood vessel development. *Nature*, **383**, 73–75.
44. Toomey, J.R., Kratzer, K.E., Lasky, N.M., Stanton, J.J., and Broze, G.J. Jr. (1996) Targeted disruption of the murine tissue factor gene results in embryonic lethality. *Blood*, **88**, 1583–1587.
45. Røttingen, J.A., Enden, T., Camerer, E., Iversen, J.G., and Prydz, H. (1995) Binding of human factor VIIa to tissue factor induces cytosolic Ca2+ signals in J82 cells, transfected COS-1 cells, Madin-Darby canine kidney cells and in human endothelial cells induced to synthesize tissue factor. *J. Biol. Chem.*, **270**, 4650–4660.
46. Nakagawa, K., Zhang, Y., Tsuji, H., Yoshizumi, M., Kasahara, T., Nishimura, H., Nawroth, P.P., and Nakagawa, M. (1998) The angiogenic effect of tissue factor on tumors and wounds. *Semin. Thromb. Haemost.*, **24**, 207–210.
47. Randolph, G.J., Luther, T., Albrecht, S., Magdolen, V., and Muller, W.A. (1998) Role of tissue factor in adhesion of mononuclear phagocytes to and trafficking through endothelium in vitro. *Blood*, **92**, 4167–4177.
48. Ott, I., Fischer, E.G., Miyagi, Y., Mueller, B.M., and Ruf, W. (1998) A role for tissue factor in cell adhesion and migration mediated by interaction with actin-binding protein 280. *J. Cell Biol.*, **140**, 1241–1253.
49. Trousseau, A. (1865) Phlegmasia alba dolens, in *Clinique Medicale de l'Hotel Dieu de Paris*, 2nd edn, vol. 3, Baillière, Paris, pp. 654–712.
50. Billroth, T. (1878) *Lectures on Surgical Pathology and Therapeutics: A Handbook for Students and Practitioners*, 8th edn, The New Sydenham Society, London, pp. 1877–1878.
51. Donati, M.B. (2007) Thrombosis and cancer: a personal view. *Thromb. Haemost.*, **98**, 126–128.
52. Donati, M.B. and Falanga, A. (2001) Pathogenetic mechanisms of thrombosis in malignancy. *Acta Haematol.*, **106**, 18–24.
53. O'Meara, R.A.Q. (1958) Coagulative properties of cancers. *Ir. J. Med. Sci.*, **394**, 474–479.
54. Dvorak, H.F., Van DeWater, L., Bitzer, A.M., Dvorak, A.M., Anderson, D., Harvey, V.S., Bach, R., Davis, G.L., DeWolf, W., and Carvalho, A.C. (1983) Procoagulant activity associated with plasma membrane vesicles shed by cultured tumor cells. *Cancer Res.*, **43**, 4434–4442.
55. Rickles, F.R., Patierno, S., and Fernandez, P.M. (2003) Tissue factor, thrombin, and cancer. *Chest*, **124** (Suppl 3), 58S–68S.
56. Donati, M.B. and Semeraro, N. (1984) Cancer cell procoagulants and their pharmacological modulation. *Haemostasis*, **14**, 422–429.
57. Versteeg, H.H., Spek, C.A., Peppelenbosch, M.P., and Richel, D.J. (2004) Tissue factor and cancer metastasis: the role of intracellular and extracellular signaling pathways. *Mol. Med.*, **10**, 6–11.
58. Ruf, W. and Mueller, B.M. (2006) Thrombin generation and the pathogenesis of cancer. *Semin. Thromb. Haemost.*, **32** (Suppl 1), 61–68.
59. Mueller, B.M. and Ruf, W. (1998) Requirement for binding of catalytically active factor VIIa in tissue factor-dependent experimental metastasis. *J. Clin. Invest.*, **101**, 1372–1378.
60. Bromberg, M.E., Konigsberg, W.H., Madison, J.F., Pawashe, A., and Garen, A. (1995) Tissue factor promotes melanoma metastasis by a pathway

independent of blood coagulation. *Proc. Natl. Acad. Sci. U.S.A.*, **92**, 8205–8209.
61. Belting, M., Dorrell, M.I., Sandgren, S. *et al.* (2004) Regulation of angiogenesis by tissue factor cytoplasmic domain signaling. *Nat. Med.*, **10**, 502–509.
62. Lorenzet, R., Peri, G., Locati, D., Allavena, P., Colucci, M., Semeraro, N., Mantovani, A., and Donati, M.B. (1983) Generation of procoagulant activity by mononuclear phagocytes: a possible mechanism contributing to blood clotting activation within malignant tissues. *Blood*, **62**, 271–273.
63. Morgan, D., Edwards, R.L., and Rickles, F.R. (1988) Monocyte procoagulant activity as a peripheral marker of clotting activation in cancer patients. *Haemostasis*, **18**, 55–65.
64. Semeraro, N. and Colucci, M. (1997) Tissue factor in health and disease. *Thromb. Haemost.*, **78**, 759–764.
65. Lwaleed, B.A., Bass, P.S., and Cooper, A.J. (2001) The biology and tumour-related properties of monocyte tissue factor. *J. Pathol.*, **193**, 3–12.
66. Contrino, J., Hair, G., Kreutzer, D.L., and Rickles, F.R. (1996) In situ detection of tissue factor in vascular endothelial cells: Correlation with the malignant phenotype of human breast disease. *Nat. Med.*, **2**, 209–215.
67. Parry, G.C. and Mackman, N. (2000) Mouse embryogenesis requires the tissue factor extracellular domain but not the cytoplasmic domain. *J. Clin. Invest.*, **105**, 1547–1554.
68. Nagy, J.A., Brown, L.F., Senger, D.R., Lanir, N., Van de Water, L., Dvorak, A.M., and Dvorak, H.F. (1989) Pathogenesis of tumor stroma generation: a critical role for leaky blood vessels and fibrin deposition. *Biochim. Biophys. Acta*, **3**, 305–326.
69. Dvorak, H.F., Nagy, J.A., Berse, B., Brown, L.F., Yeo, K.T., Yeo, T.K., Dvorak, A.M., van de Water, L., Sioussat, T.M., and Senger, D.R. (1992) Vascular permeability factor, fibrin, and the pathogenesis of tumor stroma formation. *Ann. N. Y. Acad. Sci.*, **667**, 101–111.
70. Fernandez, P.M., Patierno, S.R., and Rickles, F.R. (2004) Tissue factor and fibrin in tumor angiogenesis. *Semin. Thromb. Haemost.*, **1**, 31–44.
71. Wojtukiewicz, M.Z., Sierko, E., and Rak, J. (2004) Contribution of the hemostatic system to angiogenesis in cancer. *Semin. Thromb. Haemost.*, **1**, 5–20.
72. Verheul, H.M., Hoekman, K., Luykx-de Bakker, S., Eekman, C.A., Folman, C.C., Broxterman, H.J., and Pinedo, H.M. (1997) Platelet: transporter of vascular endothelial growth factor. *Clin. Cancer Res.*, **12**, 2187–2190.
73. Zhang, Y., Deng, Y., Luther, T., Muller, M., Ziegler, R., Waldherr, R., Stern, D.M., and Nawroth, P.P. (1994) Tissue factor controls the balance of angiogenic and antiangiogenic properties of tumor cells in mice. *J. Clin. Invest.*, **94**, 1320–1327.
74. Abdulkadir, S.A., Carvalhal, G.F., Kaleem, Z., Kisiel, W., Humphrey, P.A., Catalona, W.J., and Milbrandt, J. (2000) Tissue factor expression and angiogenesis in human prostate carcinoma. *Hum. Pathol.*, **31**, 443–447.
75. Koomagi, R. and Volm, M. (1998) Tissue-factor expression in human non-small-cell lung carcinoma measured by immunohistochemistry: correlation between tissue factor and angiogenesis. *Int. J. Cancer*, **79**, 19–22.
76. Abe, K., Shoji, M., Chen, J., Bierhaus, A., Danave, I., Micko, C., Casper, K., Dillehay, D.L., Nawroth, P.P., and Rickles, F.R. (1999) Regulation of vascular endothelial growth factor production and angiogenesis by the cytoplasmic tail of tissue factor. *Proc. Natl. Acad. Sci. U.S.A.*, **96**, 8663–8668.
77. Clauss, M., Gerlach, M., Gerlach, H., Brett, J., Wang, F., Familletti, P.C., Pan, Y.C., Olander, J.V., Connolly, D.T., and Stern, D. (1990) Vascular permeability factor: a tumor-derived polypeptide that induces endothelial cell and monocyte procoagulant activity, and promotes monocyte migration. *J. Exp. Med.*, **172**, 1535–1545.
78. Mechtcheriakova, D., Wlachos, A., Holzmüller, H., Binder, B.R., and Hofer, E. (1999) Vascular endothelial cell growth factor-induced tissue factor expression in endothelial cells is mediated by EGR-1. *Blood*, **93**, 3811–3823.

79. Ollivier, V., Bentolila, S., Chabbat, J., Hakim, J., and de Prost, D. (1998) Tissue factor-dependent vascular endothelial growth factor production by human fibroblasts in response to activated factor VII. *Blood*, **91**, 2698–2703.

80. Rak, J. and Klement, G. (2000) Impact of oncogenes and tumor suppressor genes on deregulation of hemostasis and angiogenesis in cancer. *Cancer Metastasis Rev.*, **19**, 93–96.

81. Wojtukiewicz, M.Z., Sierko, E., Klement, P., and Rak, J. (2001) The hemostatic system and angiogenesis in malignancy. *Neoplasia*, **5**, 371–384.

82. Chen, J., Bierhaus, A., Schiekofer, S., Andrassy, M., Chen, B., Stern, D.M., and Nawroth, P.P. (2001) Tissue factor–a receptor involved in the control of cellular properties, including angiogenesis. *Thromb. Haemost.*, **1**, 334–345.

83. Yan, S.F., Pinsky, D.J., and Stern, D.M. (2000) A pathway leading to hypoxia-induced vascular fibrin deposition. *Semin. Thromb. Haemost.*, **26**, 479–483.

84. Napoleone, E., Zurlo, F., Latella, M.C., Amore, C., Di Santo, A., Iacoviello, L., Donati, M.B., and Lorenzet, R. (2009) Paclitaxel downregulates tissue factor in cancer and host tumor-associated cells. *Eur. J. Cancer*, **45**, 470–477.

85. Coughlin, S.R. (2000) Thrombin signalling and protease-activated receptors. *Nature*, **407**, 258–264.

86. Yin, Y.J., Salah, Z., Grisaru-Granovsky, S., Cohen, I., Even-Ram, S.C., Maoz, M., Uziely, B., Peretz, T., and Bar-Shavit, R. (2003) Human protease-activated receptor 1 expression in malignant epithelia: a role in invasiveness. *Arterioscler. Thromb. Vasc. Biol.*, **23**, 940–944.

87. Camerer, E. (2007) Protease signaling in tumor progression. *Thromb. Res.*, **120** (Suppl 2), S75–S81.

88. Yin, Y.J., Salah, Z., Maoz, M., Ram, S.C., Ochayon, S., Neufeld, G., Katzav, S., and Bar-Shavit, R. (2003) Oncogenic transformation induces tumor angiogenesis: a role for PAR1 activation. *FASEB J.*, **17**, 163–174.

89. Palumbo, J.S. and Degen, J.L. (2007) Mechanisms linking tumor cell-associated procoagulant function to tumor metastasis. *Thromb. Res.*, **120** (Suppl 2), S22–S28.

90. Ruf, W. (2007) Tissue factor and PAR signaling in tumor progression. *Thromb. Res.*, **120** (Suppl 2), S7–S12.

Part II
Mechanisms of Angiogenesis and Lymphangiogenesis

Tumor Angiogenesis – From Molecular Mechanisms to Targeted Therapy. Edited by Francis S. Markland, Stephen Swenson, and Radu Minea
Copyright © 2010 WILEY-VCH Verlag GmbH & Co. KGaA, Weinheim
ISBN: 978-3-527-32091-2

2 Molecular Mechanisms of Angiogenesis

Andrew C. Dudley and Lena Claesson-Welsh

2.1 Vessel Formation

Blood is carried by a hierarchical network of vessels which are lined by a single layer of specialized endothelial cells (ECs). Circulation of blood allows delivery of oxygen to the tissues [1], which is essential for the generation of energy by the mitochondria. Passage of blood through thin-walled capillaries in the intestinal tract allows uptake of nutrients from food, whereas the hepatic and renal circulation allows storage of nutrients and detoxification of the blood. The ECs deposit a basement membrane composed of fibronectin, collagen IV, laminins, and heparan sulfate proteoglycans [2], which may directly or indirectly influence diverse processes such as cell differentiation, attachment, migration, polarization, guidance, and survival. Furthermore, depending on the vessel size and position relative to the heart, the ECs are surrounded by supporting cells of smooth muscle origin. Large vessels such as the aorta carry oxygenated blood from the heart and are surrounded by several layers of smooth muscle cells. Arteries branch to become finer vessels, arterioles, which in turn lead on to capillaries, the finest vessels. Capillaries are more sparsely surrounded by supporting cells called *pericytes*, which are important for the stability of the vessel [3] and which continuously exchange molecular information with the ECs. The blood flows from the capillary bed into postcapillary venules, where exchange of solutes and proteins takes place through regulated vascular leakage [4]. Venules merge to form veins, which lead the blood back to the heart and lungs for renewed oxygenation. Veins have a sparse muscle coat and are equipped with valves that aid in propagation of the blood. For a review on morphological and molecular properties of ECs in different vessels in various normal organs, see [5].

Formation of new vessels is critical in a number of processes where tissues expand. The principles of blood vessel formation differ vastly depending on whether blood vessels form during embryogenesis (denoted vasculogenesis), during physiological processes such as wound healing or growth of the endometrium (denoted regulated angiogenesis), or during pathologies such as inflammation and cancer (dysregulated angiogenesis). In addition, vessels may form through

Tumor Angiogenesis – From Molecular Mechanisms to Targeted Therapy. Edited by Francis S. Markland, Stephen Swenson, and Radu Minea
Copyright © 2010 WILEY-VCH Verlag GmbH & Co. KGaA, Weinheim
ISBN: 978-3-527-32091-2

splitting, in a process called *intussusception*, or be derived from circulating bone marrow-derived progenitors (denoted adult or postnatal vasculogenesis). Thus, mechanisms operating during embryogenesis may reappear and play a significant role during pathological angiogenesis. To successfully understand tumor angiogenesis, it is therefore essential to also recognize these embryonic processes.

2.1.1
Vasculogenesis

Stem cells committed to the vascular lineage emerge in the posterior primitive streak and migrate into the extraembryonic yolk sac. Vasculogenesis denotes the process whereby these progenitors, called *angioblasts*, differentiate and form an immature vascular plexus around embryonic day 7.5 in the mouse [6]. The exact manner in which angioblast differentiation to mature ECs occurs is not yet clear; however, there seems to be different pathways in the embryo proper compared to the yolk sac. Vasculogenesis in the yolk sac has been extensively studied and has been suggested to involve a bipotential stem cell, the hemangioblast, that differentiates to form both hematopoietic and endothelial lineages [7]. In the yolk sac, progenitor cells/hemangioblasts form a cell mass referred to as a *blood island*. Cells located peripherally in the blood island differentiate to form ECs, whereas cells located in the center form early hematopoietic cells. In the embryo, angioblasts form independently of hematopoiesis [8]. The unequivocal existence of the hemangioblast *in vivo* is still debated; however, strong evidence for the generation of hematopoietic cells through the formation of a hemogenic endothelium intermediate was recently presented [9].

2.1.2
Circulating Endothelial Progenitor Cells

Circulating bone marrow–derived endothelial precursor cells (CEPs) have been implicated in both physiological and pathological angiogenesis [10]. Following chemotactic cues, these progenitors home to sites of active angiogenesis and incorporate in the vessel wall. To what extent the progenitors fully integrate has however been controversial [11] and evidence has been presented that bone marrow-derived progenitors are myeloid cells that migrate to a position very close to the vessel wall, but do not actually become integrated. Myeloid cells may play an important "accessory" role during angiogenesis by releasing growth, chemotactic, and tissue remodeling factors [12].

2.1.3
Intussusception

Intussusception (growth within itself) involves the formation of new vessels by splitting of a preexisting vessel along its long axis. Intussusception is the most rapid mechanism for formation of new vasculature while maintaining intact circulation and it occurs in a variety of organs, especially in capillary beds close to epithelial surfaces (for a review, see [13]). The smooth muscle cell pillar that eventually extends through the EC lining, splits the vessel into two new, smaller vessels. Intussusception is a common mechanism for remodeling of the vascular bed during development and in the newborn, but occurs less frequently in the adult.

2.1.4 Sprouting Angiogenesis

Sprouting angiogenesis denotes the process whereby new vessels extend from preexisting vasculature by formation of a vascular sprout. Following exposure to a stimulating factor (see below), selected ECs acquire the capacity to invade the surrounding tissue by forming an angiogenic sprout composed of a leading tip cell and trailing stalk cells, which orientate toward the source of the stimulus [14]. Over time a lumen is formed and once the sprout anastomoses with another vessel sprout, circulation is established. The frequency of sprouting along the axis of a vessel is tightly regulated by promoting factors such as growth factors, and negatively regulated by members of the Jagged/Notch family [15], which are critical for cell differentiation in many developmental processes. For a review on signal transduction in branching morphogenesis, see [16]. During physiological angiogenesis (e.g., wound healing), vessel sprouting is tightly regulated and once the demand for oxygen is met, the process ceases. The newly formed vessel becomes stabilized by attraction of supporting smooth muscle cells and by deposition of a vascular basement membrane.

2.2 Factors That Stimulate Blood Vessel Formation

Growth factors are major stimulators of angiogenesis. A large number of second messengers may then modulate, positively or negatively, their downstream effects. Studies on gene-targeted animals have been instrumental in deducing the role of different growth factors, especially during embryonic development. We discuss below a number of key angiogenic growth factors.

2.2.1 Vascular Endothelial Growth Factors (VEGFs)

Vascular endothelial growth factor (VEGF) denotes a family of five structurally related mammalian ligands: VEGF-A, VEGF-B, VEGF-C, VEGF-D, and placental growth factor (PlGF), which consists of covalently linked homodimers [17]. The VEGF ligands bind in a distinct pattern to three related receptor tyrosine kinases: denoted VEGFR-1, VEGFR-2, and VEGFR-3 [18]. Gene targeting of VEGF-A or VEGFR-2 results in similar phenotypes with arrest in EC differentiation and embryonic death around embryonic day E8.5. Clearly, VEGF-A/VEGFR-2 function is a strict prerequisite for EC development, EC survival, and for regulation of vascular permeability in the adult. ECs in the stable vasculature weakly express VEGFR-2; but, during physiological or pathological angiogenesis (e.g., tumors), VEGFR-2 is upregulated and is expressed at high levels [19]. A distinguishing feature of VEGF-A is that its expression is regulated by hypoxia, as was originally shown by Eli Keshet and coworkers [20].

In contrast to VEGF-A, elimination of either VEGF-B or PlGF does not result in embryonic lethality. Although VEGF-B has been assigned a specific role in revascularization of the ischemic myocardium [21], PlGF has been more generally implicated in regulation of pathological angiogenesis. Thus, PlGF-deficient animals develop normally but due to a reduced angiogenic response, tumor growth and inflammatory processes are attenuated [22]. The underlying mechanism appears to involve sensitization of ECs to the stimulatory effect of VEGF-A [23]. Gene targeting of VEGFR-1, the receptor for VEGF-A as well as for VEGF-B and PlGF leads to embryonic death around E9.5 due to excess proliferation of ECs and poor organization of cells to form lumenized vessels [24]. It is noteworthy that VEGFR-1 is naturally produced not only as a full-length receptor but also as a soluble extracellular domain. This soluble form binds VEGF-A with high affinity and has been suggested to serve as a trap, immobilizing VEGF-A and preventing it from binding to VEGFR-2 [25]. The mechanism whereby the three ligands for VEGFR-1, VEGF-A, VEGF-B, and PlGF have different biological effects is unclear but could involve differential binding to additional receptors or coreceptors, such as the neuropilins [26].

VEGF-C and VEGF-D bind to VEGFR-3. VEGF-C gene inactivation results in disruption of lymphatic vessel development and prenatal death due to tissue fluid accumulation [27]. On the other hand, VEGF-D gene targeting is compatible with normal mouse development [28]. Knockout of VEGFR-3 leads to a distinct phenotype consisting of cardiovascular failure and embryonic death at E9.5 [29]. VEGFR-3 is expressed in tip cells of sprouting vessels, in agreement with an important role for VEGFR-3 in angiogenesis [30]. VEGF-C as well as VEGFR-3 have been implicated in induction of tumor angiogenesis and lymphangiogenesis [31].

2.2.2
Fibroblast Growth Factors (FGFs)

The mammalian fibroblast growth factor (FGF) family encompasses 18 proteins (FGF1–FGF10 and FGF16–FGF23) identified to date [32]. The FGFs have several distinguishing features including, for certain FGFs, the absence of a signal sequence typical for secreted proteins. The mechanisms for secretion of, for example, FGF-1 and FGF-2, involve direct translocation across the plasma membrane or heterodimerization with carrier proteins [33]. Moreover, certain FGF variants are taken up and brought to the nucleus where they may play a role in regulation of proliferation [34]. Finally, FGFs bind with very high affinity to heparan sulfate proteoglycans, which are known to act as coreceptors for the fibroblast growth factor receptors (FGFRs) [35].

There are four FGFR tyrosine kinases (FGFR1–FGFR4). FGF-2 (bFGF) was the first angiogenic growth factor to be identified and is a strong mitogen for ECs *in vitro* [36]. Although FGFs/FGFRs have essential roles in a wide spectrum of physiological processes such as skeletal development [37], accumulating data from gene targeting efforts have not supported the notion that FGFs/FGFRs act directly on ECs *in vivo*. Clearly, FGFs/FGFRs regulate expression of a wide spectrum of cytokines that in turn may affect vessel function. For example, FGFR1-deficient embryonic stem cells show an exaggerated tendency to form ECs, due to a FGFR1-dependent change in expression of cytokines such as interleukin 4 and pleiothropin [38]. It does remain possible, however, that FGFs act directly on ECs in pathologies such as cancer [39].

2.2.3 Platelet-Derived Growth Factors (PDGFs)

The Platelet-derived growth factors (PDGFs), which are structurally related to the VEGFs, occur as covalently linked homo- or heterodimers. There are five dimeric PGDF isoforms, assembled from polypeptide chains denoted A, B, C, and D (i.e., PDGF-AA, PDGF-BB, PDGF-AB, PDGF-CC, and PDGF-DD) [40]. The PDGFs interact with two structurally related receptor tyrosine kinases: denoted PDGFRα and PDGFRβ. PDGF-BB is known to be produced by ECs and is required for the development of PDGFRβ-expressing pericytes and their tight association with vessels. Thus, gene-targeted animals lacking expression of either PDGF-BB or PDGFRβ die prenatally from the consequences of pericyte-deficiency [41–43]. It was recently suggested that the effect of PDGF on pericytes appears to be negatively modulated by VEGF through formation of VEGFR2/PDGFRβ complexes [44].

2.2.4 Angiopoietins (Ang's)

The Angs bind to receptor tyrosine kinases denoted Tie1 and Tie2 (tyrosine kinase with immunoglobulin and epidermal growth factor (EGF) homology domains, also known as Tek1 and Tek2), which are expressed on ECs; in addition, Tie2 is expressed on tumor pericytes [45]. Ang–Tie signaling controls vessel maturation during embryogenesis and vessel quiescence in the adult. Additional members of the Ang family have been identified, which appear to exert either positive or negative roles during angiogenesis under different conditions. Although not required for differentiation of ECs during development, the vasculature fails to mature in the absence of either Ang1 or Tie2 [46–48]. Ang2 deficiency on the other hand is compatible with vascular development [49]. Finally, Ang2 is reported to be upregulated in tumor-specific endothelial cells (TECs) and circulating Ang2 has emerged as a biomarker of tumor progression for a number of different tumors [50].

See Figure 2.1 for a schematic outline on growth factor loops involving VEGF, FGF, PDGF, and Ang1 in the tumor vascular bed.

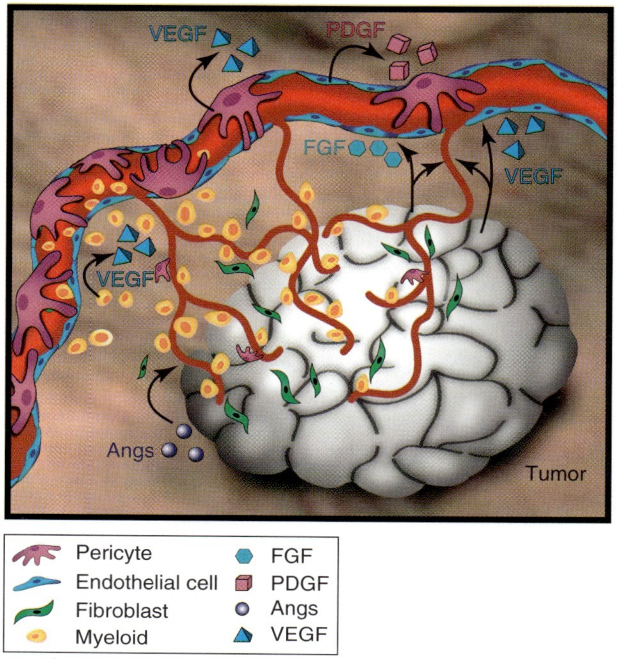

Figure 2.1 Growth factor loops in the tumor microenvironment. VEGF (alternatively denoted VEGF-A) is produced by a variety of cells in the tumor microenvironment, such as tumor cells, tumor-associated fibroblasts, inflammatory cells, and from endothelial cells themselves. VEGF acts on VEGFR2 expressed on tumor endothelial cells (growth factor loops are indicated by black arrows throughout) but also sometimes on tumor cells that may express VEGFR2 (not indicated). PDGF is produced by endothelial cells and acts on PDGFR-β expressed on pericytes. FGF may be widely produced and is retained on the producer cells or retained in the extracellular matrix through binding to heparan sulfate. FGF may act on tumor endothelial cells as well as tumor cells. Angiopoietins such as Ang1 are produced by tumor cells and acts both on endothelial cells and pericytes in tumor vessels. See text for details.

2.3
Molecular Targets in the Therapeutic Inhibition of Angiogenesis

The logic behind antiangiogenic therapies is that any solid tumor can be eradicated by targeting the associated blood vessels. While antiangiogenesis strategies hold great promise for eliminating tumors altogether, or maintaining tumors in a dormant state, this principle depends on two basic assumptions. First, in order for antiangiogenic (antiendothelial) therapies to succeed, TECs should express the receptors or other molecules that were originally singled-out to be targeted by the antiangiogenic agent. Second, TECs must not be able to acquire the adaptive mechanisms that tumor cells use to evade cytotoxic therapies. These mechanisms might include factors such as genomic instability (which underlies acquired drug resistance) characteristic for tumor cells, and the ability to upregulate compensatory or alternative pathways that can circumvent single-agent antiangiogenic therapies. As explained later, both of these assumptions have been challenged by a number of studies showing that normal ECs and TECs are indeed very different; sometimes strikingly so. The fact that the tumor endothelium may be more dynamic than once thought presents a challenge toward the design of successful antiangiogenic therapies [51]. Despite these challenges, a few successful antiangiogenic drugs have been approved for use in the clinic. In this section, we will briefly discuss some of the more successful antiangiogenic drugs that are now used to treat human cancers while focusing on inhibitors of the VEGF signaling pathway.

2.3.1
Anti-VEGF Therapies

Owing to chaotic blood flow and necrosis, many tumors are characterized by hypoxia. Hypoxia results in the stabilization of HIF (hypoxia-inducible factor), a transcription factor that switches on several hypoxia-inducible genes such as VEGF. As already discussed in detail above, VEGF then activates VEGFR-2 on the endothelium, inducing signals for migration, proliferation, and survival of EC. It was therefore logical that some of the first antiangiogenic drugs should be designed to block the activity of VEGF and its receptors (for an excellent review of VEGF-targeted therapies see Ellis *et al.* [52]). In addition to direct effects on the endothelium, VEGF blockade may also prevent the recruitment of bone marrow–derived hematopoietic or endothelial progenitor cells to sites of tumor growth [53]. These accessory cells recruited from the bone marrow are increasingly recognized to facilitate angiogenesis in solid tumors, particularly during the earliest stages of blood vessel assembly, because they secrete matrix metalloproteases (MMPs) and VEGF [54, 55].

The first anti-VEGF reagent approved for use in the clinic was the humanized anti-VEGF monoclonal antibody, bevacizumab (Avastin, Genentech), which can inhibit VEGF binding to its receptors. Avastin seems to have few side effects compared to most cytotoxic therapies. When combined with certain forms of chemotherapy, Avastin has shown benefits in patients with colorectal [56] and breast cancers [57]. For example, when combined with paclitaxel, Avastin prolongs progression-free but not overall survival in patients with breast tumors [57]. Avastin was also shown to prolong the time to disease progression in patients with renal cancer [58]. A single infusion of Avastin was shown to decrease tumor perfusion, vascular volume, and microvessel density in rectal carcinoma [53]. Because Avastin is administered in combination with cytotoxic chemotherapies, it has been difficult to tease out the direct benefit and effects of Avastin on the tumor vasculature in human subjects.

Tyrosine kinase inhibitors (TKIs) with selectivity for VEGF receptors are also approved for clinical use by the FDA. In contrast to Avastin, which binds directly to VEGF, the site of action of most TKIs is the ATP binding pocket of the receptor. For example, sorafenib (Nexavar, Bayer/Onyx) has been used for the treatment of advanced renal cell carcinoma [59] and hepatocellular carcinoma [60]. Patients on sorafenib have shown a prolonged progression-free survival compared to placebo. Similarly, sutinib (Sutent, Pfizer) is approved for the treatment of imatinib refractory or intolerant gastrointestinal stromal tumors (GISTs) and advanced renal cell carcinoma [61]. So far, patients on VEGFR-2 TKIs report very few side effects. Because both sorafenib and sutinib are TKIs, their activity is selective, but not specific, for the VEGF receptor. Some of their activity could therefore be attributed to off-target effects on other receptors such as the PDGF or FGF receptors [62], or for example, by inhibiting the growth of tumor cells, which also express VEGF receptors [63, 64].

2.3.2 Evidence for Refractoriness to Anti-VEGF Strategies

Despite the successes of benchmark therapies like Avastin, not all patients benefit from treatment. Disappointingly, overall, the clinical responses with Avastin have not been enduring. In some patients, tumors shrink only to grow again and the clinical benefit is usually measured in months, not years. Moreover, a number of phase III clinical trials have shown no benefit [52]. What are the reasons for this refractoriness? One possibility is that differences in sensitivity to antiangiogenic therapies are linked to different stages of tumor progression [65]. Another possibility is that other, compensatory growth factors such as FGF-2 can support TEC survival in tumors. For example, it was reported recently that treatment of tumor-bearing mice with an anti-VEGFR-2 antibody resulted in an initial decrease in vascularity, followed by vascular rebound and in increased expression of FGFs [66]. Increased levels of circulating FGF-2 have also been noted in human patients treated with AZD2171, a pan-VEGFR2 TKI [67]. PIGF was also reported to be increased in the circulation following treatment with sunitinib [68]. It may therefore be predicted that single-agent antiangiogenesis therapies will never succeed and combinations of different therapies, or tumor-stage- or tumor-type-specific antiangiogenic therapies will ultimately need to be developed. Interestingly, neutralizing antibodies to the angiogenic factors PIGF [69], VEGFR-3 [30], and neuropilin [70] act to suppress tumor growth in synergy with Avastin. Such combinatorial treatment can however be predicted to be very costly and not manageable by the health sector in some countries. See Figure 2.2 for a summary of mechanisms explaining refractoriness to anti-VEGF therapy.

2.4 Other Impediments to Antiangiogenic Therapies

In addition to a compensatory upregulation of alternative, proangiogenic pathways in tumor cells and tumor stromal cells in response to single-agent antiangiogenic therapies, are there other factors that might mitigate an enduring antiangiogenic

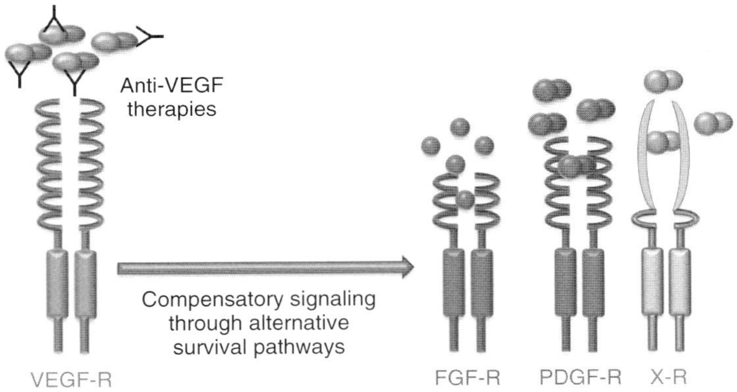

Figure 2.2 Refractoriness to antiangiogenic therapy. Treatment with, for example, neutralizing VEGF antibodies may not be successful due to escape from the dependence on VEGF signaling. The escape may rest on acquired autonomous activity downstream of VEGFR-2. Alternatively, growth factor receptors binding PDGF or FGF or other unknown receptors (XR) that normally are absent from endothelial cells, may be expressed on tumor endothelium.

response? Can intrinsic, molecular or phenotypic differences in TECs compared to normal ECs impinge on the effectiveness of antiangiogenic strategies?

2.4.1
Hallmarks of Dysregulated Tumor Angiogenesis

Vessel formation in tumors may involve several known mechanisms: (i) tumors may grow along preexisting vessels in a process denoted cooption [71], (ii) tumor-derived mitogens and chemotactic factors may stimulate the proliferation and migration of preexisiting vasculature, (iii) new vessels may appear by intussusception [72], and (iv) circulating progenitors may incorporate into the sprouting vessels although this is still being debated and appears to differ temporally and spatially between different tumor types. An important distinguishing feature between physiological and pathological angiogenesis is the continuous stimulation of ECs leading to constantly forming and dissolving vessels, which lack the hallmark of mature, stable vasculature. Thus, during conditions where tissues grow in an uncontrolled manner, such as in tumors, blood vessels also form in a dysregulated manner. Morphological abnormalities in the tumor blood vessels were noticed long ago using three-dimensional plaster casting and scanning electron microscopy [73]. Modern advances in imaging have allowed for viewing tumor blood vessel abnormalities in stunning detail [74, 75]. Abnormalities in tumor vessels can be found in the ECs themselves, the pericytes, and the basement membrane. For example, at the luminal side of capillaries, TECs may overlap and sprout across the lumen. There may be gaps in the vessel wall and leakiness. The basement membrane may have redundant layers, be discontinuous, or altogether absent [76]. There is no recognizable pattern or hierarchy in the tumor vasculature and pericytes may be loosely attached or in lower abundance compared to normal vessels. It is logical to attribute these differences in morphology between normal ECs and TECs to marked changes in production of growth factors or expression levels of the corresponding receptors, on ECs and pericytes.

2.4.2 Tumor Vessel Normalization

It has been suggested that vessel leakiness and chaotic blood flow may impair the bioavailability of chemotherapeutic drugs meant to target the tumor blood vessels or the tumor cells themselves. Some of these morphological abnormalities may be reversible or "normalized," for example, by treatment with Avastin [77, 78]. In one study, abnormal tube forming abilities of isolated TECs were reversible using the ROCK inhibitor Y27632 [79]. Normalization of tumor blood vessels can thus improve the effectiveness of chemotherapy and provides direct evidence that the morphological abnormalities in tumor blood vessels may in itself be an impediment to antiangiogenic therapies [80]. The state of normalization is probably transient, and dependent on the dose and duration of the treatment. Of note recent data show that excessive suppression of tumor vasculature by high doses of Avastin leads to EC death, which in turn may promote spread of distant metastases [81].

2.4.3 Tumor Endothelial Cell Abnormalities

The morphological differences in TECs portend an underlying change at the molecular level [82]. Indeed, in a seminal study using serial analysis of gene expression (SAGE), St. Croix demonstrated that 46 transcripts were enriched in TECs isolated from human colorectal carcinomas [83]. Several TEC-specific markers were identified, which were called *tumor endothelial markers* (*TEMs*). In a later study, it was found that some EC-specific genes were expressed only in angiogenic blood vessels of tumors and not in blood vessels during physiological angiogenesis [84]. Similar high-throughput approaches have uncovered unique gene transcriptional signatures in tumors of the breast [85] and brain [86].

Other studies have demonstrated both intra- and inter-EC heterogeneity in tumors *in vivo*. For example, the expression of TIE-2 [87], CD105, and CD31 [88] may be expressed in some TECs lining a tumor blood vessel but not in adjacent tumor ECs of the same vessel. Discontinuities in the expression of EC-specific markers and gaps between adjacent ECs result in tumor blood vessels having a mosaic appearance. Tumor blood vessels may differ temporally depending on tumor stage [89] (e.g., dysplasia versus well-differentiated carcinoma) and spatially within the tumor microenvironment (tumor border versus tumor core) [90, 91]. There may be intratumoral differences in the endothelium across different tumor types due to coevolution of tumor cells and the tumor stroma [92] or due to tumor-specific production of angiogenic growth factors. For these reasons, blood vessels in prostate cancers may be different when compared to blood vessels in breast cancers. Stage-specific and tumor-specific differences in TECs may thus contribute to the refractoriness of some tumors to antiangiogenic drugs [65].

Few studies have reported the successful isolation and long-term culture of TECs. The ability to isolate TECs and culture them for prolonged periods enables an investigation of their functional properties. In one study, TECs isolated from xenografts of melanoma and liposarcoma had multiple centrosomes and were aneuploid [93]. TEC aneuploidy has been confirmed in B-cell lymphomas in humans [94]. In another study, isolated TECs were found to overexpress EGF receptors, which made them responsive to EGF, and sensitive to epidermal growth factor receptor (EGFR) TKIs such as Iressa [95]. Additional studies have shown that TECs from renal cell carcinoma upregulate survival pathways such as Akt and are resistant to vincristine and doxorubicin [96]. As good evidence for stage-dependent differences in TECs, isolated TECs from high-grade gliomas proliferated more rapidly than TECs from low-grade gliomas [97]. Recently, TECs from prostate tumors were shown to be multipotent and have properties of adult stem cells raising new questions regarding the origins of vascular cells in tumors [98, 99].

2.5 Future Perspectives

Following Dr. Judah Folkman's suggestion that tumors could be eradicated by targeting their blood supply, interest in vascular biology has grown substantially. This "new avenue" for killing tumor cells sparked a revolution in the desire to dissect and understand the processes that regulates how blood vessels grow and die. Out of this revolution, anti-VEGF therapies evolved, which are now being used routinely in clinics. While the success of anti-VEGF strategies remains to be determined, this is definitely only the beginning. As new drugs are developed to target the receptor/ligand pathways described in detail above and as novel targets are identified, combinatorial therapies may achieve a greater efficacy than ever imagined. However, there are still obstacles in place. For example, it is clear that the ECs lining tumor blood vessels are distinct both morphologically and at the molecular level compared to their normal counterparts. Even in normal tissues, ECs are heterogeneous and can vary from organ to organ [100, 101]. Molecular and morphological heterogeneity in the vasculature may have evolved to enable the endothelium to adapt to both mechanical and microenvironmental perturbations at the blood–EC interface, but could also create a major barrier toward the goal of designing successful antiangiogenic therapies. One of the challenges will be to better characterize the biology of TECs from different tumors and during different stages of tumor progression. Isolating these cells and maintaining them for prolonged periods in culture has proven difficult. In the long-term, rigorous characterization of TECs (and other stromal cells) before and after drug treatments and from different tumors may be necessary in order to effectively design and implement antiangiogenic strategies.

References

1. Pittman, R.N. (2005) Oxygen transport and exchange in the microcirculation. *Microcirculation*, **12**, 59–70.
2. Sottile, J. (2004) Regulation of angiogenesis by extracellular matrix. *Biochim. Biophys. Acta*, **1654**, 13–22.
3. von Tell, D., Armulik, A., and Betsholtz, C. (2006) Pericytes and vascular stability. *Exp. Cell Res.*, **312**, 623–629.
4. Nagy, J.A., Benjamin, L., Zeng, H., Dvorak, A.M., and Dvorak, H.F. (2008) Vascular permeability, vascular hyperpermeability and angiogenesis. *Angiogenesis*, **11**, 109–119.
5. Aird, W.C. (2008) Endothelium in health and disease. *Pharmacol. Rep.*, **60**, 139–143.
6. Jin, S.W. and Patterson, C. (2009) The opening act: vasculogenesis and the origins of circulation. *Arterioscler. Thromb. Vasc. Biol.*, **29**, 623–629.
7. Lacaud, G., Keller, G., and Kouskoff, V. (2004) Tracking mesoderm formation and specification to the hemangioblast in vitro. *Trends Cardiovasc. Med.*, **14**, 314–317.
8. Drake, C.J. and Fleming, P.A. (2000) Vasculogenesis in the day 6.5 to 9.5 mouse embryo. *Blood*, **95**, 1671–1679.
9. Lancrin, C., Sroczynska, P., Stephenson, C., Allen, T., Kouskoff, V., and Lacaud, G. (2009) The haemangioblast generates haematopoietic cells through a haemogenic endothelium stage. *Nature*, **457**, 892–895.
10. Kopp, H.G., Ramos, C.A., and Rafii, S. (2006) Contribution of endothelial progenitors and proangiogenic hematopoietic cells to vascularization of tumor and ischemic tissue. *Curr. Opin. Hematol.*, **13**, 175–181.
11. Purhonen, S., Palm, J., Rossi, D., Kaskenpaa, N., Rajantie, I., Yla-Herttuala, S., Alitalo, K., Weissman, I.L., and Salven, P. (2008) Bone marrow-derived circulating endothelial precursors do not contribute to vascular endothelium and are not needed for tumor growth. *Proc. Natl. Acad. Sci. U.S.A.*, **105**, 6620–6625.
12. Grunewald, M., Avraham, I., Dor, Y., Bachar-Lustig, E., Itin, A., Jung, S., Chimenti, S., Landsman, L., Abramovitch, R., and Keshet, E. (2006) VEGF-induced adult neovascularization: recruitment, retention, and role of accessory cells. *Cell*, **124**, 175–189.
13. Burri, P.H., Hlushchuk, R., and Djonov, V. (2004) Intussusceptive angiogenesis: its emergence, its characteristics, and its significance. *Dev. Dyn.*, **231**, 474–488.
14. Gerhardt, H. and Betsholtz, C. (2005) How do endothelial cells orientate? *EXS*, 3–15.
15. Roca, C. and Adams, R.H. (2007) Regulation of vascular morphogenesis by Notch signaling. *Genes Dev.*, **21**, 2511–2524.
16. Horowitz, A. and Simons, M. (2008) Branching morphogenesis. *Circ. Res.*, **103**, 784–795.
17. Ferrara, N. (2005) The role of VEGF in the regulation of physiological and pathological angiogenesis. *EXS*, 209–231.
18. Olsson, A.K., Dimberg, A., Kreuger, J., and Claesson-Welsh, L. (2006) VEGF receptor signalling – in control of vascular function. *Nat. Rev. Mol. Cell Biol.*, **7**, 359–371.
19. Brekken, R.A. and Thorpe, P.E. (2001) VEGF-VEGF receptor complexes as markers of tumor vascular endothelium. *J. Control. Release*, **74**, 173–181.
20. Shweiki, D., Itin, A., Soffer, D., and Keshet, E. (1992) Vascular endothelial growth factor induced by hypoxia may mediate hypoxia-initiated angiogenesis. *Nature*, **359**, 843–845.
21. Li, X., Tjwa, M., Van Hove, I., Enholm, B., Neven, E., Paavonen, K., Jeltsch, M., Juan, T.D., Sievers, R.E., Chorianopoulos, E., Wada, H., Vanwildemeersch, M., Noel, A., Foidart, J.M., Springer, M.L., von Degenfeld, G., Dewerchin, M., Blau, H.M., Alitalo, K., Eriksson, U., Carmeliet, P., and Moons, L. (2008) Reevaluation of the role of VEGF-B suggests a restricted role in the revascularization of the ischemic

myocardium. *Arterioscler. Thromb. Vasc. Biol.*, **28**, 1614–1620.

22. Carmeliet, P., Moons, L., Luttun, A., Vincenti, V., Compernolle, V., De Mol, M., Wu, Y., Bono, F., Devy, L., Beck, H., Scholz, D., Acker, T., DiPalma, T., Dewerchin, M., Noel, A., Stalmans, I., Barra, A., Blacher, S., Vandendriessche, T., Ponten, A., Eriksson, U., Plate, K.H., Foidart, J.M., Schaper, W., Charnock-Jones, D.S., Hicklin, D.J., Herbert, J.M., Collen, D., and Persico, M.G. (2001) Synergism between vascular endothelial growth factor and placental growth factor contributes to angiogenesis and plasma extravasation in pathological conditions. *Nat. Med.*, **7**, 575–583.
23. Autiero, M., Waltenberger, J., Communi, D., Kranz, A., Moons, L., Lambrechts, D., Kroll, J., Plaisance, S., De Mol, M., Bono, F., Kliche, S., Fellbrich, G., Ballmer-Hofer, K., Maglione, D., Mayr-Beyrle, U., Dewerchin, M., Dombrowski, S., Stanimirovic, D., Van Hummelen, P., Dehio, C., Hicklin, D.J., Persico, G., Herbert, J.M., Shibuya, M., Collen, D., Conway, E.M., and Carmeliet, P. (2003) Role of PlGF in the intra- and intermolecular cross talk between the VEGF receptors Flt1 and Flk1. *Nat. Med.*, **9**, 936–943.
24. Fong, G.H., Rossant, J., Gertsenstein, M., and Breitman, M.L. (1995) Role of the Flt-1 receptor tyrosine kinase in regulating the assembly of vascular endothelium. *Nature*, **376**, 66–70.
25. Shibuya, M. (2001) Structure and dual function of vascular endothelial growth factor receptor-1 (Flt-1). *Int. J. Biochem. Cell Biol.*, **33**, 409–420.
26. Geretti, E., Shimizu, A., and Klagsbrun, M. (2008) Neuropilin structure governs VEGF and semaphorin binding and regulates angiogenesis. *Angiogenesis*, **11**, 31–39.
27. Karkkainen, M.J., Haiko, P., Sainio, K., Partanen, J., Taipale, J., Petrova, T.V., Jeltsch, M., Jackson, D.G., Talikka, M., Rauvala, H., Betsholtz, C., and Alitalo, K. (2004) Vascular endothelial growth factor C is required for sprouting of the first lymphatic vessels from embryonic veins. *Nat. Immunol.*, **5**, 74–80.
28. Baldwin, M.E., Halford, M.M., Roufail, S., Williams, R.A., Hibbs, M.L., Grail, D., Kubo, H., Stacker, S.A., and Achen, M.G. (2005) Vascular endothelial growth factor D is dispensable for development of the lymphatic system. *Mol. Cell Biol.*, **25**, 2441–2449.
29. Dumont, D.J., Jussila, L., Taipale, J., Lymboussaki, A., Mustonen, T., Pajusola, K., Breitman, M., and Alitalo, K. (1998) Cardiovascular failure in mouse embryos deficient in VEGF receptor-3. *Science*, **282**, 946–949.
30. Tammela, T., Zarkada, G., Wallgard, E., Murtomaki, A., Suchting, S., Wirzenius, M., Waltari, M., Hellstrom, M., Schomber, T., Peltonen, R., Freitas, C., Duarte, A., Isoniemi, H., Laakkonen, P., Christofori, G., Yla-Herttuala, S., Shibuya, M., Pytowski, B., Eichmann, A., Betsholtz, C., and Alitalo, K. (2008) Blocking VEGFR-3 suppresses angiogenic sprouting and vascular network formation. *Nature*, **454**, 656–660.
31. Laakkonen, P., Waltari, M., Holopainen, T., Takahashi, T., Pytowski, B., Steiner, P., Hicklin, D., Persaud, K., Tonra, J.R., Witte, L., and Alitalo, K. (2007) Vascular endothelial growth factor receptor 3 is involved in tumor angiogenesis and growth. *Cancer Res.*, **67**, 593–599.
32. Beenken, A. and Mohammadi, M. (2009) The FGF family : biology, pathophysiology and therapy. *Nat. Rev. Drug Discov.*, **8** (3), 235–253.
33. Nickel, W. (2005) Unconventional secretory routes: direct protein export across the plasma membrane of mammalian cells. *Traffic*, **6**, 607–614.
34. Olsnes, S., Klingenberg, O., and Wiedlocha, A. (2003) Transport of exogenous growth factors and cytokines to the cytosol and to the nucleus. *Physiol. Rev.*, **83**, 163–182.
35. Coombe, D.R. (2008) Biological implications of glycosaminoglycan interactions with haemopoietic cytokines. *Immunol. Cell Biol.*, **86**, 598–607.

36. Folkman, J., Klagsbrun, M., Sasse, J., Wadzinski, M., Ingber, D., and Vlodavsky, I. (1988) A heparin-binding angiogenic protein–basic fibroblast growth factor–is stored within basement membrane. *Am. J. Pathol.*, **130**, 393–400.
37. Chen, L. and Deng, C.X. (2005) Roles of FGF signaling in skeletal development and human genetic diseases. *Front. Biosci.*, **10**, 1961–1976.
38. Magnusson, P.U., Dimberg, A., Mellberg, S., Lukinius, A., and Claesson-Welsh, L. (2007) FGFR-1 regulates angiogenesis through cytokines interleukin-4 and pleiothropin. *Blood*, **110**, 4214–4222.
39. Rusnati, M. and Presta, M. (2007) Fibroblast growth factors/fibroblast growth factor receptors as targets for the development of anti-angiogenesis strategies. *Curr. Pharm. Des.*, **13**, 2025–2044.
40. Fredriksson, L., Li, H., and Eriksson, U. (2004) The PDGF family: four gene products form five dimeric isoforms. *Cytokine Growth Factor Rev.*, **15**, 197–204.
41. Soriano, P. (1994) Abnormal kidney development and hematological disorders in PDGF beta-receptor mutant mice. *Genes Dev.*, **8**, 1888–1896.
42. Lindahl, P., Johansson, B.R., Leveen, P., and Betsholtz, C. (1997) Pericyte loss and microaneurysm formation in PDGF-B-deficient mice. *Science*, **277**, 242–245.
43. Leveen, P., Pekny, M., Gebre-Medhin, S., Swolin, B., Larsson, E., and Betsholtz, C. (1994) Mice deficient for PDGF B show renal, cardiovascular, and hematological abnormalities. *Genes Dev.*, **8**, 1875–1887.
44. Greenberg, J.I., Shields, D.J., Barillas, S.G., Acevedo, L.M., Murphy, E., Huang, J., Scheppke, L., Stockmann, C., Johnson, R.S., Angle, N., and Cheresh, D.A. (2008) A role for VEGF as a negative regulator of pericyte function and vessel maturation. *Nature*, **456**, 809–813.
45. De Palma, M., Venneri, M.A., Galli, R., Sergi Sergi, L., Politi, L.S., Sampaolesi, M., and Naldini, L. (2005) Tie2 identifies a hematopoietic lineage of proangiogenic monocytes required for tumor vessel formation and a mesenchymal population of pericyte progenitors. *Cancer Cell*, **8**, 211–226.
46. Sato, T.N., Tozawa, Y., Deutsch, U., Wolburg-Buchholz, K., Fujiwara, Y., Gendron-Maguire, M., Gridley, T., Wolburg, H., Risau, W., and Qin, Y. (1995) Distinct roles of the receptor tyrosine kinases Tie-1 and Tie-2 in blood vessel formation. *Nature*, **376**, 70–74.
47. Dumont, D.J., Gradwohl, G., Fong, G.H., Puri, M.C., Gertsenstein, M., Auerbach, A., and Breitman, M.L. (1994) Dominant-negative and targeted null mutations in the endothelial receptor tyrosine kinase, tek, reveal a critical role in vasculogenesis of the embryo. *Genes Dev.*, **8**, 1897–1909.
48. Suri, C., Jones, P.F., Patan, S., Bartunkova, S., Maisonpierre, P.C., Davis, S., Sato, T.N., and Yancopoulos, G.D. (1996) Requisite role of angiopoietin-1, a ligand for the TIE2 receptor, during embryonic angiogenesis. *Cell*, **87**, 1171–1180.
49. Gale, N.W., Thurston, G., Hackett, S.F., Renard, R., Wang, Q., McClain, J., Martin, C., Witte, C., Witte, M.H., Jackson, D., Suri, C., Campochiaro, P.A., Wiegand, S.J., and Yancopoulos, G.D. (2002) Angiopoietin-2 is required for postnatal angiogenesis and lymphatic patterning, and only the latter role is rescued by Angiopoietin-1. *Dev. Cell*, **3**, 411–423.
50. Augustin, H.G., Koh, G.Y., Thurston, G., and Alitalo, K. (2009) Control of vascular morphogenesis and homeostasis through the angiopoietin-Tie system. *Nat. Rev. Mol. Cell Biol.*, **10**, 165–177.
51. Bergers, G. and Hanahan, D. (2008) Modes of resistance to anti-angiogenic therapy. *Nat. Rev. Cancer*, **8**, 592–603.
52. Ellis, L.M. and Hicklin, D.J. (2008) VEGF-targeted therapy: mechanisms of anti-tumour activity. *Nat. Rev. Cancer*, **8**, 579–591.
53. Willett, C.G., Boucher, Y., di Tomaso, E., Duda, D.G., Munn, L.L., Tong, R.T., Chung, D.C., Sahani, D.V.,

Kalva, S.P., Kozin, S.V., Mino, M., Cohen, K.S., Scadden, D.T., Hartford, A.C., Fischman, A.J., Clark, J.W., Ryan, D.P., Zhu, A.X., Blaszkowsky, L.S., Chen, H.X., Shellito, P.C., Lauwers, G.Y., and Jain, R.K. (2004) Direct evidence that the VEGF-specific antibody bevacizumab has antivascular effects in human rectal cancer. *Nat. Med.*, **10**, 145–147.

54. Murdoch, C., Muthana, M., Coffelt, S.B., and Lewis, C.E. (2008) The role of myeloid cells in the promotion of tumour angiogenesis. *Nat. Rev. Cancer*, 8, 618–631.

55. Du, R., Lu, K.V., Petritsch, C., Liu, P., Ganss, R., Passegue, E., Song, H., Vandenberg, S., Johnson, R.S., Werb, Z., and Bergers, G. (2008) HIF1alpha induces the recruitment of bone marrow-derived vascular modulatory cells to regulate tumor angiogenesis and invasion. *Cancer Cell*, **13**, 206–220.

56. Hurwitz, H., Fehrenbacher, L., Novotny, W., Cartwright, T., Hainsworth, J., Heim, W., Berlin, J., Baron, A., Griffing, S., Holmgren, E., Ferrara, N., Fyfe, G., Rogers, B., Ross, R., and Kabbinavar, F. (2004) Bevacizumab plus irinotecan, fluorouracil, and leucovorin for metastatic colorectal cancer. *N. Engl. J. Med.*, **350**, 2335–2342.

57. Miller, K., Wang, M., Gralow, J., Dickler, M., Cobleigh, M., Perez, E.A., Shenkier, T., Cella, D., and Davidson, N.E. (2007) Paclitaxel plus bevacizumab versus paclitaxel alone for metastatic breast cancer. *N. Engl. J. Med.*, **357**, 2666–2676.

58. Yang, J.C., Haworth, L., Sherry, R.M., Hwu, P., Schwartzentruber, D.J., Topalian, S.L., Steinberg, S.M., Chen, H.X., and Rosenberg, S.A. (2003) A randomized trial of bevacizumab, an anti-vascular endothelial growth factor antibody, for metastatic renal cancer. *N. Engl. J. Med.*, **349**, 427–434.

59. Escudier, B., Eisen, T., Stadler, W.M., Szczylik, C., Oudard, S., Siebels, M., Negrier, S., Chevreau, C., Solska, E., Desai, A.A., Rolland, F., Demkow, T., Hutson, T.E., Gore, M., Freeman, S., Schwartz, B., Shan, M., Simantov, R., and Bukowski, R.M. (2007) Sorafenib in advanced clear-cell renal-cell carcinoma. *N. Engl. J. Med.*, **356**, 125–134.

60. Llovet, J.M., Ricci, S., Mazzaferro, V., Hilgard, P., Gane, E., Blanc, J.F., de Oliveira, A.C., Santoro, A., Raoul, J.L., Forner, A., Schwartz, M., Porta, C., Zeuzem, S., Bolondi, L., Greten, T.F., Galle, P.R., Seitz, J.F., Borbath, I., Haussinger, D., Giannaris, T., Shan, M., Moscovici, M., Voliotis, D., and Bruix, J. (2008) Sorafenib in advanced hepatocellular carcinoma. *N. Engl. J. Med.*, **359**, 378–390.

61. Goodman, V.L., Rock, E.P., Dagher, R., Ramchandani, R.P., Abraham, S., Gobburu, J.V., Booth, B.P., Verbois, S.L., Morse, D.E., Liang, C.Y., Chidambaram, N., Jiang, J.X., Tang, S., Mahjoob, K., Justice, R., and Pazdur, R. (2007) Approval summary: sunitinib for the treatment of imatinib refractory or intolerant gastrointestinal stromal tumors and advanced renal cell carcinoma. *Clin. Cancer Res.*, **13**, 1367–1373.

62. Karaman, M.W., Herrgard, S., Treiber, D.K., Gallant, P., Atteridge, C.E., Campbell, B.T., Chan, K.W., Ciceri, P., Davis, M.I., Edeen, P.T., Faraoni, R., Floyd, M., Hunt, J.P., Lockhart, D.J., Milanov, Z.V., Morrison, M.J., Pallares, G., Patel, H.K., Pritchard, S., Wodicka, L.M., and Zarrinkar, P.P. (2008) A quantitative analysis of kinase inhibitor selectivity. *Nat. Biotechnol.*, **26**, 127–132.

63. Dallas, N.A., Fan, F., Gray, M.J., Van Buren, G. II, Lim, S.J., Xia, L., and Ellis, L.M. (2007) Functional significance of vascular endothelial growth factor receptors on gastrointestinal cancer cells. *Cancer Metastasis Rev.*, **26**, 433–441.

64. Spannuth, W.A., Nick, A.M., Jennings, N.B., Armaiz-Pena, G.N., Mangala, L.S., Danes, C.G., Lin, Y.G., Merritt, W.M., Thaker, P.H., Kamat, A.A., Han, L.Y., Tonra, J.R., Coleman, R.L., Ellis, L.M., and Sood, A.K. (2009) Functional significance of VEGFR-2 on ovarian cancer cells. *Int. J. Cancer*, **124**, 1045–1053.

65. Bergers, G., Javaherian, K., Lo, K.M., Folkman, J., and Hanahan, D. (1999) Effects of angiogenesis inhibitors on multistage carcinogenesis in mice. *Science*, **284**, 808–812.
66. Casanovas, O., Hicklin, D.J., Bergers, G., and Hanahan, D. (2005) Drug resistance by evasion of antiangiogenic targeting of VEGF signaling in late-stage pancreatic islet tumors. *Cancer Cell*, **8**, 299–309.
67. Batchelor, T.T., Sorensen, A.G., di Tomaso, E., Zhang, W.T., Duda, D.G., Cohen, K.S., Kozak, K.R., Cahill, D.P., Chen, P.J., Zhu, M., Ancukiewicz, M., Mrugala, M.M., Plotkin, S., Drappatz, J., Louis, D.N., Ivy, P., Scadden, D.T., Benner, T., Loeffler, J.S., Wen, P.Y., and Jain, R.K. (2007) AZD2171, a pan-VEGF receptor tyrosine kinase inhibitor, normalizes tumor vasculature and alleviates edema in glioblastoma patients. *Cancer Cell*, **11**, 83–95.
68. Rini, B.I., Michaelson, M.D., Rosenberg, J.E., Bukowski, R.M., Sosman, J.A., Stadler, W.M., Hutson, T.E., Margolin, K., Harmon, C.S., DePrimo, S.E., Kim, S.T., Chen, I., and George, D.J. (2008) Antitumor activity and biomarker analysis of sunitinib in patients with bevacizumab-refractory metastatic renal cell carcinoma. *J. Clin. Oncol.*, **26**, 3743–3748.
69. Fischer, C., Jonckx, B., Mazzone, M., Zacchigna, S., Loges, S., Pattarini, L., Chorianopoulos, E., Liesenborghs, L., Koch, M., De Mol, M., Autiero, M., Wyns, S., Plaisance, S., Moons, L., van Rooijen, N., Giacca, M., Stassen, J.M., Dewerchin, M., Collen, D., and Carmeliet, P. (2007) Anti-PlGF inhibits growth of VEGF(R)-inhibitor-resistant tumors without affecting healthy vessels. *Cell*, **131**, 463–475.
70. Pan, Q., Chanthery, Y., Liang, W.C., Stawicki, S., Mak, J., Rathore, N., Tong, R.K., Kowalski, J., Yee, S.F., Pacheco, G., Ross, S., Cheng, Z., Le Couter, J., Plowman, G., Peale, F., Koch, A.W., Wu, Y., Bagri, A., Tessier-Lavigne, M., and Watts, R.J. (2007) Blocking neuropilin-1 function has an additive effect with anti-VEGF to inhibit tumor growth. *Cancer Cell*, **11**, 53–67.
71. Holash, J., Maisonpierre, P.C., Compton, D., Boland, P., Alexander, C.R., Zagzag, D., Yancopoulos, G.D., and Wiegand, S.J. (1999) Vessel cooption, regression, and growth in tumors mediated by angiopoietins and VEGF. *Science*, **284**, 1994–1998.
72. Hlushchuk, R., Riesterer, O., Baum, O., Wood, J., Gruber, G., Pruschy, M., and Djonov, V. (2008) Tumor recovery by angiogenic switch from sprouting to intussusceptive angiogenesis after treatment with PTK787/ZK222584 or ionizing radiation. *Am. J. Pathol.*, **173**, 1173–1185.
73. Konerding, M.A., Malkusch, W., Klapthor, B., van Ackern, C., Fait, E., Hill, S.A., Parkins, C., Chaplin, D.J., Presta, M., and Denekamp, J. (1999) Evidence for characteristic vascular patterns in solid tumours: quantitative studies using corrosion casts. *Br. J. Cancer*, **80**, 724–732.
74. Baluk, P., Hashizume, H., and McDonald, D.M. (2005) Cellular abnormalities of blood vessels as targets in cancer. *Curr. Opin. Genet. Dev.*, **15**, 102–111.
75. McDonald, D.M. and Baluk, P. (2005) Imaging of angiogenesis in inflamed airways and tumors: newly formed blood vessels are not alike and may be wildly abnormal: Parker B. Francis lecture. *Chest*, **128**, 602S–608S.
76. Baluk, P., Morikawa, S., Haskell, A., Mancuso, M., and McDonald, D.M. (2003) Abnormalities of basement membrane on blood vessels and endothelial sprouts in tumors. *Am. J. Pathol.*, **163**, 1801–1815.
77. Fischer, C., Mazzone, M., Jonckx, B., and Carmeliet, P. (2008) FLT1 and its ligands VEGFB and PlGF: drug targets for anti-angiogenic therapy? *Nat. Rev. Cancer*, **8**, 942–956.
78. Jain, R.K. (2005) Normalization of tumor vasculature: an emerging concept in antiangiogenic therapy. *Science*, **307**, 58–62.
79. Ghosh, K., Thodeti, C.K., Dudley, A.C., Mammoto, A., Klagsbrun, M., and Ingber, D.E. (2008) Tumor-derived endothelial cells exhibit aberrant Rho-mediated mechanosensing and

abnormal angiogenesis in vitro. *Proc. Natl. Acad. Sci. U.S.A.*, **105**, 11305–11310.

80. Willett, C.G., Duda, D.G., Czito, B.G., Bendell, J.C., Clark, J.W., and Jain, R.K. (2007) Targeted therapy in rectal cancer. *Oncology (Williston Park)*, **21**, 1055–1065; discussion 1065, 1070, 1075.
81. Paez-Ribes, M., Allen, E., Hudock, J., Takeda, T., Okuyama, H., Vinals, F., Inoue, M., Bergers, G., Hanahan, D., and Casanovas, O. (2009) Antiangiogenic therapy elicits malignant progression of tumors to increased local invasion and distant metastasis. *Cancer Cell*, **15**, 220–231.
82. Aird, W.C. (2009) Molecular heterogeneity of tumor endothelium. *Cell Tissue Res.*, **335**, 271–281.
83. St Croix, B., Rago, C., Velculescu, V., Traverso, G., Romans, K.E., Montgomery, E., Lal, A., Riggins, G.J., Lengauer, C., Vogelstein, B., and Kinzler, K.W. (2000) Genes expressed in human tumor endothelium. *Science*, **289**, 1197–1202.
84. Seaman, S., Stevens, J., Yang, M.Y., Logsdon, D., Graff-Cherry, C., and St Croix, B. (2007) Genes that distinguish physiological and pathological angiogenesis. *Cancer Cell*, **11**, 539–554.
85. Parker, B.S., Argani, P., Cook, B.P., Liangfeng, H., Chartrand, S.D., Zhang, M., Saha, S., Bardelli, A., Jiang, Y., St Martin, T.B., Nacht, M., Teicher, B.A., Klinger, K.W., Sukumar, S., and Madden, S.L. (2004) Alterations in vascular gene expression in invasive breast carcinoma. *Cancer Res.*, **64**, 7857–7866.
86. Madden, S.L., Cook, B.P., Nacht, M., Weber, W.D., Callahan, M.R., Jiang, Y., Dufault, M.R., Zhang, X., Zhang, W., Walter-Yohrling, J., Rouleau, C., Akmaev, V.R., Wang, C.J., Cao, X., St Martin, T.B., Roberts, B.L., Teicher, B.A., Klinger, K.W., Stan, R.V., Lucey, B., Carson-Walter, E.B., Laterra, J., and Walter, K.A. (2004) Vascular gene expression in nonneoplastic and malignant brain. *Am. J. Pathol.*, **165**, 601–608.
87. Fathers, K.E., Stone, C.M., Minhas, K., Marriott, J.J., Greenwood, J.D., Dumont, D.J., and Coomber, B.L. (2005) Heterogeneity of Tie2 expression in tumor microcirculation: influence of cancer type, implantation site, and response to therapy. *Am. J. Pathol.*, **167**, 1753–1762.
88. di Tomaso, E., Capen, D., Haskell, A., Hart, J., Logie, J.J., Jain, R.K., McDonald, D.M., Jones, R., and Munn, L.L. (2005) Mosaic tumor vessels: cellular basis and ultrastructure of focal regions lacking endothelial cell markers. *Cancer Res.*, **65**, 5740–5749.
89. Hoffman, J.A., Giraudo, E., Singh, M., Zhang, L., Inoue, M., Porkka, K., Hanahan, D., and Ruoslahti, E. (2003) Progressive vascular changes in a transgenic mouse model of squamous cell carcinoma. *Cancer Cell*, **4**, 383–391.
90. Hellebrekers, D.M., Melotte, V., Vire, E., Langenkamp, E., Molema, G., Fuks, F., Herman, J.G., Van Criekinge, W., Griffioen, A.W., and van Engeland, M. (2007) Identification of epigenetically silenced genes in tumor endothelial cells. *Cancer Res.*, **67**, 4138–4148.
91. Langenkamp, E. and Molema, G. (2009) Microvascular endothelial cell heterogeneity: general concepts and pharmacological consequences for anti-angiogenic therapy of cancer. *Cell Tissue Res.*, **335**, 205–222.
92. Polyak, K., Haviv, I., and Campbell, I.G. (2009) Co-evolution of tumor cells and their microenvironment. *Trends Genet.*, **25**, 30–38.
93. Hida, K., Hida, Y., Amin, D.N., Flint, A.F., Panigrahy, D., Morton, C.C., and Klagsbrun, M. (2004) Tumor-associated endothelial cells with cytogenetic abnormalities. *Cancer Res.*, **64**, 8249–8255.
94. Streubel, B., Chott, A., Huber, D., Exner, M., Jager, U., Wagner, O., and Schwarzinger, I. (2004) Lymphoma-specific genetic aberrations in microvascular endothelial cells in B-cell lymphomas. *N. Engl. J. Med.*, **351**, 250–259.
95. Amin, D.N., Hida, K., Bielenberg, D.R., and Klagsbrun, M. (2006) Tumor endothelial cells express epidermal

growth factor receptor (EGFR) but not ErbB3 and are responsive to EGF and to EGFR kinase inhibitors. *Cancer Res.*, **66**, 2173–2180.
96. Bussolati, B., Deambrosis, I., Russo, S., Deregibus, M.C., and Camussi, G. (2003) Altered angiogenesis and survival in human tumor-derived endothelial cells. *FASEB J.*, **17**, 1159–1161.
97. Miebach, S., Grau, S., Hummel, V., Rieckmann, P., Tonn, J.C., and Goldbrunner, R.H. (2006) Isolation and culture of microvascular endothelial cells from gliomas of different WHO grades. *J. Neurooncol.*, **76**, 39–48.
98. Dudley, A.C., Khan, Z.A., Shih, S.C., Kang, S.Y., Zwaans, B.M., Bischoff, J., and Klagsbrun, M. (2008) Calcification of multipotent prostate tumor endothelium. *Cancer Cell*, **14**, 201–211.
99. Dudley, A.C. and Klagsbrun, M. (2009) Tumor endothelial cells have features of adult stem cells. *Cell Cycle*, **8**, 236–238.
100. Simonson, A.B. and Schnitzer, J.E. (2007) Vascular proteomic mapping in vivo. *J. Thromb. Haemost.*, **5** (Suppl 1), 183–187.
101. Aird, W.C. (2007) Phenotypic heterogeneity of the endothelium: I. Structure, function, and mechanisms. *Circ. Res.*, **100**, 158–173.

3
Proangiogenic Factors

Domenico Ribatti

3.1
Introduction

Angiogenesis is regulated, under both physiological and pathological conditions, by numerous "classic" proangiogenic factors. In recent years, evidence has been accumulated that, in addition to the "classic" factors, many other endogenous peptides or "nonclassic factors" play an important regulatory role in angiogenesis, especially under pathological conditions.

Angiogenesis is controlled by the balance between molecules that have positive and negative regulatory activity and this concept has led to the notion of the angiogenic switch, which depends on an increased production of one or more positive regulators of angiogenesis [1]. Endothelial cells (ECs) turnover in the healthy adult organism is low, the quiescence being maintained by the dominant influence of endogenous angiogenesis inhibitors over angiogenic stimuli. In pathological situations, angiogenesis may be triggered not only by the overproduction of proangiogenic factors, but also by the down-regulation of inhibitory factors.

Angiogenesis and the production of angiogenic factors are fundamental for tumor progression in the form of growth, invasion, and metastasis [2]. New vessels promote growth by conveying oxygen and nutrients, and removing catabolites. These requirements vary, however, among tumor types, and change over the course of tumor progression. ECs secrete growth factors for tumor cells and a variety of matrix-degrading proteinases that facilitate tumor invasion. An expanding endothelial surface also gives tumor cells more opportunities to enter the circulation and metastasize.

Solid tumor growth occurs by means of an avascular phase followed by a vascular phase [3]. Assuming that such growth is dependent on angiogenesis and that this depends on the release of angiogenic factors, the acquisition of an angiogenic ability can be seen as an expression of progression from neoplastic transformation to tumor growth and metastasis. Practically all solid tumors, including those of the colon, lung, breast, cervix, bladder, prostate, and pancreas, progress through these two phases. The role of angiogenesis in the growth and survival of leukemias and other hematological malignancies has become evident only since 1994, thanks to a

Tumor Angiogenesis – From Molecular Mechanisms to Targeted Therapy. Edited by Francis S. Markland, Stephen Swenson, and Radu Minea
Copyright © 2010 WILEY-VCH Verlag GmbH & Co. KGaA, Weinheim
ISBN: 978-3-527-32091-2

series of studies demonstrating that progression in several forms is clearly related to their degree of angiogenesis [4].

3.2 "Classic" Proangiogenic Factors

3.2.1 Vascular Endothelial Growth Factor

Vascular endothelial growth factor (VEGF) is an angiogenic factor *in vitro* and *in vivo*, and a mitogen for ECs with effects on vascular permeability. It plays a role in the control of blood vessel development and pathological angiogenesis; it is expressed when angiogenesis is high, and its levels are low when angiogenesis is absent [5]. VEGF and vascular endothelial growth factor receptors (VEGFRs) comprise the first EC-specific signal transduction pathway to get activated during vascular development and are critical molecules in the formation of the vascular system as evidenced in embryos, homozygous and heterozygous, for a targeted null mutation in their genes [6].

The VEGF family includes VEGF-A, VEGF-B, VEGF-C, VEGF-D, VEGF-E, and placental growth factor (PlGF) (Table 3.1). VEGF gene encodes for VEGF-A isoforms (VEGF-$A_{121-206}$) by alternative splicing that differently encodes exons 6 and 7, where the peptides responsible for the heparin-binding capacity are located [6]. The heparin-binding domains help VEGF-A to anchor to the extracellular matrix and are involved in binding to heparin sulfate and presentation to VEGFR.

Table 3.1 Main features of "classic" proangiogenic factors.

Factor	Receptors	Angiogenic activity				
		In vitro assays			*In vivo* assays	
		Proliferation	Migration	Capillary tube formation	Chorioallantoic membrane	Cornea
VEGF	VEGFR-1 VEGFR-2	S	S	S	S	S
FGF-2	FGFR-1 FGFR-2 FGFR-3 FGFR-4	S	S	S	S	S
TGF-β	TGFβR-1 TGFβR-2	I	N	S	S	S
PDGF	PDGFR	N	S	–	S	S
Angiopoietin-1	Tie 2	N	S	S	S	S

I, inhibition; N, no effect; S, stimulation; –, no findings available.

VEGF isoforms with higher heparin affinity are rapidly sequestrated by the heparan sulfate proteoglycans located at the EC's surface and in the extracellular matrix [6].

All the VEGF isoforms share common tyrosine kinase receptors [6]. VEGF-A binds with high-affinity VEGFR-1, VEGFR-2 and plays an essential role in vasculogenesis and angiogenesis. It has also been shown to induce lymphangiogenesis through VEGFR-2. VEGF-B overlaps with VEGF-A activities by activating VEGFR-1. VEGF-C and VEGF-D are both angiogenic via VEGFR-2 and VEGFR-3 and lymphangiogenic (primarily VEGF-D) via VEGFR-3 [6].

PlGF enhances angiogenesis only in pathological conditions by displacing VEGF from VEGFR-1 and thereby making more VEGF available for binding VEGFR-2, by transmitting angiogenic signals through its receptor VEGFR-1 via a novel cross talk; this causes activation of VEGFR-1 by PlGF which results in enhanced tyrosine phosphorylation of VEGFR-2, thereby amplifying VEGF-driven vessel growth [7].

The importance of VEGF in regulating tumor angiogenesis, growth, and progression has been verified by several reports showing that the inhibition of VEGF/VEGFR-2 by VEGF neutralizing antibodies, low-molecular weight VEGFR-2 inhibitors, or gene transfer of VEGFR-2 dominant–negative constructs leads to stunted tumor growth with reduced vascularization.

By using the well-established RIP1-Tag2 mouse model of multistep tumorigenesis of pancreatic islet carcinoma, it has been demonstrated that islet-specific VEGF deletion by means of the Crelox System diminishes angiogenic switching, tumor growth, and progression [8].

VEGF is expressed by numerous tumor cell lines both *in vitro* and *in vivo*, and receptors for VEGF occur only on peritumoral capillaries and not on distant endothelial cells [9]. VEGF increases vascular permeability and promotes the extravasation of plasma proteins and other circulating macromolecules from tumor vessels, leading to the formation of a fibrin gel. Fibrin provides a new provisional matrix which attracts and supports the growth of endothelial cells and fibroblasts, leading to angiogenesis and synthesis of matrix connective tissue [10].

Significantly higher serum VEGF levels in cancer patients as compared to that in normal subjects were first reported in 1994 [11] and were verified in a subsequent study of patients with various types of cancer [12].

The secretion of VEGF by tumor cells is stimulated by hypoxia [13]. As solid tumors grow in size, the cells within the expanding mass frequently become hypoxic because of the increasing distance from the nearest blood vessels, and the VEGF-mediated angiogenesis in response to hypoxia seems to be a general mechanism involved in the growth of many cancers. Hypoxia induces the expression of VEGF and its receptors via hypoxia inducible factor-1α (HIF-1α) [14, 15]. VEGF, which is secreted by the hypoxic tumor cells compartment, is distributed throughout the tumor by diffusion, thus generating a gradient.

The expression of VEGF by tumor cells is also potentiated by activation of oncogenes such as *v-ras*, *k-ras*, *v-raf*, *src*, *fos*, and *v-yes* [16], or inactivation of tumor suppressor genes such as *p53* [17] or by other cytokines such as transforming growth factor beta (TGF-β) [18] and nitric oxide [19].

3.2.2
Fibroblast Growth Factor-2

The fibroblast growth factor (FGF) family comprises a large group of about 20 polypeptides. To exert their biological activity, FGF interacts with high-affinity tyrosine kinase fibroblast growth factor receptors (FGFRs). Four members of the FGFR family (FGFR-1, FGFR-2, FGFR-3, and FGFR-4) are encoded by distinct genes and their structural variability is increased by alternative splicing [20]. FGF-2 is one of the best-characterized and investigated proangiogenic cytokines [20].

A large body of research has implicated the FGF/FGFR system as having a role in tumorigenesis [21]. Various tumor cell lines express FGF-2 [22, 23] and the appearance of an angiogenic phenotype correlates with the export of FGF-2 during the development of fibrosarcoma in transgenic mouse models [24]. Early studies showed that elevated levels of FGF-2 in urine samples collected from 950 patients having a wide variety of solid tumors as well as leukemia or lymphoma, were significantly correlated with the status and the extent of disease [25]. However, no association between increased serum level of FGF-2 and tumor type was observed in later studies on a large spectrum of metastatic carcinomas, even though two-thirds of the patients showing progressive disease had increasing serum levels of the angiogenic factor compared with less than 1/10th of the patients showing response to therapy [26]. Also, serum concentration of FGF-2 has prognostic relevance for advanced head and neck cancer [27] even though serum FGF-2 may not entirely derive from the neoplastic tissue in cancer patients [28]. Numerous studies have attempted to establish a correlation between intratumoral levels of FGF-2 mRNA or protein and intratumoral microvessel density in cancer patients. In fact, because of its pleiotropic activity that may affect both tumor vasculature and tumor parenchyma, FGF-2 may contribute to cancer progression not only by inducing neovascularization but also by acting directly on tumor cells. The capacity of tumor, stromal, and endothelial cells to express both FGF-2 and its receptors point to autocrine and paracrine functions of this growth factor in different cancers, including hematopoietic neoplasm [29].

3.2.3
Transforming Growth Factor β and Platelet-Derived Growth Factor β

Studies on targeted knockout mice have provided evidence of an essential role forTGF-β1 signaling in vascular development [30]. TGF-β1 embryos die with abnormal development of the yolk sac vasculature.

The differentiation of progenitor cells into pericytes and smooth muscle cells (SMCs) is promoted by TGF-β1 [31]. When mesenchymal cells are treated with TGF-β1, they express SMC markers, indicating differentiation toward an SMC lineage, and the differentiation can be blocked by antibodies against TGF-β1 [32].

The platelet-derived growth factor family (PDGF) comprises four family members (e.g., PFGF-A to PDGF-D) which bind, with distinct selectivity, the receptor tyrosine kinases (i.e., platelet-derived growth factor receptor) PDGFR-A and PDGFR-B expressed on EC and SMC [33]. PDGF-B is the best-characterized member in the PDGF family. Although first discovered as a secretory product of platelets during coagulation, PDGF-B is also expressed in many other cell types, such as ECs, macrophages, SMCs, fibroblasts, glial cells, neurons, and tumor cells [33]. PDGF-B is secreted by ECs, presumably in response to VEGF, and facilitates recruitment of mural cells. By releasing PDGF, ECs stimulate growth and differentiation of a PDGFR-B positive progenitor and recruit it around nascent vessels. Expression of PDGFR-B in mesenchymal progenitor cells, pericytes, and SMC is required for mural cell proliferation, migration, and incorporation in the vessel walls [33].

Animals carrying a mutated PDGF-B die perinatally of hemorrhage as a result of vascular structural defects generated by an abnormal pericyte organization process [34]. A detailed analysis of vessel development in both, PDGF-B and PDGFR-B mutant embryos, showed that SMC and pericytes initially form around vessels but, as vessels sprout and enlarge, PDGF signaling appears to be required for the comigration and proliferation of supporting cells [35].

PDGF-B expressed by tumor cells increased pericyte recruitment in several *in vivo* tumor models but failed to correct their detachment in PDGF-B deficient mice [36, 37]. Genetic abolition of the PDGF-B receptor expressed by embryonic pericytes decreased their recruitment in tumor [36]. In Lewis lung carcinoma tumors implanted in mice, inhibition by RNA interference of endothelial differentiation gene-1 (EDG-1) expression in ECs strongly reduced pericyte coverage [38].

3.2.4 Angiopoietins

The angiopoietin (Ang) family comprises at least four secreted proteins, Ang-1, Ang-2, Ang-3 and, Ang-4 all of which bind to the endothelial-specific receptor tyrosine kinase Tie-2, while Tie-1 is an orphan receptor tyrosine kinase. It is well documented that Angs play a critical role in endothelial sprouting, vessel wall remodeling, and mural cell recruitment [39].

Vasculogenesis proceeds normally in embryos lacking both Tie-1 and Tie-2, although they die early due to multiple cardiovascular defects [40, 41]. Mutation of Tie-2 does not affect initial formation of blood vessels, but embryos died in midgestation with major defects in vascular remodeling and stability [42, 43].

Ang-2 can bind to the Tie-2 receptor but does not activate it; rather it seems to act as an antagonist, counteracting the effects of Ang-1. Consistent with this, overexpression of Ang-2 in the embryos resulted in embryonic death from defects resembling those of knockouts of Ang-1 or Tie-2 [44]. The embryos showed massive vascular disruptions, which, as in the case when Ang-1 was absent, appeared to be caused by changes in both ECs and SMCs. The Ang-2 overexpressing mutant embryos had a more severe phenotype compared to the mutant embryos lacking Ang-1. Knockout embryos lacking Ang-1 display failure of EC adherence and interaction with perivascular cells and extracellular matrix [45]. Embryos lacking Ang-1 or Tie-2 develop a rather normal primary vasculature but however, this vasculature fails to undergo further remodeling. The most prominent defects are in the heart and also in the remodeling of many vascular beds into large and small vessels. Based on endothelium-restricted expression of Tie-2 and the dominant SMC expression of Angs, it appears that recruitment of SMCs or pericytes into proximity with ECs of newly formed vessels is required for Tie-2 activation.

Transgenic overexpression of Ang-1 in skin results in pronounced hypervascularization with the production of many compact, stable vessels that are resistant to leaks [46, 47]. Ang-2 seems to be the earliest marker of blood vessels that have been perturbed by invading tumor cells [48] and is overexpressed in tumor microvasculature of human glioblastoma and hepatocellular carcinoma [49, 50].

In a human glioma model developed in rat, Ang-1 led to enhanced pericyte recruitment and increased tumor growth, presumably by favoring angiogenesis [51]. On the contrary, in a colon cancer model, overexpression of Ang-1 led to smaller tumors with fewer blood vessels and a higher degree of pericyte coverage, resulting in a decreased vascular permeability and reduced hepatic metastasis [52, 53].

Many solid tumors may fail to form a well-differentiated and stable vasculature because their newly formed tumor vessels continue to overexpress Ang-2. Ang-2 induction in host vessels in the periphery of experimental C6 glioma precedes VEGF up-regulation of tumor cells, and causes regression of coopted vessels [48, 54]. Vajkoczy *et al.* [55] have demonstrated a parallel induction of Ang-2 and VEGFR-2 in quiescent host endothelial cells, suggesting that their simultaneous activity is critical for the induction of tumor angiogenesis during vascular initiation of microtumors. Consequently, the simultaneous expression of VEGFR-2 and Ang-2, rather than the expression of Ang-2 alone, may indicate the EC angiogenic phenotype and thus provide an early marker of activated host vasculature. The VEGF/Ang-2 balance may determine whether the new tumor vessels will continue to expand when the ratio of VEGF to Ang-2 is high or regress when it is low, during remodeling of the tumor microvasculature.

3.3 "Nonclassic" New Proangiogenic Factors

3.3.1 Erythropoietin

Erythropoietin (EPO) stimulates proliferation and migration of cultured, mature ECs, and lowers the apoptotic rate in ECs [56–58] and EC-expressed erythropoietin receptor (EPOR) mRNA [59]. The same effects were reported in cultured, neonatal ECs where EPO also induced capillary-like tube formation [60] and in embryonic ECs, where EPO promoted differentiation into the mature phenotype [61, 62]. These effects were mediated by the EPOR-mediated activation of JAK/STAT and PI3K/Akt pathways [63, 64].

The proangiogenic effect of EPO has been confirmed *in vivo* in the chick embryo chorioallantoic membrane (CAM) assay [64], and in experimental models of myocardium and hind limb ischemia [61, 65, 66]. The angiogenic potential of EPO has been reported to be similar to that of FGF-2 [64] and VEGF [57]. Findings suggested that EPO could stimulate angiogenesis *in vitro*, through an autocrine mechanism involving endothelin-1 (ET-1) [67, 68].

EPO enhances tumor growth by promoting angiogenesis and decreasing apoptosis [58, 69, 70].

3.3.2 Angiotensin-II

Angiotensin II (Ang II) regulates blood pressure, plasma volume, electrolyte balance, cardiovascular tissue growth, and neuronal sympathetic activity. Ang-II was angiogenic *in vivo* in the CAM and in the rabbit cornea assay [71, 72], and stimulated the growth of quiescent ECs via angiotensin-II type 1 receptors (AT_1-Rs) [73]. Sasaki *et al.* [74] provided evidence that AT_1-Rs could play an important role in ischemia-induced angiogenesis. Ang-II induced VEGF expression in SMC, which may stimulate EC proliferation, migration, and angiogenesis [75–77]. The VEGF involvement in the AT_1-R-mediated angiogenic response of ischemic tissues has been demonstrated [74]. VEGF-mediated angiogenesis was impaired by AT_1-R blockade in the cardiomyopathic hamster heart because this procedure markedly lowered VEGF mRNA expression, and capillary microvascular density [78]. In transgenic mice overexpressing Ang-II, the angiogenic response was markedly increased, the response being abolished by the AT_2-R antagonist PD123319 and unaffected by the AT_1-R antagonist losartan [79].

3.3.3 Endothelins

Endothelins (ETs) are a family of hypertensive peptides, mainly secreted by ECs. ETs and their receptors are present in a variety of tissues, where they play important physiological and pathophysiological roles, mainly concerning the cardiovascular system [80, 81].

Human umbilical vein endothelial cells (HUVECs) express high levels of ET-1 and ET_B-R mRNAs, and low levels of ET_A-R mRNA [82] and produce, secrete ET-1 [83, 84]. ET-1 and ET-3, acting via the ET_B-R, promoted *in vitro* EC proliferation [85, 86] and migration [86]. HUVECs cultured on Matrigel™ in the presence of ET-1, formed capillary-like tubular structures and the effect was inhibited by the selective ET_B-R antagonist BQ788 [82]. ET-1, in association with VEGF, showed clear proangiogenic activity in the Matrigel plug implanted into mice [82]. ET-1-producing Chinese hamster ovary cells when grafted onto CAM, induced an angiogenic effect that was prevented by an inhibitor of VEGF tyrosine kinase receptors, thereby confirming the involvement of VEGF in the ET-1 angiogenic response [87]. Moreover VEGF increased both, the expression of ET-1 mRNA in and ET-1 secretion from ECs [88]. There is also evidence that ET-1 promoted VEGF production through HIF-1α [89].

Overexpression of ET-1 and its receptors was found in tumors [90]. A potent ET receptor A antagonist, displayed antitumor activity and decreased neovascularization *in vivo* against established ovarian cancer xenografts in nude mice [91]. In primary and metastatic ovarian carcinomas, there was a highly significant correlation between ET-1 expression and microvascular density, as well as between ET-1 and VEGF expression [92]. When tested in ovarian carcinoma-derived cell lines, ET-1 increased VEGF mRNA expression and induced VEGF production in a time- and dose-dependent fashion;this occurred to a greater extent during hypoxia [89, 92].

3.3.4 Adrenomedullin

Adrenomedullin (AM), although originally isolated from human pheochromocytomas, is synthesized in several tissues and organs including the blood vessels and heart. AM possesses a proangiogenic effect under both physiological and pathophysiological conditions [93–95]. It augments vascular collateral development in response to acute ischemia [96, 97], and enhances capillary-like tube formation by HUVECs cultured on Matrigel, and blood vessel formation in the CAM assay [98]. AM gene transfer was found to induce therapeutic angiogenesis in a rabbit model of chronic hind limb ischemia [99] and Miyashita *et al.* [100] showed that AM administration improved vascular regeneration in the ischemic rat brain. AM upregulates the expression of VEGF in both *in vitro* and *in vivo* models [101, 102]. Iimuro and coworkers demonstrated that heterozygous AM knock out mice treated with AM displayed reduced capillary development, and the administration of either AM or VEGF favored blood flow recovery and capillary formation [101]. However, blocking antibodies to VEGF did not significantly inhibit AM-induced *in vitro* capillary-like tube formation by ECs [103], suggesting that AM does not act indirectly through up-regulation of VEGF.

High levels of AM expression have been detected in various types of cancer cells [104] and AM overexpressing tumors are characterized by increased vascularity [105, 106].

3.3.5 Leptin

Leptin is an adipose tissue-secreted hormone, which is involved in the regulation of satiety, metabolic rate, and thermogenesis [107].

ECs express the functionally active receptors Ob-Ra and Ob-Rb that mediate their leptin-induced proliferation through the activation of STAT-3 and extracellular signal-regulated kinases (ERKs) 1/2. Leptin also induced angiogenesis *in vivo* in the CAM and in the rat cornea assays [108, 109]. Leptin-induced new blood vessels were fenestrated, playing a critical role in the maintenance and regulation of vascular fenestration in the adipose tissue [110]. Rather contrasting findings were reported by Cohen *et al.* [111], who demonstrated that leptin induced the expression of Ang-2 in adipose tissue without concomitant increase in VEGF, thereby providing a strong angiostatic rather than angiogenic signal.

The presence of an autocrine–paracrine leptin/Ob-R system in tumors and the possible effects of leptin on tumor microenvironment and angiogenesis are consistent with a complex biological mechanism exerted by leptin/Ob-R signaling in cancer and suggest the presence of a loop in the leptin/Ob-R system. For example, leptin is secreted by hepatic tumor cells, high leptin expression was associated with an increased microvascular density and may be associated with hepatocellular carcinoma development [112–115]. Moreover, *in vivo* experiments performed in the CAM assay suggest that the use of an anti-leptin antibody as an angiostatic molecule may be a useful approach in the treatment of hepatocellular carcinoma [115] and suggest that inhibition of leptin signaling by injecting anti-leptin antibodies or a soluble form of Ob-R, could delay tumor growth.

3.3.6 Adiponectin and Neuropeptide-Y

Adiponectin, an adipose tissue–derived peptide, is a regulator of energy homeostasis and plays a role in obesity-induced insulin resistance and related complications [116].

Adiponectin-activated adenosine monophosphate kinase (AMP-K) in ECs, leads to enhanced *in vivo* angiogenesis in murine Matrigel plug and rabbit cornea assays, and inhibition of caspase 3-mediated apoptosis in HUVECs cultured *in vitro* [117, 118]. Bråkenhielm *et al.* [119] also demonstrated that adiponectin inhibited EC migration and proliferation *in vitro* and neoangiogenesis *in vivo* in the CAM and cornea assays, as well as decreased angiogenesis and induced apoptosis in tumors.

Neuropeptide-Y (NPY) is widely distributed in the nervous system, where it is thought to act as a neurotransmitter. NPY stimulated ERK 1/2 activity in rat coronary ECs in primary culture [120]. It promoted *in vitro* and *in vivo* angiogenesis in a murine Matrigel plug assay, its potency being similar to that of FGF-2. NPY has also been reported to play a major role in promoting the growth and neovascularization of neuroblastomas [121].

3.3.7 Vasoactive Intestinal Peptide and Substance-P

Vasoactive intestinal peptide (VIP) increased VEGF expression in lung cancer cells [122]. Using a rat sponge model, Hu *et al.* [123] showed that daily injections of high doses of VIP evoked intense neovascularization, as assessed by ^{133}Xe clearance technique and morphometry. Lower doses of VIP were ineffective but when administered with a subthreshold dose of interleukin-1α, evoked an angiogenic response similar to that observed with higher doses of VIP.

Substance P (SP) and a selective natural killer cell NK_1-receptor enhanced capillary growth *in vivo* in a rabbit cornea assay and stimulated proliferation and migration *in vitro* of different EC types. NK_2-Receptor and NK_3-receptor antagonists were ineffective, while SP antagonists blocked the response [124]. Fan *et al.* [125] confirmed the *in vivo* proangiogenic action of SP in a rat sponge assay, and showed that the effect was suppressed by a selective NK_1-receptor antagonist. The angiogenic effects of SP were prevented by *L-N*^{G}-nitrosamine methylester (L-NAME), suggesting the involvement of NOS/NO-dependent signaling [126]. *In vivo* experiments showed that endogenous SP could be implicated in the neoangiogenesis connected with neurogenic inflammation [127].

3.4 Conclusions and Perspectives

It is well established that the angiogenic phenotype results from the imbalance between positive and negative regulatory factors, so that the contribution of each "classic" and/or "nonclassic" angiogenic factor may play a different role in the definition of the angiogenic phenotype. Increased production of angiogenic stimuli and/or reduced production of "classic" and/or "nonclassic" angiogenic inhibitors may lead to abnormal neovascularization.

Much research effort has been concentrated on the role of angiogenesis in cancer and inhibition of angiogenesis is a major area of therapeutic development for the treatment of this disease. New pathophysiological concepts generated in the last few decades have led to the development of a large variety of new drugs that interfere with angiogenesis. Preclinical and clinical studies have made it increasingly clear that strategies that target tumor blood vessel networks will ultimately be most effective if used in conjunction with, or adjuvant to, conventional anticancer therapies.

A detailed knowledge of the mechanism of action and expression as well as the interactions of the new "nonclassic" endogenous regulators of angiogenesis with their receptors will provide new insights that are essential for the future development of chemical compounds that can modulate the activity of these new "nonclassic" endogenous regulators and may have potential for antitumor activity. In fact, tumors and other angiogenic pathologies exploit redundant mechanisms to induce angiogenesis, and neutralization of multiple factors, including both "classic" and nonclassic' regulators, may be required to suppress tumor growth.

The linkage between the laboratory and the clinic, which brought this important new development to the patient, must be maintained to further our understanding of the role of angiogenesis in normal physiology and disease in order to develop and validate intermediate and surrogate markers of benefit, and to advance to the optimal use of antiangiogenic molecules.

Acknowledgments

This work was supported by grants from the AIRC, MIUR (Italian Ministry of University and Research) (PRIN 2007), Rome, and Fondazione Cassa di Risparmio di Puglia, Bari, Italy.

References

1. Ribatti, D., Nico, B., Crivellato, E., Roccaro, A.M., and Vacca, A. (2007) The history of the angiogenic switch concept. *Leukemia*, **21**, 44–52.
2. Folkman, J. (1971) Tumor angiogenesis: therapeutic implications. *N. Engl. J. Med.*, **285**, 1182–1186.
3. Ribatti, D., Vacca, A., and Dammacco, F. (1999) The role of the vascular phase in solid tumor growth: a historical review. *Neoplasia*, **1**, 293–302.
4. Ribatti, D. and Vacca, A. (2008) Angiogenesis and antiangiogenesis in haematological diseases. *Mag. Eur. Med. Oncol.*, **1**, 31–33.
5. Ribatti, D. (2005) The crucial role of vascular permeability factor/vascular endothelial growth factor in angiogenesis: a historical review. *Br. J. Haematol.*, **128**, 303–309.
6. Ferrara, N. (2004) Vascular endothelial growth factor: basic science and clinical progress. *Endocr. Rev.*, **25**, 581–611.
7. Ribatti, D. (2008) The discovery of the placental growth factor and its role in angiogenesis: a historical review. *Angiogenesis*, **11**, 215–221.
8. Inoue, M., Hager, J.H., Ferrara, N., Gerber, H.P., and Hanahan, D. (2003) VEGF-A has a critical, nonredundant role in angiogenic switching and pancreatic β cell carcinogenesis. *Cancer Cell*, **1**, 193–202.
9. Senger, D.R., Brown, L., Claffey, K., and Dvorak, A. (1995) Vascular permeability factor, tumor angiogenesis and stroma generation. *Invasion Metastasis*, **14**, 385–394.
10. Dvorak, H.F., Harvey, V.S., Estrella, P., Brown, L.F., Mc Donagh, J., and Dvorak, A.M. (1987) Fibrin containing gels induce angiogenesis: implication for tumor stroma generation and wound healing. *Lab. Invest.*, **57**, 673–686.
11. Kondo, S., Asano, M., Matsuo, K., Ohmori, I., and Suzuki, H. (1994) Vascular endothelial growth factor/vascular permeability factor is detectable in the sera of tumor-bearing mice and cancer patients. *Biochim. Biophys. Acta*, **1221**, 211–214.
12. Yamamoto, Y., Toi, M., Kondo, S., Matsumoto, T., Suzuki, H., Kitamura, M., Tsuruta, K. *et al.* (1996) Concentrations of vascular endothelial growth factor in the sera of normal controls and cancer patients. *Clin. Cancer Res.*, **2**, 821–826.
13. Shweiki, D., Itin, A., Soffer, D., and Keshet, E. (1992) Vascular endothelial cell growth factor induced by hypoxia may mediate hypoxia-initiated angiogenesis. *Nature*, **359**, 843–845.
14. Carmeliet, P., Dor, Y., Herbert, J.M., Fukumura, D., Brusselmans, K., Dewerchin, M., Neeman, M. *et al.* (1998) Role of HIF-1α in hypoxia-mediated angiogenesis, cell proliferation and tumour angiogenesis. *Nature*, **394**, 485–490.
15. Jiang, B.H., Agani, F., Passaniti, A., and Semenza, G.L. (1997) V-SRC induces expression of hypoxia inducible factor 1 (HIF-1) and transcription of genes encoding vascular endothelial growth factor and enolase 1: involvement of HIF-1 in tumor progression. *Cancer Res.*, **57**, 5328–5335.
16. Kerbel, R.S., Viloria-Petit, A., Okada, F., and Rak, J. (1998) Establishing a link between oncogenes and tumor angiogenesis. *Mol. Med.*, **4**, 286–295.
17. Kieser, A., Welch, H.A., Brandner, G., Marme, D., and Kolck, W. (1994) Mutant p53 potentiates protein kinase C induction of vascular endothelial

growth factor expression. *Oncogene*, **9**, 963–969.

18. Pertovaara, L., Kaipainen, A., Mutsonen, T., Orpana, A., Ferrara, N., Saksela, O., and Alitalo, K. (1994) Vascular endothelial growth factor is induced in response to transforming growth factor–β in fibroblastic and epithelia cells. *J. Biol. Chem.*, **269**, 6271–6274.
19. Chin, K., Kurashima, Y., Ogura, T., Tajiri, H., Yoshida, S., and Esumi, H. (1997) Induction of vascular endothelial growth factor/vascular permeability factor by nitric oxide in human glioblastoma and hepatocellular cell lines. *Oncogene*, **15**, 437–442.
20. Presta, M., Dell'Era, P., Mitola, S., Moroni, E., Ronca, R., and Rusnati, M. (2005) Fibroblast growth factor/fibroblast growth factor receptor system in angiogenesis. *Cytokine Growth Factor Rev.*, **16**, 159–178.
21. Gross, R. and Dickson, C. (2005) Fibroblast growth factor signaling in tumorigenesis. *Cytokine Growth Factor Rev.*, **16**, 179–186.
22. Presta, M., Moscatelli, D., Joseph-Silverstein, J., and Rifkin, D.B. (1986) Purification from a human hepatoma cell line of a basic fibroblast growth factor-like molecule that stimulates capillary endothelial cell plasminogen activator production, DNA synthesis, and migration. *Mol. Cell Biol.*, **6**, 4060–4066.
23. Moscatelli, D., Presta, M., and Rifkin, D.B. (1986) Purification of a factor from human placenta that stimulates capillary endothelial cell protease production, DNA synthesis, and migration. *Proc. Natl. Acad. Sci. U.S.A.*, **83**, 2091–2095.
24. Kandel, J., Bossy-Wetzel, E., Radvanyi, F., Klagsbrun, M., Folkman, J., and Hanahan, D. (1991) Neovascularization is associated with a switch to the export of bFGF in the multistep development of fibrosarcoma. *Cell*, **66**, 1095–1104.
25. Nguyen, M., Watanabe, H., Budson, A.E., Richie, J.P., Hayes, D.F., and Folkman, J. (1994) Elevated levels of an angiogenic peptide, basic fibroblast growth factor, in the urine of patients with a wide spectrum of cancers. *J. Natl. Cancer Inst.*, **86**, 356–361.
26. Dirix, L.Y., Vermeulen, P.B., Pawinski, A., Prové, A., Benoy, I., De Pooter, C. *et al.* (1997) Elevated levels of the angiogenic cytokines basic fibroblast growth factor and vascular endothelial growth factor in sera of cancer patients. *Br. J. Cancer*, **76**, 238–243.
27. Dietz, A., Rudat, V., Conradt, C., Wiedauer, H., Ho, A., and Moehler, T. (2000) Prognostic relevance of serum levels of the angiogenic peptide bFGF in advanced carcinoma of the head and neck treated by primary radiochemotherapy. *Head Neck*, **22**, 666–673.
28. Salgado, R., Benoy, I., Vermeulen, P., van Dam, P., Van Marck, E., and Dirix, L. (2004) Circulating basic fibroblast growth factor is partly derived from the tumour in patients with colon, cervical and ovarian cancer. *Angiogenesis*, **7**, 29–32.
29. Moroni, E., Dell'Era, P., Rusnati, M., and Presta, M. (2002) Fibroblast growth factors and their receptors in hematopoiesis and hematological tumors. *J. Hematother. Stem Cell Res.*, **11**, 19–32.
30. Dickson, M.C., Martin, J.S., Cousins, F.M., Kulkarni, A.B., Karlsson, S., and Akhurst, R.J. (1995) Defective haematopoiesis and vasculogenesis in transforming growth factor-β1 knock out mice. *Development*, **121**, 1845–1854.
31. Armulik, A., Abramsson, A., and Betsholtz, C. (2005) Endothelial/pericyte interactions. *Circ. Res.*, **97**, 512–523.
32. Hirschi, K.K., Rohovsky, S.A., and D'Amore, P.A. (1998) PDGF, TGF-β, and heterotypic cell-cell interactions mediate endothelial cell-induced recruitment of 10T1/2 cells and their differentiation to a smooth muscle fate. *J. Cell Biol.*, **141**, 805–814.
33. Kazlauskas, A. (2008) Platelet-derived growth factor, in *Angiogenesis. An Integrative Approach from Science to Medicine* (eds W.D.Figg J. Folkman), Springer Science, New York, pp. 99–102.

34. Lindhal, P., Johannson, B.R., Leeven, P., and Betsholtz, C. (1997) Pericyte loss and microaneurysm formation in PDGF-B deficient mice. *Science*, **277**, 242–245.
35. Hellstrom, M., Kalen, M., Lindahl, P., Abramson, A., and Betsholtz, C. (1999) Role of PDGF-B and PDGFR-β in recruitment of vascular smooth muscle cells and pericytes during embryonic blood vessel formation in the mouse. *Development*, **126**, 3047–3055.
36. Abramsson, A., Lindblom, P., and Betshowz, C. (2003) Endothelial and nonendothelial sources of PDGF-B regulate pericyte recruitment and influence vascular pattern formation in tumors. *J. Clin. Invest.*, **112**, 1142–1151.
37. Guo, P., Hu, B., Gu, W., Xu, L., Wang, D., Hunag, H.J., Cavanee, W.K., and Cheng, S.Y. (2003) Platelet-derived growth factor-B enhances glioma angiogenesis by stimulating vascular endothelial growth factor expression in tumor endothelia and by promoting pericyte recruitment. *Am. J. Pathol.*, **162**, 1083–1093.
38. Chae, S.S., Paik, J.H., Furneaux, H., and Hla, T. (2004) Requirement for sphingosine 1-phosphate receptor-1 in tumor angiogenesis demonstrated by in vivo RNA interference. *J. Clin. Invest.*, **114**, 1082–1089.
39. Thurston, G. (2003) Role of angiopoietins and Tie receptor tyrosine kinases in angiogenesis and lymphangiogenesis. *Cell Tissue Res.*, **314**, 61–68.
40. Suri, C., Jones, P.F., Patan, S., Bartunkova, S., Maisonpierre, P.C., Davis, S., Sato, T.N., and Yancopoulos, G.D. (1996) Requisite role of angiopoietin-1, a ligand for the Tie-2 receptor, during embryonic angiogenesis. *Cell*, **87**, 1171–1180.
41. Puri, M.C., Partanen, J., Rossant, J., and Bernstein, A. (1999) Interaction of the TEK and TIE receptor tyrosine kinases during cardiovascular development. *Development*, **126**, 4569–4580.
42. Dumont, D.J., Fong, G.H., Puri, M.C., Gradwohl, G., Alitalo, K., and Breitman, M.L. (1995) Vascularization of the mouse embryo: study of flk-1, tek, tie and vascular endothelial growth factor expression during development. *Dev. Dyn.*, **203**, 80–92.
43. Sato, T.N., Tozawa, Y., Deutsch, U., Wolburg-Buchholz, K., Fujiwara, Y., Gendron Maguire, M., Gridley, T. *et al.* (1995) Distinct roles of the receptor tyrosine kinases Tie-1 and Tie-2 in blood vessel formation. *Nature*, **376**, 70–74.
44. Maisonpierre, P.C., Suri, C., Jones, P.F., Bartunkova, S., Wiegand, S.J., Radziejewski, C., Compton, D. *et al.* (1997) Angiopoietin-2, a natural antagonist for Tie2 that disrupts in vivo angiogenesis. *Science*, **277**, 55–60.
45. Davis, S. and Yancopouolos, G.F. (1999) The angiopoietins: yin and yang in angiogenesis. *Curr. Top. Microbiol. Immunol.*, **273**, 173–185.
46. Thurston, G., Suri, C., Smith, K., McClain, J., Stao, T.N., Yancoupolos, G.D., and McDonald, D.M. (1999) Leakage-resistant blood vessels in mice transgenically overexpression angiopoietin 1. *Science*, **286**, 2511–2514.
47. Suri, C., McLain, J., Thurston, G., McDonald, D., Oldmixon, E.H., Sato, T.N., and Yancopouolos, G.D. (1998) Increased vascularization in mice overexpressing angiopoietin-1. *Science*, **282**, 468–471.
48. Holash, J., Wiegand, S.J., and Yancopoulos, G.F. (1999) New model of tumor angiogenesis: dynamic balance between vessel regression and growth mediated by angiopoietins and VEGF. *Oncogene*, **18**, 5356–5362.
49. Stratmann, A., Risau, W., and Plate, K.H. (1998) Cell-type specific expression of angiopoietin-1 and angiopoietin-2 suggests a role in glioblastoma angiogenesis. *Am. J. Pathol.*, **153**, 1459–1466.
50. Tanaka, S., Mori, M., Sakamoto, Y., Makunchi, M., Sugimachi, K., and Wands, J.R. (1999) Biologic significance of angiopoietin-2 expression in human hepatocelluar carcinoma. *J. Clin. Invest.*, **103**, 341–345.
51. Machein, M.R., Knedla, A., Knoth, R., Wagner, S., Neusche, E., and Plate, K.H. (2004) Angiopoietin-1 promotes tumor angiogenesis in a rat glioma model. *Am. J. Pathol.*, **165**, 1557–1570.

52. Ahmad, S.A., Liu, W., Jung, Y.D., Fan, F., Wilson, M., Reinmuth, H., Shaheen, R.M. *et al.* (2001) The effects of angiopoietin-1 and -2 on tumor growth and angiogenesis in human colon cancer. *Cancer Res.*, **61**, 1255–1259.
53. Stoeltzing, O., Ahmad, S.A., Liu, W., McCarty, M.F., Wey, J.S., Parikh, A.A., Fan, F. *et al.* (2003) Angiopoietin-1 inhibits vascular permeability, angiogenesis, and growth of hepatic colon cancer tumors. *Cancer Res.*, **63**, 3370–3377.
54. Yancopoulos, G.D., Davis, S., Gale, N.W., Rudge, J.S., Wiegand, S.J., and Holash, J. (2000) Vascular specific growth factors and blood vessels formation. *Nature*, **407**, 242–248.
55. Vajkoczy, P., Farhadi, M., Gaumann, A., Heidenreich, R., Erber, R., Wunder, A., Tonn, J.C. *et al.* (2002) Microtumor growth initiates angiogenic sprouting with simultaneous expression of VEGF, VEGF receptor-2, and angiopoietin-2. *J. Clin. Invest.*, **109**, 777–785.
56. Anagnostou, A., Lee, E.S., Kessimian, N., Levinson, R., and Steiner, M. (1990) Erythropoietin has a mitogenic and positive chemotactic effect on endothelial cells. *Proc. Natl. Acad. Sci. U.S.A.*, **87**, 5978–5982.
57. Jaquet, K., Krause, K., Tawakol-Khodai, M., Geidel, S., and Kuck, K.H. (2002) Erythropoietin and VEGF exhibit equal angiogenic potential. *Microvasc. Res.*, **64**, 326–333.
58. Ribatti, D., Vacca, A., Roccaro, A.M., Crivellato, E., and Presta, M. (2003) Erythropoietin as an angiogenic factor. *Eur. J. Clin. Invest.*, **33**, 891–896.
59. Anagnostou, A., Liu, Z., Steiner, M., Chin, K., Lee, E.S., Kessimian, N., and Noguchi, C.T. (1994) Erythropoietin receptor mRNA expression in human endothelial cells. *Proc. Natl. Acad. Sci. U.S.A.*, **91**, 3974–3978.
60. Ashley, R.A., Dubuque, S.H., Dvorak, B., Woodward, S.S., William, S.K., and Kling, P.T. (2002) Erythropoietin stimulates vasculogenesis in neonatal rat mesenteric microvascular endothelial cells. *Pediatr. Res.*, **51**, 472–478.
61. Heeschen, C., Aicher, A., Lehmann, R., Fichtlscherer, S., Vasa, M., Urbich, C., Mildner Rihm, C. *et al.* (2003) Erythropoietin is a potent physiologic stimulus for endothelial progenitor cell mobilization. *Blood*, **102**, 1340–1346.
62. Müller-Ehmsen, J., Schmidt, A., Krausgrill, B., Schwinger, R.H.G., and Bloch, W. (2006) Role of erythropoietin for angiogenesis and vasculogenesis: from embryonic development through adulthood. *Am. J. Physiol.*, **290**, H331–H340.
63. Haller, H., Christel, C., Donnenberg, L., Thiele, P., Lindschau, C., and Luft, F.C. (1996) Signal transduction of erythropoietin in endothelial cells. *Kidney Int.*, **50**, 481–488.
64. Ribatti, D., Presta, M., Vacca, A., Ria, R., Giuliani, R., Dell'Era, P., Nico, B. *et al.* (1999) Human erythropoietin induces a proangiogenic phenotype in cultured endothelial cells and stimulates neovascularization in vivo. *Blood*, **93**, 2627–2636.
65. Calvillo, L., Latini, R., Kajstura, J., Leri, A., Anversa, P., Ghezzi, P., Salio, M., Cerami, A., and Brines, M. (2003) Recombinant human erythropoietin protects the myocardium from ischemia reperfusion injury and promotes beneficial remodeling. *Proc. Natl. Acad. Sci. U.S.A.*, **100**, 4802–4806.
66. Parsa, C.J., Matsumoto, A., Kim, J., Riel, R.U., Pascal, L.S., Walton, G.B., Thompson, R.B. *et al.* (2003) A novel protective effect of erythropoietin in the infarcted heart. *J. Clin. Invest.*, **112**, 999–1007.
67. Carlini, R.G., Dusso, A.S., Obialo, C.I., Alvarez, U.M., and Rothstein, M. (1993) Recombinant human erythropoietin (rHuEPO) increases endothelin-1 release by endothelial cells. *Kidney Int.*, **43**, 1010–1014.
68. Carlini, R.G., Reyes, A.A., and Rothstein, M. (1995) Recombinant human erythropoietin stimulates angiogenesis in vitro. *Kidney Int.*, **47**, 740–745.
69. Hardee, M.E., Arcasoy, M.O., Blackwell, K.L., Kirkpatrick, J.P., and Dewhirst, M.W. (2006) Erythropoietin

biology in cancer. *Clin. Cancer Res.*, **12**, 332–339.

70. Ribatti, D. (2008) Erythropoietin and cancer, a double-edged sword. *Leukemia Res.*, **33**, 1–4.

71. Fernandez, L.A., Twickler, J., and Mead, A. (1985) Neovascularization produced by angiotensin II. *J. Lab. Clin. Med.*, **105**, 141–145.

72. Le Noble, F.A.C., Hekking, J.W.M., Van Straaten, H.W.M., Slaaf, D.W., and Struyker Boudier, H.A.J. (1991) Angiotensin II stimulates angiogenesis in the chorioallantoic membrane of the chick embryo. *Eur. J. Pharmacol.*, **195**, 3005–3006.

73. Stoll, M., Steckelings, U.M., Paul, M., Bottari, S.P., Metzger, R., and Unger, T. (1995) The angiotensin AT2-receptor mediates inhibition of cell proliferation in coronary endothelial cells. *J. Clin. Invest.*, **95**, 651–657.

74. Sasaki, K., Murohara, T., Ikeda, H., Sugaya, T., Shimada, T., Shintani, S., and Imaizumi, T. (2002) Evidence for the importance of angiotensin II type 1 receptor in ischemia-induced angiogenesis. *J. Clin. Invest.*, **109**, 603–611.

75. Williams, B., Baker, A.Q., Gallacher, B., and Lodwick, D. (1995) Angiotensin II increases vascular permeability factor gene expression by human vascular smooth muscle cells. *Hypertension*, **25**, 913–917.

76. Chua, C.C., Hamdy, R.C., and Chua, B.H. (1998) Upregulation of vascular endothelial growth factor by angiotensin II in rat heart endothelial cells. *Biochim. Biophys. Acta*, **1401**, 187–194.

77. Otani, A., Takagi, H., Suzuma, K., and Honda, Y. (1998) Angiotensin II potentiates vascular endothelial growth factor-induced angiogenic activity in retinal microcapillary endothelial cells. *Circ. Res.*, **82**, 619–628.

78. Shimizu, T., Okamoto, H., Chiba, S., Matsui, Y., Sugawara, T., Akino, M., Nan, J. *et al.* (2003) VEGF-mediated angiogenesis is impaired by angiotensin type 1 receptor blockade in cardiomyopathic hamster hearts. *Cardiovasc. Res.*, **58**, 203–212.

79. Walther, T., Menrad, A., Orzechowski, H.D., Siemeister, G., Paul, M., and Schirner, M. (2003) Differential regulation of in vivo angiogenesis by angiotensin II receptors. *FASEB J.*, **17**, 2061–2067.

80. Kedzierski, R.M. and Yanagisawa, M. (2001) Endothelin system: the double-edged sword in health and disease. *Annu. Rev. Pharmacol.*, **41**, 851–876.

81. Rossi, G.P., Seccia, T.M., and Nussdorfer, G.G. (2001) Reciprocal regulation of endothelin-1 and nitric oxide: relevance in the physiology and pathology of the cardiovascular system. *Int. Rev. Cytol.*, **209**, 241–272.

82. Salani, D., Taraboletti, G., Rosanò, L., Di Castro, V., Borsotti, P., Giavazzi, R., and Bagnato, A. (2000) Endothelin-1 induces an angiogenic phenotype in cultured endothelial cells and stimulates neovascularization in vivo. *Am. J. Pathol.*, **157**, 1703–1711.

83. Fujitani, Y., Oda, K., Takimoto, M., Inui, T., Okada, T., and Urade, Y. (1992) Autocrine receptors for endothelins in the primary culture of endothelial cells of human umbilical vein. *FEBS Lett.*, **298**, 79–83.

84. Flynn, M.A., Haleen, S.J., Welch, K . M., Cheng, X . M., and Reynolds, E.E. (1998) Endothelin B receptors on human endothelial and smooth muscle cells show equivalent binding pharmacology. *J. Cardiovasc. Pharmacol.*, **32**, 106–116.

85. Morbidelli, L., Orlando, C., Maggi, C.A., Ledda, F., and Ziche, M. (1995) Proliferation and migration of endothelial cells is promoted by endothelins via activation of ETB receptors. *Am. J. Physiol.*, **269**, H685–H695.

86. Noiri, E., Hu, Y., Bahou, W.F., Keese, C.R., Giaever, I., and Goligorsky, M.S. (1997) Permissive role of nitric oxide in endothelin-induced migration of endothelial cells. *J. Biol. Chem.*, **272**, 1747–1752.

87. Cruz, A., Parnot, C., Ribatti, D., Corvol, P., and Gasc, J.M. (2001) Endothelin-1, a regulator of angiogenesis in the chick chorioallantoic membrane. *J. Vasc. Res.*, **38**, 536–545.

88. Matsuura, A., Yamochi, W., Hirata, K., Kawashima, S., and Yokoyama, M. (1998) Stimulatory interaction between vascular endothelial growth factor and endothelin-1 on each gene expression. *Hypertension*, **32**, 89–95.
89. Spinella, F., Rosanò, L., Di Castro, V., Natali, P.G., and Bagnato, A. (2002) Endothelin-1 induces vascular endothelial growth factor by increasing hypoxia-inducible factor-1α in ovarian carcinoma cells. *J. Biol. Chem.*, **277**, 27850–27855.
90. Bagnato, A. and Spinella, F. (2002) Emerging role of endothelin-1 in tumor angiogenesis. *Trends Endocrinol. Metab.*, **14**, 44–50.
91. Rosanò, L., Varmi, M., Salani, D., Di Castro, V., Spinella, F., Natali, P.G., and Bagnato, A. (2001) Endothelin-1 induces proteinase activation and invasiveness of ovarian carcinoma cells. *Cancer Res.*, **61**, 8340–8346.
92. Salani, D., Di Castro, V., Nicotra, M.R., Rosanò, L., Tecce, R., Venuti, A., Natali, P.G., and Bagnato, A. (2000) Role of endothelin-1 in neovascularization of ovarian carcinoma. *Am. J. Pathol.*, **157**, 1537–1547.
93. Nikitenko, L.L., Smith, D.M., Hague, S., Wilson, C.R., Bicknell, R., and Rees, M.C.P. (2002) Adrenomedullin and the microvasculature. *Trends Pharmacol. Sci.*, **23**, 101–103.
94. Nagaya, N., Mori, H., Mutokami, S., Kangawa, K., and Kitamura, S. (2005) Adrenomedullin: angiogenesis and gene therapy. *Am. J. Physiol.*, **288**, R1432–R1437.
95. Ribatti, D., Nico, B., Spinazzi, R., Vacca, A., and Nussdorfer, G.G. (2005) The role of adrenomedullin in angiogenesis. *Peptides*, **26**, 1670–1675.
96. Abe, M., Sata, M., Nishimatsu, H., Nagata, D., Suzuki, E., Terauchi, Y., Kadowaki, T. *et al.* (2003) Adrenomedullin augments collateral development in response to acute ischemia. *Biochem. Biophys. Res. Commun.*, **306**, 10–15.
97. Iwase, T., Nagaya, N., Fujii, T., Itoh, T., Ishibashi-Ueda, H., Yamagishi, M., Miyatake, K. *et al.* (2005) Adrenomedullin enhances angiogenic potency of bone marrow transplantation in a rat model of hindlimb ischemia. *Circulation*, **111**, 356–362.
98. Ribatti, D., Guidolin, D., Conconi, M.T., Nico, B., Baiguera, S., Parnigotto, P.P., Vacca, A., and Nussdorfer, G.G. (2003) Vinblastine inhibits the angiogenic response induced by adrenomedullin in vitro and in vivo. *Oncogene*, **22**, 6458–6461.
99. Tokunaga, N., Nagaya, N., Shirai, M., Tanaka, E., Ishibashi-Ueda, H., Harada-Shiba, M., Kanda, M. *et al.* (2004) Adrenomedullin gene transfer induces therapeutic angiogenesis in a rabbit model of chronic hind limb ischemia. Benefits of a novel nonviral vector, gelatin. *Circulation*, **109**, 526–531.
100. Miyashita, K., Itoh, H., Arai, H., Suganami, T., Sawada, N., Fukunaga, Y., Sone, M. *et al.* (2006) The neuroprotective and vasculo-neuro-regenerative roles of adrenomedullin in ischemic brain and its therapeutic potential. *Endocrinology*, **147**, 1642–1653.
101. Iimuro, S., Shindo, T., Moriyama, N., Amaki, T., Niu, P., Takeda, N., Iwata, H. *et al.* (2004) Angiogenic effects of adrenomedullin in ischemia and tumor growth. *Circ. Res.*, **95**, 415–423.
102. Albertin, G., Rucinski, M., Carraro, G., Forneris, M., Andreis, P.G., Malendowicz, L.K., and Nussdorfer, G.G. (2005) Adrenomedullin and vascular endothelium growth factor genes are overexpressed in the regenerating rat adrenal cortex, and AM and VEGF reciprocally enhance their mRNA expression in cultured rat adrenocortical cells. *Int. J. Mol. Med.*, **16**, 431–435.
103. Fernandez-Sauze, S., Delfino, C., Mabrouk, K., Dussert, C., Chinot, O., Martin, P.M., Grisoli, F. *et al.* (2004) Effects of adrenomedullin on endothelial cells in the multistep process of angiogenesis: involvement of CRLR/RAMP2 and CRLR/RAMP3 receptors. *Int. J. Cancer*, **108**, 797–804.
104. Nikitenko, L.L., Fox, S.B., Kekoe, S., Rees, M.C.P., and Bicknell, R. (2006) Adrenomedullin and tumor angiogenesis. *Br. J. Cancer*, **94**, 1–7.

105. Oehler, M.K., Fischer, D.C., Orlowska-Volk, M., Herrie, F., Kiebach, D.G., Rees, M.C., and Bicknell, R. (2003) Tissue and plasma expression of the angiogenic peptide adrenomedullin in the breast cancer. *Br. J. Cancer*, **89**, 1927–1933.
106. Oehler, M.K., Hague, S., Reese, M.C., and Bicknell, R. (2002) Adrenomedullin promotes formation of xenografted endometrial tumors by stimulation of autocrine growth and angiogenesis. *Oncogene*, **21**, 2815–2821.
107. Ahima, R.S. and Flier, J.S. (2000) Leptin. *Annu. Rev. Physiol.*, **62**, 413–437.
108. Bouloumie, A., Drexler, H.C.A., Lafontan, M., and Busse, R. (1998) Leptin, the product of ob gene, promotes angiogenesis. *Circ. Res.*, **83**, 1059–1066.
109. Sierra-Honigmann, M.R., Nath, A.K., Murakami, C., Garcia-Cardena, G., Papapetropoulos, A., Sessa, W.C., Madge, L.A. *et al.* (1998) Biological action of leptin as an angiogenic factor. *Science*, **281**, 1683–1686.
110. Cao, R., Bråkenhielm, E., Wahlestedt, C., Thyberg, J., and Cao, Y. (2001) Leptin induces vascular permeability and synergistically stimulates angiogenesis with FGF-2 and VEGF. *Proc. Natl. Acad. Sci. U.S.A.*, **98**, 6390–6395.
111. Cohen, B., Barkan, D., Levy, Y., Goldberg, I., Fridman, E., Kopolovic, J., and Rubinstein, M. (2001) Leptin induces angiopoietin-2 expression in adipose tissues. *J. Biol. Chem.*, **276**, 7697–7700.
112. Rose, D.P., Komminou, D., and Stephenson, G.D. (2004) Obesity, adipocytokines, and insulin resistance in breast cancer. *Obes. Rev.*, **5**, 153–165.
113. Wang, S.N., Yeh, Y.T., Yang, S.F., Chai, C.Y., and Lee, K.T. (2006) Potential role of leptin expression in hepatocellular carcinoma. *J. Clin. Pathol.*, **59**, 930–934.
114. Wang, S.N., Chuang, S.C., Yeh, Y.T., Yang, S.F., Chai, C.Y., Chen, W.T., Kuo, K.K. *et al.* (2006) Potential prognostic value of leptin receptor in hepatocellular carcinoma. *J. Clin. Pathol.*, **59**, 1267–1271.
115. Ribatti, D., Belloni, A.S., Nico, B., Di Comite, M., Crivellato, E., and Vacca, A. (2008) Leptin-leptin receptor are involved in angiogenesis in human hepatocellular carcinoma. *Peptides*, **29**, 1596–1602.
116. Kadowaki, T. and Yamauchi, T. (2005) Adiponectin and adiponectin receptors. *Endocr. Rev.*, **26**, 439–451.
117. Kobayashi, H., Ouchi, N., Kihara, S., Walsh, K., Kumada, M., Abe, Y., Funahashi, T., and Matsuzawa, Y. (2004) Selective suppression of endothelial cell apoptosis by the high molecular weight form of adiponectin. *Circ. Res.*, **94**, 27–31.
118. Ouchi, N., Kobayashi, H., Kihara, S., Kumada, M., Sato, K., Inoue, T., Funahashi, T., and Walsh, K. (2004) Adiponectin stimulates angiogenesis by promoting cross talk between AMP-activated protein kinase and Akt signaling in endothelial cells. *J. Biol. Chem.*, **279**, 1304–1309.
119. Bråkenhielm, E., Veitonmäki, N., Cao, R., Kihara, S., Matsuzawa, Y., Zhivotovsky, B., Funahashi, T., and Cao, Y. (2004) Adiponectin-induced antiangiogenesis and antitumor activity involve caspase-mediated endothelial cell apoptosis. *Proc. Natl. Acad. Sci. U.S.A.*, **101**, 2476–2481.
120. Zukowska-Grojec, Z., Karwatowska-Prokopczuk, E., Fisher, T.A., and Ji, H. (1998) Mechanisms of vascular growth promoting effects of neuropeptide- Y: role of its inducible receptors. *Regul. Pept.*, **75/76**, 231–238.
121. Kitlinska, J. (2007) Neuropeptide-Y (NPY) in neuroblastoma: effect on growth and vascularization. *Peptides*, **28**, 405–412.
122. Casibang, M., Purdom, S., Jacowlew, S., Neckers, L., Zia, F., Ben-Av, P., Hla, T. *et al.* (2001) Prostaglandin E2 and vasoactive intestinal peptide increase vascular endothelial cell growth factor mRNAs in lung cancer cells. *Lung Cancer*, **31**, 203–212.
123. Hu, D., Hiley, C.R., and Fan, T.P. (1996) Comparative studies of the angiogenic activity of vasoactive intestinal

peptide, endothelins-1 and – 3 and angiotensin II in a rat sponge model. *Br. J. Pharmacol.*, **117**, 545–551.

124. Ziche, M., Morbidelli, L., Pacini, M., Geppetti, P., Alessandri, G., and Maggi, C.A. (1990) Substance P stimulates neovascularization in vivo and proliferation of cultured endothelial cells. *Microvasc. Res.*, **40**, 264–278.

125. Fan, T.P., Hu, D.E., Guard, S., Gresham, G.A., and Watling, K.J. (1993) Stimulation of angiogenesis by substance P and interleukin-1 in the rat and its inhibition by NK1 or interleukin-1 receptor antagonists. *Br. J. Pharmacol.*, **110**, 43–49.

126. Ziche, M., Morbidelli, L., Masini, E., Amerini, S., Granger, H.J., Maggi, C.A., Geppeetti, P., and Ledda, F. (1994) Nitric oxide mediates angiogenesis in vivo and endothelial cell growth and migration in vitro promoted by substance P. *J. Clin. Invest.*, **94**, 2036–2044.

127. Seegers, H.C., Hood, V.C., Kidd, B.L., Cruwys, S.C., and Walsh, D.A. (2003) Enhancement of angiogenesis by endogenous substance P release and neurokinin-1 receptors during neurogenic inflammation. *J. Pharmacol. Exp. Ther.*, **306**, 8–12.

4
The Role of Accessory Cells in Tumor Angiogenesis

Nobuyuki Takakura

4.1
Introduction

Vascular development is orchestrated by two processes, vasculogenesis and angio genesis. Vasculogenesis is the process by which mesodermal cells become committed to angioblast differentiation and form a primitive vascular plexus and larger organized vessels in the embryo, whereas angiogenesis is the process of new blood vessel formation by sprouting and remodeling of preexisting vessels [1]. In both processes, interactions between endothelial cells (ECs) and the surrounding mesenchymal cells, termed *mural cells* (such as pericytes or vascular smooth muscle cells) are critical [2]. Many molecules and factors regulating the processes of vasculogenesis and angiogenesis have been identified, and found to be involved in maintaining the integrity of vessels by recruitment and formation of the periendothelial layer or by mediating interactions between arteries and veins [2, 4, 5]. Of these factors, two receptor tyrosine kinase subfamilies are characterized by their largely endothelial cell-specific expression. One is the vascular endothelial growth factor (VEGF) family and its cognate receptors (Flt-1/VEGFR1, Flk-1/KDR/VEGFR2, and Flt-4/VEGFR3), for which crucial roles have been demonstrated in genetically engineered mouse mutants [6, 8, 9]. The other family includes Tie1 and Tie2; onset of expression of these receptors in the embryo seems to follow VEGFR expression [10]. These receptors, like VEGFRs, play a critical role in embryonic vascular formation [11–13]. Embryos lacking Tie2 or Tie1 fail to develop a structurally stable vasculature, resulting in hemorrhage at E9.5 and 13.5, respectively. Compared with the early defect in vasculogenesis seen in *VEGF* or *VEGFR* mutant embryos, mice lacking *Tie1* or *Tie2* thus exhibit later defects in angiogenesis and vascular remodeling, as well as in vascular integrity.

During blood vessel maturation processes, mural cells need to adhere to ECs. It has been suggested that this is promoted by angiopoietin-1 (Ang1), a ligand for Tie2 on ECs, which is produced by mural cells [13–17] after their recruitment

Tumor Angiogenesis – From Molecular Mechanisms to Targeted Therapy. Edited by Francis S. Markland, Stephen Swenson, and Radu Minea
Copyright © 2010 WILEY-VCH Verlag GmbH & Co. KGaA, Weinheim
ISBN: 978-3-527-32091-2

to the neighborhood of ECs by platelet-derived growth factors (PDGFs) and their cognate receptors, especially PDGFRβ [18, 19]. Under normoxic conditions, Ang1 is continuously produced by mural cells, resulting in the maintenance of stability of the blood vessels. However under hypoxic conditions, Ang2, an antagonist of Ang1, is produced and released from ECs [20]. In this situation, Tie2 is inactivated, resulting in dissociation of mural cells from ECs, after which the ECs can migrate into the hypoxic region. This initiates sprouting angiogenesis.

Angiogenesis occurs under physiological conditions such as during embryonic development, wound healing, ovulation, and so on, and is also very closely associated with progression of different pathological states [21, 22]. Abnormally rapid proliferation of blood vessels is implicated in over 20 diseases such as cancer, psoriasis, inflammatory diseases, diabetic retinopathy, and so on. In pathological angiogenesis, it was widely believed that ECs derived from those resident in preexisting vessels are responsible for forming neovasculature. However, it has been demonstrated that cells derived from the bone marrow may also contribute to postnatal angiogenesis [23]. Most studies have focused initially on the contribution of endothelial progenitor cells (EPCs) to this process. In tumor angiogenesis, it has been suggested that VEGFR2 + EPCs directly contribute to the tumor ECs in a murine B6RV2 lymphoma model, and in a Lewis lung carcinoma (LLC) model in Id-mutant mice, which show impaired angiogenesis [24]. However, several reports have argued against the direct contribution of bone marrow–derived cells to vascular ECs [25, 26]. One line of evidence suggested that cells expressing VE-cadherin but not hematopoietic lineage markers could represent a significant source of ECs for new vessels in the tumor during early stages of tumorigenesis, although resident ECs were the major contributors to neovascularization at later stages [27, 28]. As time course dependency of EPCs was observed during tumor development, it is possible that their relative contribution to neovascularization may be different for different tumor types. Further precise analysis will be required to clarify the contribution of EPCs to tumor angiogenesis.

On the other hand, in contrast to the concept that bone marrow-derived cells contribute to vascular cells directly in the tumor, many reports suggest that cells of the hematopoietic lineage are mobilized and then entrapped in tumor tissues. They then function as accessory cells promoting the sprouting of resident ECs by releasing angiogenic signals, such as VEGF and angiopoietins, and matrix metalloproteinases (MMPs) [29]. It is widely accepted that tumor cells are the major producers of proangiogenic factors [30]. However, not only cancer cells themselves but also the tumor stroma, composed of tumor-associated fibroblasts, pericytes, mesenchymal stem cells, as well as hematopoietic cells, produce proangiogenic factors and function as accessory cell components. Of all these potential accessory cells for tumor angiogenesis, in this chapter we focus on the complex role of hematopoietic lineage cells for angiogenesis, with an emphasis on the differentiation capacity of cells of the hematopoietic lineage to vascular cells such as ECs and mural cells [31, 32].

4.2 Developmental Association of Vascular Cells and Hematopoietic Cells

There is no doubt that the evolution of blood vessel formation is closely associated with increasing body size. In small animals, diffusion is sufficient to supply the body and heart's pumping activity is not required. However, as body size increases, diffusion is not sufficient to transport materials such as blood cells and fluid throughout the whole body. Therefore, heartlike tubes with diastalsis developed to facilitate blood flow. Moreover as body size increases, stromal cells and extracellular matrix fill the interstitium for structural maintenance of each tissue/organ, resulting in increased interstitial pressure. To compensate for this interstitial hypertension, intraluminal pressure in the blood vessel must be greater than in the interstitium to ensure transport of materials into the parenchyma, hence the requirement for blood vessels.

A close evolutionary association between blood vessels (blood flow) and hematopoietic cells is suggested from data on blood cell migration in *Drosophila* [33]. There are no blood vessels in the *Drosophila* embryo; however, hemocytes (mainly phagocytes) have to migrate to phagocytose the dead cells induced by programmed cell death. These cells develop in the head mesoderm and extensively migrate during development [34], but it has not been elucidated how developmental migrations are coordinated with hemocyte homing toward dead cells along the migration pathway. Cho *et al.* found that VEGF, usually utilized for the development, proliferation, and migration of ECs in vertebrates, is involved in hemocyte migration. (Figure 4.1a) [33]. The VEGF receptor homolog is expressed in hemocytes and VEGFs are expressed along hemocyte migration routes. This VEGF pathway guides developmental migration of blood cells. Thus, VEGF is utilized in evolution for blood flow in the absence of blood vessels.

The spaces between blocks of organs/tissues represents the initial space for development of tubes which eventually formed an open circulatory system. Gradually, the intraluminal space became coated with extracellular matrix material, to which some hematopoietic cells could adhere. Indeed, in *Halocynthia roretzi* (common sea squirt), there are no ECs in the blood vessels but some hemocytes adhere to matrices in the intraluminal part of the blood vessels (Figure 4.1b, unpublished data). At present, it is not clear whether hemocytes adhering to matrices actually represent the evolutionary origin of ECs, but in vertebrates the lining of the intraluminal part of blood vessels is almost exclusively by ECs which are covered with mural cells such as pericytes and vascular smooth muscle cells (Figure 4.1c), thus forming the closed circulatory system. Depending on the demands of the tissues, the closed circulatory systems supply materials in a controlled manner by permeability through intracellular and intercellular pathways of the ECs.

In vertebrates, close association between hematopoietic cells and ECs is first observed in the yolk sac, during early embryogenesis. Here, mesodermal cells form clusters termed *blood islands,* in the centers of which fetal nucleated erythrocytes are produced and where the outer cells differentiate into ECs, resulting in the formation

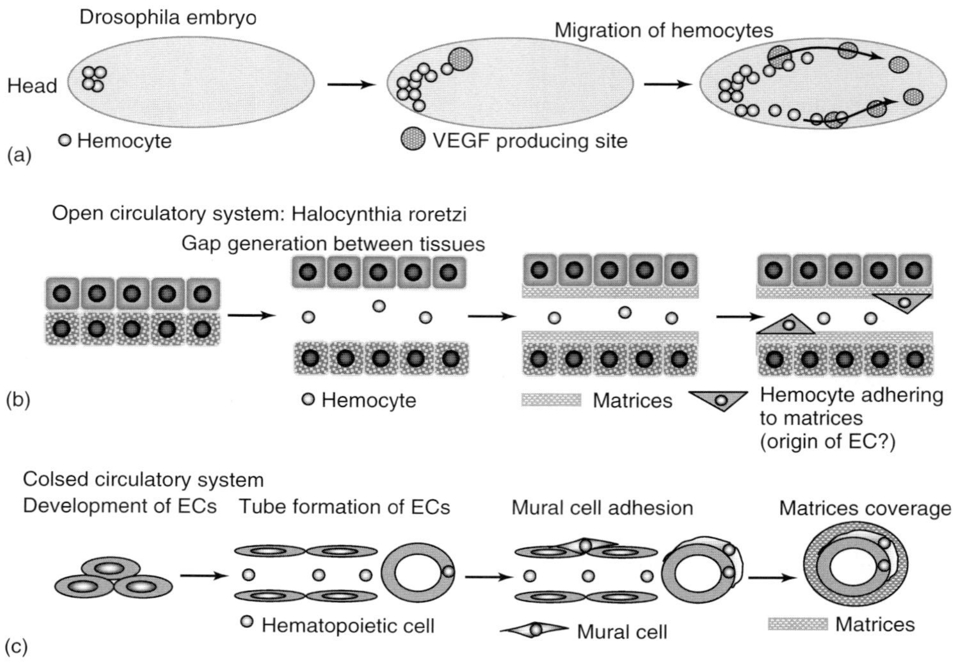

Figure 4.1 Evolution of blood vessels. (a) Migration of hematocytes caused by VEGF produced from cells in the apoptotic area of the *Drosophila* embryo. (b) Gaps between tissues/organs provide paths for blood flow in lower species with open circulatory systems; example of *Halocynthia roretzi* (unpublished data). (c) For closed circulatory systems, endothelial cells (ECs) form tubes and then mural cells adhere from outside the lumen. Finally matrices cover blood vessels for structural stabilization. VEGF, vascular endothelial growth factor; EC, endothelial cell.

of blood vessels. This type of blood vessel formation is termed *cavitation* and is not usually utilized in the intraembryonic tissues/organs. Thus, cells forming blood islands generate both hematopoietic cells and ECs, and are known as *hemangioblasts* [35]. Growth factors, transcription factors, signaling molecules for differentiation of hemangioblasts to ECs or hematopoietic stem cells (HSCs) have been extensively analyzed [36] and are summarized in Figure 4.2.

It was widely believed that hemangioblasts were no longer present after birth, but recently they have been shown to exist in adults as well, although their numbers are extremely low [37, 38]. There are several possibilities to explain the existence of hemangioblasts in adults. One is that they are remnant minor populations surviving from embryogenesis; this is consistent with the presence of hemangioblasts in human cord blood [39]. Alternatively, they may be derived from dedifferentiated cells by epigenetic gene modification. Interestingly, hemangioblastic activity has been observed in chronic myeloid leukemia tumor masses and is suggested to be involved in blood vessel formation therein [40, 41].

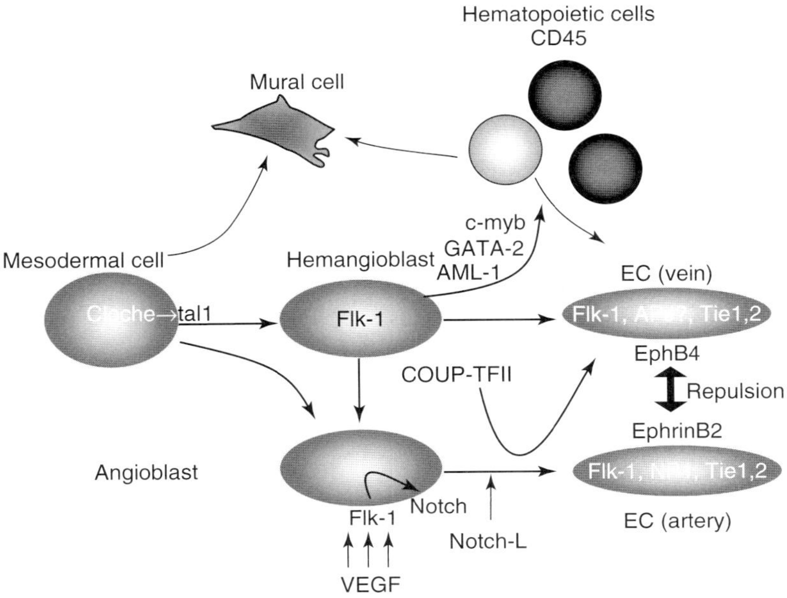

Figure 4.2 Existence of a common progenitor, termed a *hemangioblast*, differentiating into hematopoietic cells and endothelial cells (ECs). Receptors and transcription factors for specific development of each cell type are described in the schema. Differentiation potency of hematopoietic cells to vascular cells (mural cells and EC) has been reported. VEGF, vascular endothelial growth factor; EC, endothelial cell; tal1, T-cell acute lymphocytic leukemia 1; AML-1, acute myeloid leukemia-1; GATA-2, GATA binding protein-2; c-myb, transcription factor c-myb; COUP-TFII, chicken ovalbumin upstream promoter-transcription factor II; Flk-1; VEGFR2, VEGF receptor; Tie; TEK; angiopoietin receptor, NP-1, neuropilin-1; APJ, apelin receptor.

Due to the existence of a common ancestor, the hemangioblast, hematopoietic system, and vascular system are closely associated with each other during very early stages of embryogenesis. It has been suggested that ECs can support hematopoietic activity because HSCs adhere tightly to ECs at several sites in the embryo including the yolk sac [42], omphalomesenteric and vitelline artery, and dorsal aorta [16, 35, 42–45]. In addition, some stromal cell lines that are able to support hematopoiesis have been characterized as ECs [46]. These observations suggest a close interaction between hematopoiesis and vascular development. In addition to this evidence that formation of an endothelial network precedes hematopoiesis, a common molecular basis utilized in vascular development and hematopoiesis has been elucidated. Flk1/VEGFR2 mediates fundamental developmental steps in both endothelial angiogenesis and hematopoiesis [47–51]. Lack of Tie2, which is critical for interactions between mural cells and ECs, leads to failure of endothelial cell support for hematopoietic stem cell proliferation [16]. Thus, the vascular and hematopoietic systems are evolutionarily and developmentally closely associated and are both coordinately regulated under pathological as well as physiological conditions.

4.3 Inflammation and Cancer

Cancer has been called an *incurable inflammation*. In tissue repair accompanied by inflammation, granulation tissue formation is initiated by activation of fibroblasts and remodeling of the extracellular matrix, together with infiltration of inflammatory cells such as granulocytes, monocytes, lymphocytes, and so on. Simultaneously, angiogenesis is induced in the granulation tissue. Once inflammation is attenuated, and remodeled tissue is restored, the new blood vessels are reabsorbed. However, the tumor environment continues to contain many inflammatory cells such as macrophages, granulocytes, eosinophils, dendritic cells, and mast cells (MCs), as well as lymphocytes; under these conditions, there is no subsidence of angiogenic activity and tissue remodeling with extracellular matrices and myofibroblasts also continues.

Persistent infections by microbial and viral agents or prolonged exposure to environmental carcinogens can induce chronic inflammation. It is now recognized that cancers can develop from "subthreshold neoplastic states" by induction of somatic changes [52, 53]. Inflammatory conditions such as cigarette smoke-induced bronchitis, HBV-, and HCV-induced hepatitis, *Helicobacter pylori*-induced gastritis, inflammatory bowel disease, and papilloma virus-induced cervicitis are all closely associated with the ontogeny of cancer in the lung, stomach, liver, colon, and cervix, respectively. It has been estimated that a significant proportion of these chronic inflammatory diseases progresses to cancer in predisposed patients [54–56]. Caused by chronic inflammation, chromosomal genetic instability in susceptible cell populations can be induced by different factors such as reactive oxygen species released from inflammatory cells causing accelerated accumulation of genetic and epigenetic alterations that affect the expression or function of protooncogenes and tumor suppressor genes in somatic cells [54, 56]. Moreover, inflammatory cells produce different cytokines, chemokines, and matrix metalloproteinases which have been suggested to affect proliferation, survival, adhesion, and migration of transformed cells [56].

In acute inflammatory responses, neutrophils are the first recruited effectors. Subsequently monocytes, which finally differentiate into macrophages, migrate into the inflammatory foci. Activated macrophages produce various growth factors which promote proliferation and activation of ECs and fibroblasts in the inflammatory foci. MCs also participate in acute inflammation by releasing inflammatory mediators such as cytokines, histamine, and proteases. These factors secreted by inflammatory cells are important for generating granulation tissue; however, they also induce angiogenic switching during the premalignant stage of carcinogenesis and act to promote uncontrolled and disorganized blood vessel formation in the tumor environment [56–58].

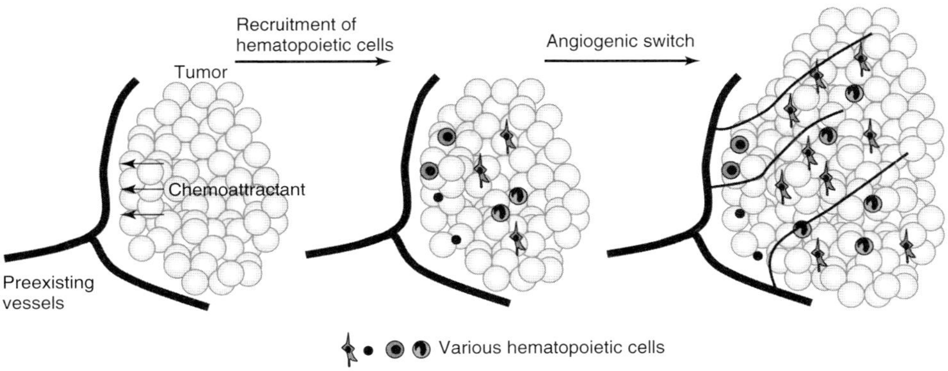

Figure 4.3 Hematopoietic cells promote angiogenesis as accessory cell components. Chemoattractants released in the tumor environment trigger migration of various hematopoietic cells into the tumor. Such hematopoietic cells produce many proangiogenic factors which result in promoting the angiogenic switch and abnormally rapid proliferation of blood vessels supporting tumor growth.

4.4 Hematopoietic Cells Promote Angiogenesis as an Accessory Cell Component

Hematopoietic lineage cells contributing to angiogenesis have been extensively analyzed under different physiological and pathological conditions. They are found to promote angiogenesis by secreting a variety of proangiogenic factors, including growth factors and MMPs. Moreover, direct differentiation of hematopoietic cells to vascular cells, such as ECs and mural-like cells, has also been reported. In the next section, the function of several hematopoietic cell types in promoting tumor angiogenesis is described (Figure 4.3 and Table 4.1).

4.4.1 Mast Cells (MC)

4.4.1.1 Physiological Function

MCs are found in tissues forming an interface between the host and the external environment, such as the skin, or respiratory and gastrointestinal mucosa. MCs are also located in muscular organs such as the heart and uterus, and around blood vessels in general, and form a "barrier" between host tissue and the bloodstream. The density of MCs may be substantially increased at sites of inflammation, wound healing, or tissue fibrosis. MCs are recruited from the bone marrow as progenitors circulating in blood and entering the tissue where their final maturation takes place. Thereafter, MCs may retain the capacity to proliferate on cytokine stimulation, suggesting that their density in tissue is dependent on local MC proliferation and MC progenitor influx.

Well-known roles of MCs reflect their involvement in allergy. When high-affinity IgE receptors on their surface are activated, de-granulation of effector molecules

Table 4.1 Summary of hematopoietic cells involved in tumor angiogenesis.

Cell types	Chemoattractants[a]	Proangiogenic factors	Differentiation[b]
Mast cell (MC)	SCF, IL3 Ligands for CXCR4, CCR3, CCR4	VEGF, bFGF, IL8, CCL2, CXCL12, MMP9, MCP-4, 6	Probably no
Monocyte/Mϕ	MCP-1(CCL2), CSF1, PlGF, CCL3, CCL4, CCL5, VEGF, S100A8, S100A9	VEGF, bFGF, TNFα, IL1β, CXCL8, PDGF, MMPs	ECs? Mural cells?
Tie2-expressing monocyte (TEM)	Ang1? 2?	bFGF	Not examined?
Myeloid-derived suppressor cells (MDSC)	CCL2, CXCL12, CXCL5, SCF, Bv8	MMP2, 9, 13, 12, 14, Bv8	ECs?
Neutrophil	CXCL1, 6, 8	MMP9, VEGF, oncostatin M, CXCL1, 9	No
Eosinophil	CCL11	VEGF, CCL1, bFGF, IL6, CXCL8, GM-CSF,	No
DC	VEGF, CXCL8, 12, HGF β-defensin	TNFα, osteopontin, CXCL1, 2, 3, 5, 8	ECs
HSC	SDF1, and probably by Ang1, VEGF	MMP2, 9	ECs, mural cells

[a] For migration of hematopoietic cells into the tumor.
[b] Differentiation of hematopoietic lineage cells into vascular cells.

(histamine, lipid mediators, cytokines, and proteases) is induced. MCs also have an important role in generating and maintaining innate and adaptive immune responses as well as in the development of autoimmune disorders and tolerance [59].

4.4.1.2 Localization in the Tumor Environment and Recruitment to Tumors

MCs are found in the tumor environment from premalignant stages through to malignant tumor progression, and especially at the periphery of invasive tumors, suggesting a host–tumor barrier function. Many studies have demonstrated the localization of MCs in tumors such as those of the lung [60], larynx [61], kidney [62], breast [63], stomach [64], esophagus [65], oral cavity [66], colon [67], and endometrium [68] as well as in melanoma [69] and multiple myeloma [70]. The density of MC in tumors correlates with the density of blood vessels, suggesting an association of MC with angiogenesis [71, 72].

There are many ways to induce recruitment of MCs into tumors by factors released by tumor constituents. Stem cell factor (SCF), a ligand for c-Kit expressed on MCs, is a major chemoattractant for these cells [73]. Inhibition of SCF expression in mammary tumor cells by antisense DNA attenuated infiltration of MC into the tumor environment, and resulted in a reduction of microvessel density in the tumor and decreased tumor growth [74]. Interleukin-3 (IL-3) is a growth factor for MC *in vitro* and also acts as a chemoattractant for murine MC (*in vitro*) [75]. It has been reported that expression of IL-3 in melanocytic lesions correlates with an increased number of MCs in the perilesional stroma, suggesting recruitment of MC by IL-3 [76]. Moreover, it has been suggested that B lymphocytes are required for establishing chronic inflammatory states that promote *de novo* carcinogenesis. Although the factors responsible have not yet been identified, mediators derived from B lymphocytes have been suggested to induce recruitment of MC in a *de novo* epithelial carcinogenesis model using K14-HPV16 mice [77]. There are certainly additional chemoattractants for MCs, because these cells express a variety of chemokine receptors such as CXCR4, CCR3, and CCR5 [78], and are ligands for these receptors to be expressed by tumors; they could also induce recruitment of MCs.

4.4.1.3 **Functions in Angiogenesis**

Important functions of MCs in promoting angiogenesis have been reported in several murine tumor models. Firstly, the role of host MC in tumor-associated angiogenesis was investigated by comparing the angiogenic response of genetically MC-deficient W/Wv c-Kit mutant mice and MC-sufficient +/+ littermates in a model of s.c. B16-BL6 melanoma. The angiogenic response was found to be slower and initially less intense in W/Wv mice than in +/+ mice. Furthermore, transplantation of normal bone marrow into W/Wv mice to repair this MC deficiency restored the full angiogenic response [79]. This strongly suggests that a function of MC is promoting angiogenesis. Similarly, using W/Wv mice, MC deficiency was reported to decrease premalignant angiogenesis in HPV16-induced cutaneous squamous cell carcinoma [80]. Using an MC inhibitor, Soucek *et al.* recently reported that MC recruitment is required for angiogenesis and macroscopic expansion of Myc-induced pancreatic islet tumors [81].

A number of reports indicate that MCs produce a variety of proangiogenic cytokines and growth factors, including VEGF, basic fibroblast growth factor (bFGF, also known as FGF2) IL-8, MMP9, tumor necrosis factor α (TNFα), transforming growth factor β (TGFβ), CCL2, CXCL8, heparin, and histamine. They also contain angiogenic proteases in their secretory granules [82]. In addition, MCs indirectly stimulate angiogenesis by secreting MC-specific serine proteases (monocyte chemoattractant protein-4 (MCP-4 and MCP-6) that activate pro-MMPs and stimulate fibroblasts to synthesize collagen [71, 80]. These factors must affect proliferation, migration, and tube formation of ECs in the tumor environment [71, 72]. It has been reported that MCs are observed around tumors before the formation of new blood vessels and then accumulate near blood vessels in tumors [74, 80]. From the intimate colocalization of MCs and ECs, transdifferentiation of MCs to ECs has even been suggested; however, such transdifferentiation capacity is controversial [83]. De-granulation and secretion of proangiogenic factors has been suggested to be induced by hypoxia depending on hypoxia-inducible factor (HIF) [84–86].

4.4.2 Monocyte/Macrophage Lineage Cells

4.4.2.1 Physiological Function

Macrophages develop from monocytes and are involved in the host response to implanted materials. They are activated in response to tissue damage or infection, causing an increase in the production of cytokines, chemokines, and other inflammatory molecules. The macrophage phenotype is characterized according to distinct functional properties, surface markers, and the cytokine profile. M1, classically activated proinflammatory macrophages, are known to be induced by interferon γ (IFN-γ) alone or in combination with LPS, TNF, and GM-CSF. In general, M1-activated macrophages then become IL-12^{high}, IL-23^{high}, IL-10^{low} secretors, metabolize arginine, produce high levels of inducible nitric oxide synthetase (iNOS), secrete toxic reactive oxygen and nitric oxide intermediates, as well as inflammatory cytokines such as IL-1β, IL-6, and TNF. They act as inducers and effector cells in Th1-type inflammatory responses. In contrast, M2, alternatively activated macrophages, are induced by exposure to a variety of signals including the cytokines IL-4, IL-13, and IL-10, immune complexes, and glucocorticoid or secosteroid (vitamin D3) hormones. M2-activated macrophages are IL-12^{low}, IL-23^{low}, and IL-10^{high} and possess the ability to facilitate tissue repair and regeneration [87]. They also have high levels of scavenger, mannose, and galactose receptors. Moreover, they produce arginase in place of arginine, subsequently generating ornithine and polyamines, and are involved in polarized Th2 reactions.

4.4.2.2 Localization in the Tumor Environment and Recruitment to Tumors

Macrophages infiltrating tumors are usually termed tumor-associated macrophages (TAMs). Typically, they are similar to M2 phenotype macrophages [88, 89]. However, growing evidence suggests that their phenotype is dependent on the stage of tumor development, with the display of a tumorigenic, M1-like phenotype in sites of chronic inflammation where tumors develop and a tumor-promoting, M2-like phenotype in established tumors [90].

Massive macrophage infiltrates are observed in many human tumors [91] but their biological significance is still not established. Historically, it was believed that activated macrophages are effector cells that phagocytose microorganisms and kill tumor cells. However, many recent clinical reports have highlighted a correlation between the degree of macrophage infiltration in tumors and a poor prognosis [92]. TAMs initially extravasate across the tumor vasculature as monocytes from the blood and then differentiate into TAMs *in situ* [93]. Recruitment of monocytes into the tumor environment is induced by various chemoattractants derived from constituents of the tumor, such as tumor cells themselves, and stromal cells [94]. MCP-1, also known as CCL2 [95–97] and colony stimulating factor-1 (CSF-1, also known as M-CSF) [92, 98] are well-known chemoattractants for monocytes/macrophages in tumors. The extent of MCP-1 expression in human cancers correlates with both TAM infiltration and tumor malignancy [99]. Using mice containing a recessive null mutation in the CSF-1 gene (op/op mice), a significant role of CSF-1 for recruitment of macrophages, tumor growth, and metastasis has been reported in the Polyoma Middle T antigen-expressing mouse model (PyMT mice) of breast cancer [92, 98]. Placental growth factor (PlGF, PGF) [100], CCL3 (MIP-1α), CCL4 (MIP-1β), CCL5 (RANTES) [101], VEGF [102], and S100A8 and S100A9 [103] have also been reported as chemoattractants for macrophages.

4.4.2.3 **Functions in Angiogenesis**

A correlation between a high number of TAMs and increased density of blood vessels has been reported in human tumors [104–107]. This strongly suggests that macrophages might induce tumor angiogenesis. Using a transgenic mouse susceptible to mammary cancer, the PyMT mouse mentioned above, it has been suggested that the angiogenic switch, identified as the formation of a high-density vessel network, and progression to malignancy, are both regulated by macrophages infiltrating the primary mammary tumors. Depletion of macrophages using genetically CSF-1-deficient mice resulted in a 50% reduction in vascular density causing delayed tumor progression and metastasis, whereas genetic restoration of the macrophage population specifically in these tumors rescued the vessel phenotype. Furthermore, premature induction of macrophage infiltration into premalignant lesions promoted early onset of the angiogenic switch, independent of tumor progression [108]. Moreover, when macrophages were depleted using clodronate liposomes, both neovascularization and tumor growth were reported to be reduced in the LL cell tumor xenograft model [109].

TAMs express a variety of proangiogenic and matrix remodeling factors such as VEGF, bFGF, TNFα, interleukin 1β (IL-1β) CXCL8, PDGF, MMPs, and urokinase-type plasminogen activator (uPA) [57]. They frequently infiltrate necrotic areas of tumors to scavenge cellular debris. In addition, they accumulate in hypoxic areas of the tumor. It has been reported that hypoxia stimulates the expression of several proangiogenic factors by stabilizing HIFs in TAMs [110]. Interestingly, upregulation of NFκB in macrophages is induced under tissue hypoxia and it has been suggested that high levels of NFκB activation in murine TAMs skews them toward a protumor M2 phenotype [111]. In contrast, a previous report showed that reduced levels of NFκB activity promote an M2 phenotype in TAMs in murine fibrosarcoma [89]. Therefore, it is unclear whether hypoxia relating to the NFκB pathway actually affects macrophage differentiation or not.

In terms of transdifferentiation of TAMs to vascular cells, recent studies have also suggested that monocyte/macrophage lineage cells can indeed transdifferentiate into ECs and/or mural-like cells when exposed to sustained stimulation by angiogenic growth factors [112–114]. It remains to be determined whether monocytes newly recruited into tumors or differentiated into TAMs directly contribute to the formation of new tumor blood vessels in this way.

4.4.3 Tie 2–Expressing Monocytes (TEM)

4.4.3.1 Physiological Function

Tie2-expressing monocytes (TEMs), as the name suggests, are a subset of circulating and tumor-infiltrating monocytes that express Tie2, a receptor for angiopoietins. It appears that TEMs are a subpopulation of TAMs, possibly overlapping with M2 macrophages [115]. However, TEMs can be distinguished from other TAM populations by their surface marker profile [116–119]. TEMs have been identified in the peripheral blood of both humans and mice [116–119] and are distinct from the previously described circulating endothelial cells (CEPs) or EPCs expressing Tie2 [118]. However, the physiology of TEMs is not yet well understood.

4.4.3.2 Localization in the Tumor Environment and Recruitment to Tumors

TEMs have been observed in several mouse tumor models such as subcutaneous tumor grafts, orthotopically growing gliomas and spontaneous pancreatic tumors [116]. TEMs represent 1–15% of total $CD11b^{+}$ myeloid cells in tumors but are infrequently observed in normal organs [116–119]. In humans, TEMs are found in tumors including those of the kidney, colon, pancreas, and lung, as well as in soft tissue sarcomas [118]. TEMs preferentially localize around angiogenic tumor vessels and are absent from necrotic tumor regions.

Targeted elimination of TEMs by means of a suicide gene reduced tumor angiogenesis of human gliomas grafted into the mouse brain and induced substantial tumor regression (see Section 4.5) [116]. As TEM elimination did not affect the recruitment of TAMs to necrotic tumor areas, it is unlikely that TEMs comprise precursors of other TAM populations. Rather, it would appear that TEMs represent a distinct lineage of monocytes/macrophages.

Unlike monocytes, TEMs seem to not express chemokine (C–C motif) receptor 2 (CCR2), suggesting that they are recruited to tumors by a mechanism different from that of monocytes, which are a source of TAMs. It has been suggested that an immature HSC population expressing Tie2 is recruited into the tumor by molecules secreted from cells in the tumor environment [120]. Ang1 induces recruitment of hematopoietic stem/progenitor cells from bone marrow [121], suggesting that ligands for Tie2 expressed on TEMs may induce recruitment of TEMs into the tumor. Indeed, several lines of evidence have shown that TEMs could be recruited by Ang2 [117, 118], a cytokine abundantly expressed by both hypoxic tumor cells and ECs in tumor blood vessels [122, 123]. However, colocalization of TEM and Ang2 has not been shown in tumors.

4.4.3.3 Functions in Angiogenesis

Several lines of evidence suggest a close association of TEM with tumor angiogenesis in human and mouse tumor models [116, 118]. However, at present, how TEMs regulate tumor angiogenesis is not precisely determined, but secretion of the proangiogenic cytokine bFGF from TEMs has been suggested [116].

4.4.4 Myeloid-Derived Suppressor Cells (MDSCs)

4.4.4.1 Physiological Function

$CD11b^+/Gr1^+$ myeloid cells are a heterogeneous population including cells of the granulocytic lineage (e.g., neutrophils), the monocytic lineage, immature dendritic cells (iDCs), and progenitor populations capable of differentiating into any of the other three subpopulations [124]. In addition, as the name myeloid-derived suppressor cells (MDSCs) suggests, a fraction of $CD11b^+/Gr1^+$ cells has been associated with suppression of immune function. MDSCs can be broadly characterized by their expression of Gr1. $CD11b^+Gr1^{high}$ cells have an immature neutrophil phenotype, and $CD11b^+Gr1^{low}$ cells resemble monocytes [125, 126].

In contrast to mature neutrophils, MDSCs are potently immunosuppressive and express low levels of major histocompatibility complex (MHC) class II molecules and CD80 required for T-cell activation and survival [125, 127, 128]. Phenotypically, MDSCs are quite similar to immature myeloid cells; however, recent studies have demonstrated clear differences in the biology of normal immature myeloid cells and the MDSCs that accumulate in tumor-bearing hosts [129]. Indeed, a possible role of MDSC in suppressing the antitumor function of T- and natural killer (NK) cells has been suggested [129].

4.4.4.2 Localization in the Tumor Environment and Recruitment to Tumors

The presence of MDSCs has been observed in the peripheral blood of cancer patients and in tumors [130–132]. Correlation of the number of circulating MDSCs with the clinical stage has been reported in cancer patients [131]. In several tumor models in mice, increased numbers of MDSCs were observed in bone marrow, peripheral blood, and spleen [128, 133–135]. Especially in the tumor environment from colorectal carcinoma or LLC, it has been estimated that the tumor contains around 5% of MDSCs [136].

It has been suggested that recruitment of the MDSC is induced by a variety of chemokines and cytokines such as CCL2, CXCL12, CXCL5, and SCF through their respective cognate receptors CCR2, CXCR4, CXCR2, and c-Kit on MDSC [126, 135, 137]. Recently, Bv8 was characterized as a crucial regulator of MDSC [134]. Both Bv8 and EG-VEGF bind two highly homologous G-protein-coupled receptors termed PKR-1 (also known as EG-VEGFR1) and PKR-2 (EG-VEGFR2) [138]. Bv8 expression is upregulated in $CD11b^+/Gr1^+$ cells following implantation of tumor cells [137] and induces mobilization of hematopoietic cells into the peripheral blood [139]. Thus, it appears that in tumor-bearing mice, increased levels of Bv8 mobilize MDSCs from the bone marrow and increase their number in the circulation.

4.4.4.3 Functions in Angiogenesis

Originally, MDSCs were thought to play a role in immunosuppression of antitumor effectors, exerted by their production of arginase 1 and iNOS (nitric oxide synthase 2A) which suppress the antitumor functions of T- and NK cells [140–142]. However, MDSCs also induce tumor angiogenesis significantly. Coinjection of MDSCs and colorectal cancer cells enhanced tumor angiogenesis, and resulted in rapid growth of tumors [136]. Upon inhibition of MDSC infiltration using anti-Bv8 neutralizing antibody, tumor angiogenesis was attenuated [139].

MDSCs produce a variety of proangiogenic factors. For matrix remodeling, MDSCs in the tumor environment produce several MMPs (MMP2, MMP9, MMP13, and MMP14) [135, 136]. Hypoxia induces expression of Bv8 and CXCL12 [134, 143], suggesting that MDSCs are recruited to the site of hypoxia and can function effectively under these conditions. Upon exposure to tissue hypoxia, immature myeloid cell populations can differentiate into vascular cells such as ECs and mural cells [114]. It has been reported that some MDSCs change their morphology from round to bipolar shape and also express endothelial markers such as CD31 and VEGFR2 [136], suggesting that some have the potential to differentiate and be incorporated into the tumor endothelium. Taken together, these data suggest that MDSCs contribute to tumor angiogenesis mainly by secreting MMPs and differentiating into vascular cells.

4.4.5 Neutrophils

4.4.5.1 Physiological Function

Neutrophils are terminally differentiated cells. Differentiation from myeloblastic and myelocytic progenitors involves tightly regulated sequential gene expression that leads to the formation of a granulocyte with specific protein contents. Hematopoietic cytokines promote neutrophil progenitor proliferation and differentiation; the major cytokine for neutrophil proliferation and survival is granulocyte-colony stimulating factor (G-CSF) [144]. Mice and humans deficient in either G-CSF or its receptor suffer from profound neutropenia. These phagocytic, polymorphonuclear neutrophils are the most abundant but also very short-lived human white blood cells, acting as first defenders in acute inflammatory responses to invading microorganisms. Neutrophil turnover is rapid, $\sim 10^9$ cells per kg of body weight leave the bone marrow per day in healthy humans. Human neutrophils express many cell surface markers commonly expressed on other subsets of leukocytes. CD66B (also known as CEACAM8) is suggested to be uniquely expressed by neutrophils. Identification of neutrophils within human tumors has to date, mainly relied on staining with the cytoplasmic marker myeloperoxidase (MPO). In murine tumors, $Gr1^+$ cells are usually considered to be neutrophils or cells derived from a neutrophil precursor. However, it should be remembered that murine TEM and MDSC also express Gr1 [58].

4.4.5.2 Localization in the Tumor Environment and Recruitment to Tumors

Neutrophils are abundantly present in most tumor types [145, 146]. A correlation between the number of neutrophils and clinical prognosis has been reported [147]. Recruitment of neutrophils is mainly mediated by CXC chemokines overexpressed in various human cancer and tumor cell lines [148] that bind to CXCR1 (also known as IL8RA) and/or CXCR2 [145, 149–151]. The level of expression of CXC8 correlates with the number of neutrophils infiltrating the tumor environment [147, 152]. Moreover, it has been reported that the number of neutrophils also correlates with microvessel density in tumors of patients with myxofibrosarcoma [153]. Neutrophil depletion using anti-Gr1 antibody inhibited CXCL1- or CXCL8-mediated tumor angiogenesis significantly [154]. In contrast, enhancement of neutrophil infiltration by overexpression of CXCL6 in human melanoma cells induced marked angiogenesis in nude mouse xenograft models [155]. In RIP1-Tag2 transgenic mice, neutrophils were predominantly found inside angiogenic islet dysplasias and tumors in the pancreas. Transient depletion of neutrophils using anti-Gr1 antibody significantly reduced the frequency of initial angiogenic switching in dysplasias [156], suggesting that infiltrating neutrophils can play a crucial role in activating angiogenesis in a previously quiescent tissue vasculature during the early stages of carcinogenesis.

4.4.5.3 Functions in Angiogenesis

Neutrophils infiltrating tumors produce MMP9 to degrade extracellular matrices and induce angiogenesis [157, 158]. MMP9 activity is suppressed by the formation of MMP9–TIMP1 complexes. However, neutrophils seem not to produce TIMP1 themselves [159], suggesting that MMP9 produced from neutrophils may rapidly stimulate angiogenesis [160]. Upon activation, neutrophils secrete several proangiogenic factors. When they are stimulated with TNFα, they release VEGF, resulting in promotion of angiogenesis [161]. Cell-to-cell contact of neutrophils with breast cancer cells induces oncostatin M production from the former, which in turn induces VEGF production from the latter [162]. Moreover, neutrophils produce CXCL8 and CXCL1 [163]. CXCL8 seems to act on neutrophils via an autocrine loop, resulting in release of MMP9 [164]. Neutrophils seem not to express hypoxic sensors because they do not produce VEGF upon exposure to hypoxia [165].

4.4.6 Eosinophils

4.4.6.1 Physiological Function

It is widely accepted that eosinophils are terminally differentiated, nonreplicating effector cells playing a central, potentially beneficial role in the clearance of parasitic infections, primarily through the exocytotic release of eosinophil granule-derived cytotoxic proteins. Eosinophils play potentially maladaptive roles in asthma and other atopic diseases by releasing specific eicosanoid lipids, that is, leukotriene C4, and cationic granule-derived proteins. However, accumulating evidence has suggested additional roles for eosinophils in immunoregulatory systems involved in both, the adaptive and innate arms of immunity. Specifically, eosinophils have key immunoregulatory roles as antigen-presenting cells and as modulators of $CD4^+$ T-cell, dendritic cell, B cell, MC, neutrophil, and basophil functions [166]. Phenotypically, eosinophils are characterized by the expression of CD125, a low affinity receptor for IL-5 as well as CCR3, and specific intracellular granules containing cationic proteins. They also express receptors for IgE [167].

4.4.6.2 Localization in the Tumor Environment and Recruitment to Tumors

It has been reported that eosinophils are present in various tumors, such as nasopharyngeal [168] and oral squamosal carcinomas [169], gastrointestinal tumors [170] and Hodgkin lymphoma [171]. Correlations between the number of eosinophils and clinical stage have not yet been established. Because of the expression of CCR3 on eosinophils, CCL11, a ligand for CCR3, has been suggested to recruit these cells [172, 173]. Indeed, a correlation between increased numbers of eosinophils and expression of CCL11 by tumor components of Hodgkin lymphomas has been reported [172]. However, the precise molecular mechanism of eosinophil recruitment to tumors has not been elucidated.

4.4.6.3 Functions in Angiogenesis

It has been suggested that eosinophils produce a variety of proangiogenic factors, such as VEGF [174], CCL11 [175], bFGF, IL-6, CXCL8, GM-CSF, PDGF, TGFβ [176], and MMP9 [177]. Of these, using *in vivo* models such as aorta ring assays and CAM assays, eosinophil-derived VEGF and CCL11 have been shown to induce angiogenesis [174, 175]. However, the function of these molecules in tumorigenesis, especially in angiogenesis, has not been clarified. In some tumor types, it seems that eosinophilia is associated with tumor suppression rather than angiogenesis [178]. Therefore, further precise analysis of the eosinophil function in tumor angiogenesis is required.

4.4.7 Dendritic Cells (DCs)

4.4.7.1 Physiological Function

Populations of DCs are heterogeneous and reside in most peripheral tissues, particularly at sites of the interface with the environment; for example, skin and mucosa. DCs patrol through the blood, peripheral tissues, lymph, and secondary lymphoid organs. In the periphery, immature DCs (iDCs) can capture and process antigens. Thereafter, they migrate toward T-cell-rich areas of the secondary lymphoid organs through the afferent lymphatics. During this migration, DCs lose their capacity to internalize further antigens and acquire the capacity to present their captured antigens to naïve T-cells, a process referred to as *maturation of DC*. Antigens are presented in such a way that antigen-specific naïve T-cells become activated and start to proliferate (T-cell priming). Moreover, when DCs initiate a T-cell-mediated adaptive immune response, they also play an important role in polarization of T-cell reactivity toward type-1 and/or type-2 responses [179].

4.4.7.2 Localization in the Tumor Environment and Recruitment to Tumors

Immature rather than mature DCs have been reported to be located in a variety of tumors [167, 180–182]. VEGF [183, 184], β-defensin [185], CXCL12 [186], HGF [187], and CXCL8 [188] produced in the tumor environment have been suggested to induce recruitment of iDCs from the peripheral blood [58].

4.4.7.3 Functions in Angiogenesis

iDCs produce a variety of proangiogenic factors and induce angiogenesis directly and indirectly. TNFα and CXCL8 produced by iDCs in ovarian cancer ascites have been shown to promote angiogenesis *in vivo* [186, 188]. Osteopontin released from iDCs has been suggested to act as a trigger for the production of the proangiogenic factor IL-1β from monocytes [189]. Various chemokines such as CXCL1, CXCL2, CXCL3, and CXCL5 produced by DCs have been suggested to recruit cells other proangiogenic myeloid cells [190]; however, the *in vivo* function of chemokines derived from DCs has not been investigated in detail [58]. Interestingly, it was recently reported that a population of $CD45^{+}CD11c^{+}MHC\text{-}II^{+}$ DC precursors that massively infiltrate human ovarian carcinomas also express the EC-specific marker VE-cadherin. Those cells could assemble into blood vessels *in vivo* following implantation of tumor-derived $CD45^{+}$VE-cadherin^{+} cells into the flanks of immunodeficient mice [191]. Moreover, differentiation of tumor-derived $CD11c^{+}$ DCs into endothelial-like cells has been induced by culture with VEGF and oncostatin M [185, 192]. Therefore, it may be possible that such iDCs directly contribute to blood vessel formation in tumors.

That DCs may possess proangiogenic properties is supported by the finding that VEGF produced from tumors [184, 193] sustains the immature phenotype of iDCs and their proangiogenic activity induced by HGF [187], TGF β [194], prostaglandin E2 [195], lactate [196], and osteopontin [197] produced by tumors [58].

4.4.8
Hematopoietic Stem Cells (HSCs)

4.4.8.1 Physiological Function

HSCs are responsible for production and maintenance of mature blood cells. They are defined by two functional properties: vast proliferative self-renewal ability, and multilineage hematopoietic differentiation potential. In adults, localization of HSCs has traditionally been thought to be restricted to the bone marrow microenvironment. However, a novel view has emerged that HSCs present in adult tissue may be recruited from the bone marrow through the peripheral circulation, into tissues, becoming part of a regenerative and/or inflammatory process at these "distal" sites. This process has been referred to as the *plasticity of stem cells.* Once within the tissue, the fate of these primitive HSCs is determined by locally elaborated growth factors which control their development into hematopoietic cells or tissue-specific cells, including vascular cells such as ECs and mural cells [114].

4.4.8.2 Localization in the Tumor Environment and Recruitment to Tumors

There are no reports of the presence of HSC populations in the human tumor environment. In a mouse tumor xenograft model, using the PC3 prostate cancer cell line inoculated into nude mice, abundant $CD45^{+}$c-kit^{+}SCA-1^{+} HSCs are observed in the periphery of tumors, especially at the interface between normal tissue and tumor tissue in the early stage of tumorigenesis [120]. In the later stage of tumorigenesis, similar to TEMs and MDSCs, HSCs locate in the near vicinity of

ECs. Inhibition of HSC migration into the tumor site, using neutralizing antibody to c-Kit-suppressed maturation of blood vessels, especially in the periphery of the tumor suggests a function of HSCs in furthering the maturation process of blood vessels and the importance of SCF for migration of HSCs into the tumor [120]. Possible molecular cues for migration of HSCs into tumor sites are SDF1 [198], VEGF, and Ang1 [121].

4.4.8.3 Functions in Angiogenesis

In vivo functions of HSCs in tumor angiogenesis have not been established; however, HSCs induce migration of ECs mediated by the production of Ang1, suggesting that Ang1 from HSCs induces sprouting angiogenesis from preexisting blood vessels [199]. It has been reported that peripheral CD34-positive HSCs/progenitors express high levels of MMP-2 and -9 [200]. Our own preliminary data also show that embryonic HSCs (all positive for CD45, c-Kit, CD34, and negative for Lineage markers) prominently express MMP-9 (data not shown). These findings suggest that HSCs digest extracellular matrices by means of MMPs and induce chemotaxis of ECs by the Ang1 that they secrete.

AML1 (acute myeloid leukemia-1)/Runx1-deficient mice lack HSC development and show lethality during embryogenesis because of massive hemorrhages in the ventricles of the central nervous system, in the vertebral canal, within the pericardial space, and in the peritoneal cavity. Using this model, it has been reported that HSCs support newly developed blood vessels structurally by their differentiation into mural cells/pericytes [114]. The differentiation capacity of HSCs from adult bone marrow, and not only from embryonal cells, into mural cells has now been documented by several groups [31, 114]. It has been suggested that tumor pericytes are recruited from the bone marrow and contribute to vascular stabilization and survival of the vascular niche in pancreatic tumors [201]. In addition, detection of bone marrow HSC-derived pericytes within the newly formed vascular architecture has been reported in central nervous system tumors [202].

4.5 Clinical Therapeutic Applications

Based on the evidence of important roles played by cells of the hematopoietic lineage in promoting angiogenesis, blocking the infiltration of hematopoietic cells into tumors is a promising approach for managing cancer in patients. When tumor cells were inoculated into mice after bone marrow suppression using anti-c-kit antibody, tumor growth was severely suppressed [120]. However, once the tumor mass is established, this method is not effective and inhibition of tumor growth is limited (unpublished data). This suggested that hematopoietic cells are actually important for initiating tumor angiogenesis, the so-called angiogenic switch. In human cancer patients, because the tumor mass is already established at diagnosis and tumor angiogenesis has started, inhibition of hematopoietic migration alone may not effectively suppress tumor angiogenesis although progression of metastasizing lesions before angiogenic switching occurs might be suppressed.

To suppress the growth of the original tumor as well as the metastases, depletion of hematopoietic cells or silencing of proangiogenic activity in already-migrating hematopoietic cells should result in regression of the blood vessels in the tumor environment. De Palma *et al.* used *tk* suicide gene transfection for depletion of TEM in the tumor environment [116]. They generated transgenic mice harboring the *tk* gene under the transcriptional control of the Tie2 promoter (Tie2-tk mice). In theory, proliferating cells expressing Tie2 could then be selectively killed by the administration of ganciclovir, using gene-directed enzyme–prodrug therapy. When tumor-bearing nude mice that had been previously transplanted with bone marrow-derived from the Tie-2-tk mice were treated with ganciclovir, TEMs in the tumor environment were completely eliminated, resulting in a marked reduction in tumor angiogenesis and growth [116]. As in this experiment, inhibition of maintenance, including migration of monocytes into murine tumors using neutralizing antibodies that block the activity of specific molecules such as CSF1, PlGF, and Bv8, tumor growth as well as tumor angiogenesis could be markedly suppressed [97, 134, 203, 204]. Moreover, the angiotensin type 1 receptor has been suggested to be involved in macrophage infiltration into tumors and the usefulness of blockade of this receptor (ARB) for suppression of tumor growth has been investigated in clinical trials [205, 206].

Because it seems that regulation of hematopoietic cells in tumors is a promising approach for managing cancer patients, in order to develop an optimal strategy for inhibition of tumor angiogenesis, it is important to understand how responses of the inflammatory system as a whole are affected when one chemokine, myeloid cell type or proangiogenic factor is ablated.

4.6 Conclusions

In this chapter, the function of hematopoietic cells as accessory cell constituents supporting tumor angiogenesis has been described. As accessory cells, stromal cells such as fibroblasts and mural cells also act to promote angiogenesis. It has been reported that carcinoma-associated fibroblasts (CAFs) extracted from human breast carcinomas promote the growth of admixed breast carcinoma cells significantly more than do normal mammary fibroblasts derived from the same patients [207]. In that study, it was reported that CAFs promote angiogenesis by recruiting EPCs into carcinomas, an effect mediated in part by SDF-1/CLCL12. In addition, several papers suggested that CAFs secrete a variety of proangiogenic factors such as VEGF [208, 209], PDGF-C [210], and MMPs [211].

Although tumor hypoxia and VEGF form the main axis for the induction of tumor angiogenesis, it has emerged that VEGF production can be triggered by several components in the tumor microenvironment, such as hematopoietic cells and stromal cells, perhaps recruited from the bone marrow. At present, the function of lymphocytes and erythrocytes in tumor angiogenesis is not known; however, they may also induce angiogenesis as observed for other hematopoietic components. Including these hematopoietic lineage cells, precise molecular analysis of how hematopoietic cells and other stromal cells such as CAFs act as accessory cell components to induce tumor angiogenesis will shed light on identifying new molecular targets to efficiently inhibit tumor angiogenesis.

References

1. Risau, W. (1997) Mechanisms of angiogenesis. *Nature*, **386**, 671–674.
2. Folkman, J. and D'Amore, P.A. (1996) Blood vessel formation: what is its molecular basis? *Cell*, **87**, 1153–1155.
3. Hanahan, D. (1997) Signaling vascular morphogenesis and maintenance. *Science*, **277**, 48–50.
4. Wang, H.U., Chen, Z.-F., and Anderson, D.J. (1998) Molecular distinction and angiogenic interaction between embryonic arteries and veins revealed by ephrin-B2 and its receptor Eph-B4. *Cell*, **93**, 741–753.
5. Gale, N.W. and Yancopoulos, G.D. (1999) Growth factors acting via endothelial cell-specific receptor tyrosine kinases: VEGFs, Angiopoietins, and ephrins in vascular development. *Genes Dev.*, **13**, 1055–1066.
6. Shalaby, F., Rossant, J., Tamaguchi, T.P., Gertsenstein, M., Wu, X.-F., Breitman, M.L., and Schuh, A.C. (1995) Failure of blood-island formation and vasculogenesis in Flk-1-deficient mice. *Nature*, **376**, 62–66.
7. Fong, G.-H., Rossant, J., Gertsenstein, M., and Breitman, M.L. (1995) Role of the Flt-1 receptor tyrosine kinase in regulating the assembly of vascular endothelium. *Nature*, **376**, 66–70.
8. Dumont, D.J., Jussila, L., Taipale, J., Lymboussaki, T., Mustonen, T., Pajusola, K., Breitman, M., and Alitalo, K. (1998) Cardiovascular failure in mouse embryos deficient in VEGF receptor-3. *Science*, **282**, 946–949.
9. Ferrara, N., Carver-Moore, K., Chen, H., Dowd, M., Lu, L., O'Shea, K.S., Powell-Braxton, L. Hillan, K.J. *et al.* (1996) Heterozygous embryonic lethality induced by targeted inactivation of the VEGF gene. *Nature*, **380**, 439–442.
10. Dumont, D.J., Fong, G.-H., Puri, M.C., Gradwohl, G., Alitalo, K., and Breitman, M.L. (1995) Vascularization of the mouse embryo: a study of flk-1, tek, tie, and vascular endothelial growth factor expression during development. *Dev. Dyn.*, **203**, 80–92.
11. Dumont, D.J., Gradwohl, G., Fong, G.-H., Puri, M.C., Gerstenstein, M., Auerbach, A., and Breitman, M.L. (1994) Dominant-negative and targeted null mutations in the endothelial receptor tyrosine kinase, tek, reveal a critical role in vasculogenesis of the embryo. *Genes Dev.*, **8**, 1897–1909.
12. Puri, M.C., Rossant, J., Alitalo, K., Bernstein, A., and Partanen, J. (1995) The receptor tyrosine kinase TIE is required for integrity and survival of vascular endothelial cells. *EMBO J.*, **14**, 5884–5891.
13. Sato, T.N., Tozawa, Y., Deutsch, U., Wolburg-Buchholz, K., Fujiwara, Y., Gendron-Maguire, M., Gridley T. *et al.* (1995) Distinct roles of the receptor tyrosine kinases Tie-1 and Tie-2 in blood vessel formation. *Nature*, **376**, 70–74.
14. Davis, S., Aldrich, T.H., Jones, P.F., Acheson, A., Compton, D.L., Jain, V., Ryan, T.E. *et al.* (1996) Isolation of angiopoietin-1, a ligand for the TIE2 receptor, by secretion-trap expression cloning. *Cell*, **87**, 1161–1169.
15. Maisonpierre, P.C., Suri, C., Jones, P.F., Bartunkova, S., Wiegand, S.J., Radziejewski, C., Compton, D. *et al.* (1997) Angiopoietin-2, a natural antagonist for Tie2 that disrupts in vivo angiogenesis. *Science*, **277**, 55–61.
16. Takakura, N., Huang, X.-L., Naruse, T., Hamaguchi, I., Dumont, D.J., Yancopoulos, G.D., and Suda, T. (1998) Critical role of the TIE2 endothelial

cell receptor in the development of definitive hematopoiesis. *Immunity*, **9**, 677–686.

17. Suri, C., Jones, P.F., Patan, S., Bartunkova, S., Maisonpierre, P.C., Davis, S., Sato, T.N., and Yancopoulos, G.D. (1996) Requisite role of angiopoietin-1, a ligand for the Tie-2 receptor, during embryonic angiogenesis. *Cell*, **87**, 1171–1180.
18. Hellström, M., Kalén, M., Lindahl, P., Abramsson, A., and Betsholtz, C. (1999) Role of PDGF-B and PDGFR-beta in recruitment of vascular smooth muscle cells and pericytes during embryonic blood vessel formation in the mouse. *Development*, **126**, 3047–3055.
19. Lindblom, P., Gerhardt, H., Liebner, S., Abramsson, A., Enge, M., Hellstrom, M., Backstrom, G. *et al.* (2003) Endothelial PDGF-B retention is required for proper investment of pericytes in the microvessel wall. *Genes Dev.*, **17**, 1835–1840.
20. Oh, H., Takagi, H., Suzuma, K., Otani, A., Matsumura, M., and Honda, Y. (1999) Hypoxia and vascular endothelial growth factor selectively up-regulate angiopoietin-2 in bovine microvascular endothelial cells. *J. Biol. Chem.*, **274**, 15732–15739.
21. Risau, W. and Flamme, I. (1995) Vasculogenesis. *Annu. Rev. Cell Dev. Biol.*, **11**, 73–91.
22. Red-Horse, K., Crawford, Y., Shojaei, F., and Ferrara, N. (2007) Endothelium-microenvironment interactions in the developing embryo and in the adult. *Dev. Cell*, **12**, 181–194.
23. Asahara, T., Murohara, T., Sullivan, A., Silver, M., van der Zee, R., Li, T., Witzenbichler, B. *et al.* (1997) Isolation of putative progenitor endothelial cells for angiogenesis. *Science*, **275**, 964–967.
24. Lyden, D., Hattori, K., Dias, S., Costa, C., Blaikie, P., Butros, L., Chadburn, A. *et al.* (2001) Impaired recruitment of bone-marrow-derived endothelial and hematopoietic precursor cells blocks tumor angiogenesis and growth. *Nat. Med.*, **7**, 1194–1201.
25. Göthert, J.R., Gustin, S.E., van Eekelen, J.A., Schmidt, U., Hall, M.A., Jane, S.M., Green, A.R. *et al.* (2004) Genetically tagging endothelial cells in vivo: bone marrow-derived cells do not contribute to tumor endothelium. *Blood*, **104**, 1769–1777.
26. Purhonen, S., Palm, J., Rossi, D., Kaskenpää, N., Rajantie, I., Ylä-Herttuala, S., Alitalo, K. *et al.* (2008) Bone marrow-derived circulating endothelial precursors do not contribute to vascular endothelium and are not needed for tumor growth. *Proc. Natl. Acad. Sci. U.S.A.*, **105**, 6620–6625.
27. Nolan, D.J., Ciarrocchi, A., Mellick, A.S., Jaggi, J.S., Bambino, K., Gupta, S., Heikamp, E. *et al.* (2007) Bone marrow-derived endothelial progenitor cells are a major determinant of nascent tumor neovascularization. *Genes Dev.*, **21**, 1546–1558.
28. Shaked, Y., Ciarrocchi, A., Franco, M., Lee, C.R., Man, S., Cheung, A.M., Hicklin, D.J. *et al.* (2006) Therapy-induced acute recruitment of circulating endothelial progenitor cells to tumors. *Science*, **313**, 1785–1787.
29. Palma, M.D. and Naldini, L. (2006) Role of haematopoietic cells and endothelial progenitors in tumour angiogenesis. *Biochim. Biophys. Acta*, **1766**, 159–166.
30. Ferrara, N. (2004) Vascular endothelial growth factor: basic science and clinical progress. *Endocr. Rev.*, **25**, 581–611.
31. Sata, M., Saiura, A., Kunisato, A., Tojo, S., Okada, T., Tokuhisa, H., Hirai, M. *et al.* (2002) Hematopoietic stem cells differentiate into vascular cells that participate in the pathogenesis of atherosclerosis. *Nat. Med.*, **8**, 403–409.
32. Rafii, S. and Lyden, D. (2003) Therapeutic stem and progenitor cell transplantation for organ vascularization and regeneration. *Nat. Med.*, **9**, 702–712.
33. Cho, N.K., Keyes, L., Johnson, E., Heller, J., Ryner, L., Karim, F., and Krasnow, M.A. (2002) Developmental control of blood cell migration by the Drosophila VEGF pathway. *Cell*, **108**, 865–876.

34. Tepass, U., Fessler, L.I., Aziz, A., and Hartenstein, V. (1994) Embryonic origin of hemocytes and their relationship to cell death in Drosophila. *Development*, **120**, 1829–1837.
35. Eichmann, A., Corbel, C., Nataf, V., Vaigot, P., Breant, C., and Le Douarin, N.M. (1997) Ligand-dependent development of the endothelial and hemopoietic lineages from embryonic mesodermal cells expressing vascular endothelial growth factor receptor 2. *Proc. Natl. Acad. Sci. U.S.A.*, **94**, 5141–5146.
36. Oettgen, P. (2001) Transcriptional regulation of vascular development. *Circ. Res.*, **89**, 380–388.
37. Prindull, G. (2005) Hemangioblasts representing a functional endothelio-hematopoietic entity in ontogeny, postnatal life, and CML neovasculogenesis. *Stem Cell Rev.*, **1**, 277–284.
38. Grant, M.B., May, W.S., Caballero, S., Brown, G.A., Guthrie, S.M., Mames, R.N., Byrne, B.J. *et al.* (2002) Adult hematopoietic stem cells provide functional hemangioblast activity during retinal neovascularization. *Nat. Med.*, **8**, 607–612.
39. Cogle, C.R., Wainman, D.A., Jorgensen, M.L., Guthrie, S.M., Mames, R.N., and Scott, E.W. (2004) Adult human hematopoietic cells provide functional hemangioblast activity. *Blood*, **103**, 133–135.
40. Gunsilius, E., Duba, H.C., Petzer, A.L., Kähler, C.M., Grünewald, K., Stockhammer, G., Gabl, C. *et al.* (2000) Evidence from a leukaemia model for maintenance of vascular endothelium by bone-marrow-derived endothelial cells. *Lancet*, **355**, 1688–1691.
41. Fang, B., Zheng, C., Liao, L., Han, Q., Sun, Z., Jiang, X., and Zhao, R.C. (2005) Identification of human chronic myelogenous leukemia progenitor cells with hemangioblastic characteristics. *Blood*, **105**, 2733–2740.
42. Moore, M.A.S. and Metcalf, D. (1970) Ontogeny of the hematopoietic system: yolk sac of in vivo and in vitro colony forming cells in the developing mouse embryo. *Br. J. Haematol.*, **18**, 279–296.
43. Dieterlen-Lièvre, F. and Martin, C. (1981) Diffuse intraembryonic hematopoiesis in normal and chimeric avian development. *Dev. Biol.*, **88**, 180–191.
44. Medvinsky, A. and Dierzak, E. (1996) Definitive hematopoiesis is autonomously initiated by the AGM region. *Cell*, **86**, 897–906.
45. Cumano, A., Dieterlen-Lièvre, F., and Godin, I. (1996) Lymphoid potential, probed before circulation in mouse, is restricted to caudal intraembryonic splanchnopleura. *Cell*, **86**, 907–916.
46. Lu, L.S., Wang, S.J., and Auerbach, R. (1996) In vitro and in vivo differentiation into B cells, T cells, and myeloid cells of primitive yolk sac hematopoietic precursor cells expanded > 100-fold by coculture with a clonal yolk sac endothelial cell line. *Proc. Natl. Acad. Sci. U.S.A.*, **93**, 14782–14787.
47. Choi, K., Kennedy, M., Kazarov, A., Papadimitriou, J.C., and Keller, G. (1998) A common precursor for hematopoietic and endothelial cells. *Development*, **125**, 725–732.
48. Fraser, S.T., Ogawa, M., Yu, R.T., Nishikawa, S., Yoder, M.C., and Nishikawa, S. (2002) Definitive hematopoietic commitment within the embryonic vascular endothelial-cadherin(+) population. *Exp. Hematol.*, **30**, 1070–1078.
49. Miyagi, T., Takeno, M., Nagafuchi, H., Takahashi, M., and Suzuki, N. (2002) Flk1+ cells derived from mouse embryonic stem cells reconstitute hematopoiesis in vivo in SCID mice. *Exp. Hematol.*, **30**, 1444–1453.
50. Damert, A., Miquerol, L., Gertsenstein, M., Risau, W., and Nagy, A. (2002) Insufficient VEGFA activity in yolk sac endoderm compromises haematopoietic and endothelial differentiation. *Development*, **129**, 1881–1892.
51. Zippo, A., De Robertis, A., Bardelli, M., Galvagni, F., and Oliviero, S. (2004) Identification of Flk-1 target genes in vasculogenesis: Pim-1 is required for endothelial and mural cell differentiation in vitro. *Blood*, **103**, 4536–4544.
52. Rous, P. and Kidd, J. (1941) Conditional neoplasms and subthreshold

neoplastic states: a study of the tar tumors of rabbits. *J. Exp. Med.*, **73**, 365–389.

53. Mackenzie, I.C. and Rous, P. (1941) The experimental disclosure of latent neoplastic changes in tarred skin. *J. Exp. Med.*, **73**, 391–415.

54. Vakkila, J. and Lotze, M.T. (2004) Inflammation and necrosis promote tumour growth. *Nat. Rev. Immunol.*, **4**, 641–648.

55. Balkwill, F., Charles, K.A., and Mantovani, A. (2005) Smoldering and polarized inflammation in the initiation and promotion of malignant disease. *Cancer Cell*, **7**, 211–217.

56. Coussens, L.M. and Werb, Z. (2002) Inflammation and cancer. *Nature*, **420**, 860–867.

57. Bergers, G. and Benjamin, L.E. (2003) Tumorigenesis and the angiogenic switch. *Nat. Rev. Cancer*, **3**, 401–410.

58. Murdoch, C., Muthana, M., Coffelt, S.B., and Lewis, C.E. (2008) The role of myeloid cells in the promotion of tumour angiogenesis. *Nat. Rev. Cancer*, **8**, 618–631.

59. Metz, M. and Maurer, M. (2007) Mast cells – key effector cells in immune responses. *Trends Immunol.*, **28**, 234–241.

60. Dundar, E., Oner, U., Peker, B.C., Metintas, M., Isiksoy, S., and Ak, G. (2008) The significance and relationship between mast cells and tumour angiogenesis in nonsmall cell lung carcinoma. *J. Int. Med. Res.*, **36**, 88–95.

61. Sawatsubashi, M., Yamada, T., Fukushima, N., Mizokami, H., Tokunaga, O., and Shin, T. (2000) Association of vascular endothelial growth factor and mast cells with angiogenesis in laryngeal squamous cell carcinoma. *Virchows Arch.*, **436**, 243–248.

62. Tuna, B., Yorukoglu, K., Unlu, M., Mungan, M.U., and Kirkali, Z. (2006) Association of mast cells with microvessel density in renal cell carcinomas. *Eur. Urol.*, **50**, 530–534.

63. Ribatti, D., Finato, N., Crivellato, E., Guidolin, D., Longo, V., Mangieri, D., Nico, B. *et al.* (2007) Angiogenesis and mast cells in human breast cancer sentinel lymph nodes with and without micrometastases. *Histopathology*, **51**, 837–842.

64. Yano, H., Kinuta, M., Tateishi, H., Nakano, Y., Matsui, S., Monden, T., Okamura, J. *et al.* (1999) Mast cell infiltration around gastric cancer cells correlates with tumor angiogenesis and metastasis. *Gastric Cancer*, **2**, 26–32.

65. Elpek, G.O., Gelen, T., Aksoy, N.H., Erdogan, A., Dertsiz, L., Demircan, A., and Keles, N. (2001) The prognostic relevance of angiogenesis and mast cells in squamous cell carcinoma of the oesophagus. *J. Clin. Pathol.*, **54**, 940–944.

66. Iamaroon, A., Pongsiriwet, S., Jittidecharaks, S., Pattanaporn, K., Prapayasatok, S., and Wanachantararak, S. (2003) Increase of mast cells and tumor angiogenesis in oral squamous cell carcinoma. *J. Oral Pathol. Med.*, **32**, 195–199.

67. Acikalin, M.F., Onera, U., Topcu, I., Yasar, B., Kiper, H., and Colak, E. (2005) Tumour angiogenesis and mast cell density in the prognostic assessment of colorectal carcinomas. *Dig. Liver Dis.*, **37**, 162–169.

68. Ribatti, D., Finato, N., Crivellato, E., Marzullo, A., Mangieri, D., Nico, B., Vacca, A., and Beltrami, C.A. (2005) Neovascularization and mast cells with tryptase activity increase simultaneously with pathologic progression in human endometrial cancer. *Am. J. Obstet. Gynecol.*, **193**, 1961–1965.

69. Ribatti, D., Ennas, M.G., Vacca, A., Ferreli, F., Nico, B., Orru, S., and Sirigu, P. (2003) Tumor vascularity and tryptasepositive mast cells correlate with a poor prognosis in melanoma. *Eur. J. Clin. Invest.*, **33**, 420–425.

70. Ribatti, D., Vacca, A., Nico, B., Quondamatteo, F., Ria, R., Minischetti, M., Marzullo, A. *et al.* (1999) Bone marrow angiogenesis and mast cell density increase simultaneously with progression of human multiple myeloma. *Br. J. Cancer*, **79**, 451–455.

71. Ribatti, D., Vacca, A., Nico, B., Crivellato, E., Roncali, L., and Dammacco, F. (2001) The role of

mast cells in tumour angiogenesis. *Br. J. Haematol*, **115**, 514–521.

72. Crivellato, E., Nico, B., and Ribatti, D. (2008) Mast cells and tumour angiogenesis: new insight from experimental carcinogenesis. *Cancer Lett.*, **269**, 1–6.
73. Wershil, B.K., Tsai, M., Geissler, E.N., Zsebo, K.M., and Galli, S.J. (1992) The rat c-kit ligand, stem cell factor, induces c-kit receptor-dependent mouse mast cell activation in vivo. Evidence that signaling through the c-kit receptor can induce expression of cellular function. *J. Exp. Med.*, **175**, 245–255.
74. Zhang, W., Stoica, G., Tasca, S.I., Kelly, K.A., and Meininger, C.J. (2000) Modulation of tumor angiogenesis by stem cell factor. *Cancer Res.*, **60**, 6757–6762.
75. Matsuura, N. and Zetter, B.R. (1989) Stimulation of mast cell chemotaxis by interleukin 3. *J. Exp. Med.*, **170**, 1421–1426.
76. Reed, J.A., McNutt, N.S., Bogdany, J.K., and Albino, A.P. (1996) Expression of the mast cell growth factor interleukin-3 in melanocytic lesions correlates with an increased number of mast cells in the perilesional stroma: implications for melanoma progression. *J. Cutan. Pathol.*, **23**, 495–505.
77. de Visser, K.E., Korets, L.V., and Coussens, L.M. (2005) De novo carcinogenesis promoted by chronic inflammation is B lymphocyte dependent. *Cancer Cell*, **7**, 411–423.
78. Juremalm, M. and Nilsson, G. (2005) Chemokine receptor expression by mast cells. *Chem. Immunol. Allergy*, **87**, 130–144.
79. Starkey, J.R., Crowle, P.K., and Taubenberger, S. (1988) Mastcell-deficient W/Wv mice exhibit a decreased rate of tumor angiogenesis. *Int. J. Cancer*, **42**, 48–52.
80. Coussens, L.M., Raymond, W.W., Bergers, G., Laig-Webster, M., Behrendtsen, O., Werb, Z., Caughey, G.H., and Hanahan, D. (1999) Inflammatory mast cells upregulate angiogenesis during squamous epithelial carcinogenesis. *Genes Dev.*, **13**, 1382–1397.
81. Soucek, L., Lawlor, E.R., Soto, D., Shchors, K., Swigart, L.B., and Evan, G.I. (2007) Mast cells are required for angiogenesis and macroscopic expansion of Mycinduced pancreatic islet tumors. *Nat. Med.*, **13**, 1211–1218.
82. Theoharides, T.C., Kempuraj, D., Tagen, M., Conti, P., and Kalogeromitros, D. (2007) Differential release of mast cell mediators and the pathogenesis of inflammation. *Immunol. Rev.*, **217**, 65–78.
83. Nico, B., Mangieri, D., Crivellato, E., Vacca, A., and Ribatti, D. (2008) Mast cells contribute to vasculogenic mimicry in multiple myeloma. *Stem Cells Dev.*, **17**, 19–22.
84. Jeong, H.J., Chung, H.S., Lee, B.R., Kim, S.J., Yoo, S.J., Hong, S.H., and Kim, H.M. (2003) Expression of proinflammatory cytokines via HIF-1α and NFκB activation on desferrioxamine-stimulated HMC-1 cells. *Biochem. Biophys. Res. Commun.*, **306**, 805–811.
85. Fujita, Y., Mimata, H., Nasu, N., Nomura, T., Nomura, Y., and Nakagawa, M. (2002) Involvement of adrenomedullin induced by hypoxia in angiogenesis in human renal cell carcinoma. *Int. J. Urol.*, **9**, 285–295.
86. Garayoa, M., Martínez, A., Lee, S., Pío, R., An, W.G., Neckers, L., Trepel, J. *et al.* (2000) Hypoxia-inducible factor-1 (HIF-1) up-regulates adrenomedullin expression in human tumor cell lines during oxygen deprivation: a possible promotion mechanism of carcinogenesis. *Mol. Endocrinol.*, **14**, 848–862.
87. Mantovani, A., Sozzani, S., Locati, M., Allavena, P., and Sica, A. (2002) Macrophage polarization: tumor-associated macrophages as a paradigm for polarized M2 mononuclear phagocytes. *Trends Immunol.*, **23**, 549–555.
88. Biswas, S.K., Gangi, L., Paul, S., Schioppa, T., Saccani, A., Sironi, M., Bottazzi, B. *et al.* (2006) A distinct and unique transcriptional program expressed by tumor-associated macrophages (defective NF-κB and enhanced IRF-3/STAT1 activation). *Blood*, **107**, 2112–2122.

89. Saccani, A., Schioppa, T., Porta, C., Biswas, S.K., Nebuloni, M., Vago, L., Bottazzi, B. *et al.* (2006) p50 nuclear factor-k B overexpression in tumor-associated macrophages inhibits M1 inflammatory responses and antitumor resistance. *Cancer Res.*, **66**, 11432–11440.
90. Biswas, S.K., Sica, A., and Lewis, C.E. (2008) Plasticity of macrophage function during tumor progression: regulation by distinct molecular mechanisms. *J. Immunol.*, **180**, 2011–2017.
91. Bingle, L., Brown, N.J., and Lewis, C.E. (2002) The role of tumour-associated macrophages in tumour progression: implications for new anticancer therapies. *J. Pathol.*, **196**, 254–265.
92. Lin, E.Y. and Pollard, J.W. (2004) Role of infiltrated leucocytes in tumour growth and spread. *Br. J. Cancer*, **90**, 2053–2058.
93. Yamashiro, S., Takeya, M., Nishi, T., Kuratsu, J., Yoshimura, T., Ushio, Y., and Takahashi, K. (1994) Tumor-derived monocyte chemoattractant protein-1 induces intratumoral infiltration of monocyte-derived macrophage subpopulation in transplanted rat tumors. *Am. J. Pathol.*, **145**, 856–867.
94. Murdoch, C., Giannoudis, A., and Lewis, C.E. (2004) Mechanisms regulating the recruitment of macrophages into hypoxic areas of tumors and other ischemic tissues. *Blood*, **104**, 2224–2234.
95. Bottazzi, B., Walter, S., Govoni, D., Colotta, F., and Mantovani, A. (1992) Monocyte chemotactic cytokine gene transfer modulates macrophage infiltration, growth, and susceptibility to IL-2 therapy of a murine melanoma. *J. Immunol.*, **148**, 1280–1285.
96. Loberg, R.D., Ying, C., Craig, M., Yan, L., Snyder, L.A., and Pienta, K.J. (2007) CCL2 as an important mediator of prostate cancer growth in vivo through the regulation of macrophage infiltration. *Neoplasia*, **9**, 556–562.
97. Gazzaniga, S., Bravo, A.I., Guglielmotti, A., van Rooijen, N., Maschi, F., Vecchi, A., Mantovani, A. *et al.* (2007) Targeting tumor-associated macrophages and inhibition of MCP-1 reduce angiogenesis and tumor growth in a human melanoma xenograft. *J. Invest. Dermatol.*, **127**, 2031–2041.
98. Lin, E.Y., Nguyen, A.V., Russell, R.G., and Pollard, J.W. (2001) Colony-stimulating factor 1 promotes progression of mammary tumors to malignancy. *J. Exp. Med.*, **193**, 727–740.
99. Ueno, T., Toi, M., Saji, H., Muta, M., Bando, H., Kuroi, K., Koike, M. *et al.* (2000) Significance of macrophage chemoattractant protein-1 in macrophage recruitment, angiogenesis, and survival in human breast cancer. *Clin. Cancer Res.*, **6**, 3282–3289.
100. Fischer, C., Jonckx, B., Mazzone, M., Zacchigna, S., Loges, S., Pattarini, L., Chorianopoulos, E. *et al.* (2007) Anti-PlGF inhibits growth of VEGF(R)-inhibitor-resistant tumors without affecting healthy vessels. *Cell*, **131**, 463–475.
101. Scotton, C., Milliken, D., Wilson, J., Raju, S., and Balkwill, F. (2001) Analysis of CC chemokine and chemokine receptor expression in solid ovarian tumours. *Br. J. Cancer*, **85**, 891–897.
102. Dineen, S.P., Lynn, K.D., Holloway, S.E., Miller, A.F., Sullivan, J.P., Shames, D.S., Beck, A.W. *et al.* (2008) Vascular endothelial growth factor receptor 2 mediates macrophage infiltration into orthotopic pancreatic tumors in mice. *Cancer Res.*, **68**, 4340–4346.
103. Hiratsuka, S., Watanabe, A., Aburatani, H., and Maru, Y. (2006) Tumour-mediated upregulation of chemoattractants and recruitment of myeloid cells predetermines lung metastasis. *Nat. Cell Biol.*, **8**, 1369–1375.
104. Leek, R.D., Lewis, C.E., Whitehouse, R., Greenall, M., Clarke, J., and Harris, A.L. (1996) Association of macrophage infiltration with angiogenesis and prognosis in invasive breast carcinoma. *Cancer Res.*, **56**, 4625–4629.
105. Onita, T., Ji, P.G., Xuan, J.W., Sakai, H., Kanetake, H., Maxwell, P.H., Fong, G.H. *et al.* (2002) Hypoxia-induced,

perinecrotic expression of endothelial Per-ARNT-Sim domain protein-1/hypoxia-inducible factor-2α correlates with tumor progression, vascularization, and focal macrophage infiltration in bladder cancer. *Clin. Cancer Res.*, **8**, 471–480.

106. Takanami, I., Takeuchi, K., and Kodaira, S. (1999) Tumor-associated macrophage infiltration in pulmonary adenocarcinoma: association with angiogenesis and poor prognosis. *Oncology*, **57**, 138–142.

107. Valkovic, T., Dobrila, F., Melato, M., Sasso, F., Rizzardi, C., and Jonjic, N. (2002) Correlation between vascular endothelial growth factor, angiogenesis, and tumor associated macrophages in invasive ductal breast carcinoma. *Virchows Arch.*, **440**, 583–588.

108. Lin, E.Y., Li, J.F., Gnatovskiy, L., Deng, Y., Zhu, L., Grzesik, D.A., Qian, H. *et al.* (2006) Macrophages regulate the angiogenic switch in a mouse model of breast cancer. *Cancer Res.*, **66**, 11238–11246.

109. Kimura, Y.N., Watari, K., Fotovati, A., Hosoi, F., Yasumoto, K., Izumi, H., Kohno, K. *et al.* (2007) Inflammatory stimuli from macrophages and cancer cells synergistically promote tumor growth and angiogenesis. *Cancer Sci.*, **98**, 2009–2018.

110. Lewis, C.E. and Pollard, J.W. (2006) Distinct role of macrophages in different tumor microenvironments. *Cancer Res.*, **66**, 605–612.

111. Hagemann, T., Lawrence, T., McNeish, I., Charles, K.A., Kulbe, H., Thompson, R.G., Robinson, S.C., and Balkwill, F.R. (2008) "Re-educating" tumor-associated macrophages by targeting NF-κ B. *J. Exp. Med.*, **205**, 1261–1268.

112. Fernandez Pujol, B., Lucibello, F.C., Gehling, U.M., Lindemann, K., Weidner, N., Zuzarte, M.L., Adamkiewicz, J. *et al.* (2000) Endothelial-like cells derived from human CD14 positive monocytes. *Differentiation*, **65**, 287–300.

113. Kuwana, M., Okazaki, Y., Kodama, H., Satoh, T., Kawakami, Y., and Ikeda, Y. (2006) Endothelial differentiation potential of human monocyte-derived multipotential cells. *Stem Cells*, **24**, 2733–2743.

114. Yamada, Y. and Takakura, N. (2006) Physiological pathway of differentiation of hematopoietic stem cell population into mural cells. *J. Exp. Med.*, **203**, 1055–1065.

115. Balkwill, F., Charles, K.A., and Mantovani, A. (2005) Smoldering and polarized inflammation in the initiation and promotion of malignant disease. *Cancer Cell*, **7**, 211–217.

116. De Palma, M., Venneri, M.A., Galli, R., Sergi Sergi, L., Politi, L.S., Sampaolesi, M., and Naldini, L. (2005) Tie-2 identifies a hematopoietic lineage of proangiogenic monocytes required for tumor vessel formation and a mesenchymal population of pericyte progenitors. *Cancer Cell*, **8**, 211–226.

117. Murdoch, C., Tazzyman, S., Webster, S., and Lewis, C.E. (2007) Expression of Tie-2 by human monocytes and their responses to angiopoietin-2. *J. Immunol.*, **178**, 7405–7411.

118. Venneri, M.A., De Palma, M., Ponzoni, M., Pucci, F., Scielzo, C., Zonari, E., Mazzieri, R. *et al.* (2007) Identification of proangiogenic Tie-2-expressing monocytes (TEMs) in human peripheral blood and cancer. *Blood*, **109**, 5276–5285.

119. Nowak, G., Karrar, A., Holmén, C., Nava, S., Uzunel, M., Hultenby, K., and Sumitran-Holgersson, S. (2004) Expression of vascular endothelial growth factor receptor-2 or Tie-2 on peripheral blood cells defines functionally competent cell populations capable of reendothelialization. *Circulation*, **110**, 3699–3707.

120. Okamoto, R., Ueno, M., Yamada, Y., Takahashi, N., Sano, H., Suda, T., and Takakura, N. (2005) Hematopoietic cells regulate the angiogenic switch during tumorigenesis. *Blood*, **105**, 2757–2763.

121. Hattori, K., Dias, S., Heissig, B., Hackett, N.R., Lyden, D., Tateno, M., Hicklin, D.J. *et al.* (2001) Vascular endothelial growth factor and angiopoietin-1 stimulate postnatal

hematopoiesis by recruitment of vasculogenic and hematopoietic stem cells. *J. Exp. Med.*, **193**, 1005–1014.

122. Gu, J., Yamamoto, H., Ogawa, M., Ngan, C.Y., Danno, K., Hemmi, H., Kyo, N. *et al.* (2006) Hypoxia-induced up-regulation of angiopoietin-2 in colorectal cancer. *Oncol. Rep.*, **15**, 779–783.

123. Stratmann, A., Risau, W., and Plate, K.H. (1998) Cell type specific expression of angiopoietin-1 and angiopoietin-2 suggests a role in glioblastoma angiogenesis. *Am. J. Pathol.*, **153**, 1459–1466.

124. Gabrilovich, D. (2004) Mechanisms and functional significance of tumour-induced dendritic-cell defects. *Nat. Rev. Immunol.*, **4**, 941–952.

125. Movahedi, K., Guilliams, M., Van den Bossche, J., Van den Bergh, R., Gysemans, C., Beschin, A., De Baetselier, P., and Van Ginderachter, J.A. (2008) Identification of discrete tumor induced myeloid-derived suppressor cell subpopulations with distinct T cell-suppressive activity. *Blood*, **111**, 4233–4244.

126. Sawanobori, Y., Ueha, S., Kurachi, M., Shimaoka, T., Talmadge, J.E., Abe, J., Shono, Y. *et al.* (2008) Chemokine-mediated rapid turnover of myeloid-derived suppressor cells in tumor bearing mice. *Blood*, **111**, 5457–5466.

127. Bronte, V., Apolloni, E., Cabrelle, A., Ronca, R., Serafini, P., Zamboni, P., Restifo, N.P., and Zanovello, P. (2000) Identification of a CD11b+/Gr-1+/CD31+ myeloid progenitor capable of activating or suppressing CD8+ T cells. *Blood*, **96**, 3838–3846.

128. Kusmartsev, S. and Gabrilovich, D.I. (2002) Immature myeloid cells and cancer-associated immune suppression. *Cancer Immunol. Immunother.*, **51**, 293–298.

129. Gabrilovich, D.I., Bronte, V., Chen, S.H., Colombo, M.P., Ochoa, A., Ostrand-Rosenberg, S., and Schreiber, H. (2007) The terminology issue for myeloid-derived suppressor cells. *Cancer Res.*, **67**, 425.

130. Almand, B., Clark, J.I., Nikitina, E., van Beynen, J., English, N.R., Knight, S.C., Carbone, D.P., and Gabrilovich, D.I. (2001) Increased production of immature myeloid cells in cancer patients: a mechanism of immunosuppression in cancer. *J. Immunol.*, **166**, 678–689.

131. Diaz-Montero, C.M., Salem, M.L., Nishimura, M.I., Garrett-Mayer, E., Cole, D.J., and Montero, A.J. (2009) Increased circulating myeloid-derived suppressor cells correlate with clinical cancer stage, metastatic tumor burden, and doxorubicin-cyclophosphamide chemotherapy. *Cancer Immunol. Immunother.*, **58**, 49–59.

132. Hoechst, B., Ormandy, L.A., Ballmaier, M., Lehner, F., Krüger, C., Manns, M.P., Greten, T.F., and Korangy, F. (2008) A new population of myeloid-derived suppressor cells in hepatocellular carcinoma patients induces CD4+CD25+Foxp3+ T cells. *Gastroenterology*, **35**, 234–243.

133. Melani, C., Chiodoni, C., Forni, G., and Colombo, M.P. (2003) Myeloid cell expansion elicited by the progression of spontaneous mammary carcinomas in c-erbB-2 transgenic BALB/c mice suppresses immune reactivity. *Blood*, **102**, 2138–2145.

134. Shojaei, F., Wu, X., Zhong, C., Yu, L., Liang, X.H., Yao, J., Blanchard, D. *et al.* (2007) Bv8 regulates myeloid-cell-dependent tumour angiogenesis. *Nature*, **450**, 825–831.

135. Yang, L., Huang, J., Ren, X., Gorska, A.E., Chytil, A., Aakre, M., Carbone, D.P. *et al.* (2008) Abrogation of TGFβ signaling in mammary carcinomas recruits Gr-1+CD11b+ myeloid cells that promote metastasis. *Cancer Cell*, **13**, 23–35.

136. Yang, L., DeBusk, L.M., Fukuda, K., Fingleton, B., Green-Jarvis, B., Shyr, Y., Matrisian, L.M. *et al.* (2004) Expansion of myeloid immune suppressor Gr+CD11b+ cells in tumor-bearing host directly promotes tumor angiogenesis. *Cancer Cell*, **6**, 409–421.

137. Pan, P.Y., Wang, G.X., Yin, B., Ozao, J., Ku, T., Divino, C.M., and

Chen, S.H. (2008) Reversion of immune tolerance in advanced malignancy: modulation of myeloid-derived suppressor cell development by blockade of stem-cell factor function. *Blood*, **111**, 219–228.

138. Masuda, Y., Takatsu, Y., Terao, Y., Kumano, S., Ishibashi, Y., Suenaga, M., Abe, M. *et al.* (2002) Isolation and identification of EG-VEGF/prokineticins as cognate ligands for two orphan G-protein-coupled receptors. *Biochem. Biophys. Res. Commun.*, **293**, 396–402.
139. LeCouter, J., Zlot, C., Tejada, M., Peale, F., and Ferrara, N. (2004) Bv8 and endocrine gland-derived vascular endothelial growth factor stimulate hematopoiesis and hematopoietic cell mobilization. *Proc. Natl. Acad. Sci. U.S.A.*, **101**, 16813–16818.
140. Dolcetti, L., Marigo, I., Mantelli, B., Peranzoni, E., Zanovello, P., and Bronte, V. (2008) Myeloid-derived suppressor cell role in tumor-related inflammation. *Cancer Lett.*, **267**, 216–225.
141. Sinha, P., Clements, V.K., Miller, S., and Ostrand-Rosenberg, S. (2005) Tumor immunity: a balancing act between T cell activation, macrophage activation and tumor-induced immune suppression. *Cancer Immunol. Immunother.*, **54**, 1137–1142.
142. Nagaraj, S. and Gabrilovich, D.I. (2007) Myeloid-derived suppressor cells. *Adv. Exp. Med. Biol.*, **601**, 213–223.
143. Du, R., Lu, K.V., Petritsch, C., Liu, P., Ganss, R., Passegué, E., Song, H. *et al.* (2008) HIF1α induces the recruitment of bone marrow-derived vascular modulatory cells to regulate tumor angiogenesis and invasion. *Cancer Cell*, **13**, 206–220.
144. Furze, R.C. and Rankin, S.M. (2008) Neutrophil mobilization and clearance in the bone marrow. *Immunology*, **125**, 281–288.
145. Eck, M., Schmausser, B., Scheller, K., Brandlein, S., and Muller-Hermelink, H.K. (2003) Pleiotropic effects of CXC chemokines in gastric carcinoma: differences in CXCL8 and CXCL1 expression between diffuse and intestinal types of gastric carcinoma. *Clin. Exp. Immunol.*, **134**, 508–515.
146. Nielsen, B.S., Timshel, S., Kjeldsen, L., Sehested, M., Pyke, C., Borregaard, N., and Danø, K. (1996) 92 kDa type IV collagenase (MMP-9) is expressed in neutrophils and macrophages but not in malignant epithelial cells in human colon cancer. *Int. J. Cancer*, **65**, 57–62.
147. Bellocq, A., Antoine, M., Flahault, A., Philippe, C., Crestani, B., Bernaudin, J.F., Mayaud, C. *et al.* (1998) Neutrophil alveolitis in bronchioloalveolar carcinoma: induction by tumor-derived interleukin-8 and relation to clinical outcome. *Am. J. Pathol.*, **152**, 83–92.
148. Xie, K. (2001) Interleukin-8 and human cancer biology. *Cytokine Growth Factor Rev.*, **12**, 375–391.
149. Gijsbers, K., Gouwy, M., Struyf, S., Wuyts, A., Proost, P., Opdenakker, G., Penninckx, F. *et al.* (2005) GCP-2/CXCL6 synergizes with other endothelial cell-derived chemokines in neutrophil mobilization and is associated with angiogenesis in gastrointestinal tumors. *Exp. Cell Res.*, **303**, 331–342.
150. Luan, J., Shattuck-Brandt, R., Haghnegahdar, H., Owen, J.D., Strieter, R., Burdick, M., Nirodi, C. *et al.* (1997) Mechanism and biological significance of constitutive expression of MGSA/GRO chemokines in malignant melanoma tumor progression. *J. Leukoc. Biol.*, **62**, 588–597.
151. Arenberg, D.A., Keane, M.P., DiGiovine, B., Kunkel, S.L., Morris, S.B., Xue, Y.Y., Burdick, M.D. *et al.* (1998) Epithelial-neutrophil activating peptide (ENA-78) is an important angiogenic factor in non-small cell lung cancer. *J. Clin. Invest.*, **102**, 465–472.
152. Lee, L.F., Hellendall, R.P., Wang, Y., Haskill, J.S., Mukaida, N., Matsushima, K., and Ting, J.P. (2000) IL-8 reduced tumorigenicity of human ovarian cancer in vivo due to neutrophil infiltration. *J. Immunol.*, **164**, 2769–2775.
153. Mentzel, T., Brown, L.F., Dvorak, H.F., Kuhnen, C., Stiller, K.J., Katenkamp, D., and Fletcher, C.D. (2001) The association between tumour progression and vascularity in myxofibrosarcoma

and myxoid/round cell liposarcoma. *Virchows Arch.*, **438**, 13–22.

154. Benelli, R., Morini, M., Carrozzino, F., Ferrari, N., Minghelli, S., Santi, L., Cassatella, M. *et al.* (2002) Neutrophils as a key cellular target for angiostatin: implications for regulation of angiogenesis and inflammation. *FASEB J.*, **16**, 267–269.

155. Van Coillie, E., Van Aelst, I., Wuyts, A., Vercauteren, R., Devos, R., De Wolf-Peeters, C., Van Damme, J., and Opdenakker, G. (2001) Tumor angiogenesis induced by granulocyte chemotactic protein-2 as a countercurrent principle. *Am. J. Pathol.*, **159**, 1405–1414.

156. Nozawa, H., Chiu, C., and Hanahan, D. (2006) Infiltrating neutrophils mediate the initial angiogenic switch in a mouse model of multistage carcinogenesis. *Proc. Natl. Acad. Sci. U.S.A.*, **103**, 12493–12498.

157. Coussens, L.M., Tinkle, C.L., Hanahan, D., and Werb, Z. (2000) MMP-9 supplied by bone marrow-derived cells contributes to skin carcinogenesis. *Cell*, **103**, 481–490.

158. Bergers, G., Brekken, R., McMahon, G., Vu, T.H., Itoh, T., Tamaki, K., Tanzawa, K. *et al.* (2000) Matrix metalloproteinase-9 triggers the angiogenic switch during carcinogenesis. *Nat. Cell Biol.*, **2**, 737–744.

159. Masure, S., Proost, P., Van, D.J., and Opdenakker, G. (1991) Purification and identification of 91-kDa neutrophil gelatinase. Release by the activating peptide interleukin-8. *Eur. J. Biochem.*, **198**, 391–398.

160. Ardi, V.C., Kupriyanova, T.A., Deryugina, E.I., and Quigley, J.P. (2007) Human neutrophils uniquely release TIMP-free MMP-9 to provide a potent catalytic stimulator of angiogenesis. *Proc. Natl. Acad. Sci. U.S.A.*, **104**, 20262–20267.

161. McCourt, M., Wang, J.H., Sookhai, S., and Redmond, H.P. (1999) Proinflammatory mediators stimulate neutrophil-directed angiogenesis. *Arch. Surg.*, **134**, 1325–1331.

162. Queen, M.M., Ryan, R.E., Holzer, R.G., Keller-Peck, C.R., and Jorcyk, C.L. (2005) Breast cancer cells stimulate neutrophils to produce oncostatin M: potential implications for tumor progression. *Cancer Res.*, **65**, 8896–8904.

163. Cassatella, M.A. (1999) Neutrophil-derived proteins: selling cytokines by the pound. *Adv. Immunol.*, **73**, 369–509.

164. Van den Steen, P.E., Proost, P., Wuyts, A., Van Damme, J., and Opdenakker, G. (2000) Neutrophil gelatinase B potentiates interleukin-8 tenfold by aminoterminal processing, whereas it degrades CTAP-III, PF-4, and GRO-alpha and leaves RANTES and MCP-2 intact. *Blood*, **96**, 2673–2681.

165. Koehne, P., Willam, C., Strauss, E., Schindler, R., Eckardt, K.U., and Bührer, C. (2000) Lack of hypoxic stimulation of VEGF secretion from neutrophils and platelets. *Am. J. Physiol. Heart Circ. Physiol.*, **279**, H817–H824.

166. Akuthota, P., Wang, H.B., Spencer, L.A., and Weller, P.F. (2008) Immunoregulatory roles of eosinophils: a new look at a familiar cell. *Clin. Exp. Allergy*, **38**, 1254–1263.

167. Troy, A.J., Summers, K.L., Davidson, P.J., Atkinson, C.H., and Hart, D.N. (1998) Minimal recruitment and activation of dendritic cells within renal cell carcinoma. *Clin. Cancer Res.*, **4**, 585–593.

168. Looi, L.M. (1987) Tumor-associated tissue eosinophilia in nasopharyngeal carcinoma. A pathologic study of 422 primary and 138 metastatic tumors. *Cancer*, **59**, 466–470.

169. Dorta, R.G., Landman, G., Kowalski, L.P., Lauris, J.R., Latorre, M.R., and Oliveira, D.T. (2002) Tumour-associated tissue eosinophilia as a prognostic factor in oral squamous cell carcinomas. *Histopathology*, **41**, 152–157.

170. Nielsen, H.J., Hansen, U., Christensen, I.J., Reimert, C.M., Brünner, N., and Moesgaard, F. (1999) Independent prognostic value of eosinophil and mast cell infiltration in colorectal cancer tissue. *J. Pathol.*, **189**, 487–495.

171. Teruya-Feldstein, J., Jaffe, E.S., Burd, P.R., Kingma, D.W., Setsuda, J.E., and Tosato, G. (1999) Differential

chemokine expression in tissues involved by Hodgkin's disease: direct correlation of eotaxin expression and tissue eosinophilia. *Blood*, **93**, 2463–2470.

172. Jose, P.J., Adcock, I.M., Griffiths-Johnson, D.A., Berkman, N., Wells, T.N., Williams, T.J., and Power, C.A. (1994) Eotaxin: cloning of an eosinophil chemoattractant cytokine and increased mRNA expression in allergen-challenged guinea-pig lungs. *Biochem. Biophys. Res. Commun.*, **205**, 788–794.
173. Daugherty, B.L., Siciliano, S.J., DeMartino, J.A., Malkowitz, L., Sirotina, A., and Springer, M.S. (1996) Cloning, expression, and characterization of the human eosinophil eotaxin receptor. *J. Exp. Med.*, **183**, 2349–2354.
174. Puxeddu, I., Alian, A., Piliponsky, A.M., Ribatti, D., Panet, A., and Levi-Schaffer, F. (2005) Human peripheral blood eosinophils induce angiogenesis. *Int. J. Biochem. Cell Biol.*, **37**, 628–636.
175. Salcedo, R., Young, H.A., Ponce, M.L., Ward, J.M., Kleinman, H.K., Murphy, W.J., and Oppenheim, J.J. (2001) Eotaxin (CCL11) induces in vivo angiogenic responses by human CCR3+ endothelial cells. *J. Immunol.*, **166**, 7571–7578.
176. Munitz, A. and Levi-Schaffer, F. (2004) Eosinophils: 'new' roles for 'old' cells. *Allergy*, **59**, 268–275.
177. Ohno, I., Ohtani, H., Nitta, Y., Suzuki, J., Hoshi, H., Honma, M., Isoyama, S. *et al.* (1997) Eosinophils as a source of matrix metalloproteinase-9 in asthmatic airway inflammation. *Am. J. Respir. Cell Mol. Biol.*, **16**, 212–219.
178. Simson, L., Ellyard, J.I., Dent, L.A., Matthaei, K.I., Rothenberg, M.E., Foster, P.S., Smyth, M.J., and Parish, C.R. (2007) Regulation of carcinogenesis by IL-5 and CCL11: a potential role for eosinophils in tumor immune surveillance. *J. Immunol.*, **178**, 4222–4229.
179. Joffre, O., Nolte, M.A., Spörri, R., and Reise Sousa, C. (2009) Inflammatory signals in dendritic cell activation and the induction of adaptive immunity. *Immunol. Rev.*, **227**, 234–247.
180. Gabrilovich, D., Ishida, T., Oyama, T., Ran, S., Kravtsov, V., Nadaf, S., and Carbone, D.P. (1998) Vascular endothelial growth factor inhibits the development of dendritic cells and dramatically affects the differentiation of multiple hematopoietic lineages in vivo. *Blood*, **92**, 4150–4166.
181. Ghiringhelli, F., Puig, P.E., Roux, S., Parcellier, A., Schmitt, E., Solary, E., Kroemer, G. *et al.* (2005) Tumor cells convert immature myeloid dendritic cells into TGFβ-secreting cells inducing CD4+CD25+ regulatory T cell proliferation. *J. Exp. Med.*, **202**, 919–929.
182. Fricke, I. and Gabrilovich, D.I. (2006) Dendritic cells and tumor microenvironment: a dangerous liaison. *Immunol. Invest.*, **35**, 459–483.
183. Gabrilovich, D.I., Chen, H.L., Girgis, K.R., Cunningham, H.T., Meny, G.M., Nadaf, S., Kavanaugh, D., and Carbone, D.P. (1996) Production of vascular endothelial growth factor by human tumors inhibits the functional maturation of dendritic cells. *Nat. Med.*, **2**, 1096–1103.
184. Dikov, M.M., Oyama, T., Cheng, P., Takahashi, T., Takahashi, K., Sepetavec, T., Edwards, B. *et al.* (2001) Vascular endothelial growth factor effects on nuclear factor-κ B activation in hematopoietic progenitor cells. *Cancer Res.*, **61**, 2015–2021.
185. Conejo-Garcia, J.R., Benencia, F., Courreges, M.C., Kang, E., Mohamed-Hadley, A., Buckanovich, R.J., Holtz, D.O. *et al.* (2004) Tumor-infiltrating dendritic cell precursors recruited by a β-defensin contribute to vasculogenesis under the influence of VEGF-A. *Nat. Med.*, **10**, 950–958.
186. Curiel, T.J., Cheng, P., Mottram, P., Alvarez, X., Moons, L., Evdemon-Hogan, M., Wei, S. *et al.* (2004) Dendritic cell subsets differentially regulate angiogenesis in human ovarian cancer. *Cancer Res.*, **64**, 5535–5538.

187. Okunishi, K., Dohi, M., Nakagome, K., Tanaka, R., Mizuno, S., Matsumoto, K., Miyazaki, J. *et al.* (2005) A novel role of hepatocyte growth factor as an immune regulator through suppressing dendritic cell function. *J. Immunol.*, **175**, 4745–4753.

188. Feijoó, E., Alfaro, C., Mazzolini, G., Serra, P., Peñuelas, I., Arina, A., Huarte, E. *et al.* (2005) Dendritic cells delivered inside human carcinomas are sequestered by interleukin-8. *Int. J. Cancer*, **116**, 275–281.

189. Naldini, A., Leali, D., Pucci, A., Morena, E., Carraro, F., Nico, B., Ribatti, D., and Presta, M. (2006) Cutting edge: IL-1β mediates the proangiogenic activity of osteopontin-activated human monocytes. *J. Immunol.*, **177**, 4267–4270.

190. Scimone, M.L., Lutzky, V.P., Zittermann, S.I., Maffia, P., Jancic, C., Buzzola, F., Issekutz, A.C., and Chuluyan, H.E. (2005) Migration of polymorphonuclear leucocytes is influenced by dendritic cells. *Immunology*, **114**, 375–385.

191. Conejo-Garcia, J.R., Buckanovich, R.J., Benencia, F., Courreges, M.C., Rubin, S.C., Carroll, R.G., and Coukos, G. (2005) Vascular leukocytes contribute to tumor vascularization. *Blood*, **105**, 679–681.

192. Gottfried, E., Kunz-Schughart, L.A., Weber, A., Rehli, M., Peuker, A., Müller, A., Kastenberger, M. *et al.* (2007) Differentiation of human tumour associated dendritic cells into endothelial-like cells: an alternative pathway of tumour angiogenesis. *Scand. J. Immunol.*, **65**, 329–335.

193. Laxmanan, S., Robertson, S.W., Wang, E., Lau, J.S., Briscoe, D.M., and Mukhopadhyay, D. (2005) Vascular endothelial growth factor impairs the functional ability of dendritic cells through Id pathways. *Biochem. Biophys. Res. Commun.*, **334**, 193–198.

194. Alard, P., Clark, S.L., and Kosiewicz, M.M. (2004) Mechanisms of tolerance induced by TGF beta-treated APC: CD4 regulatory T cells prevent the induction of the immune response possibly through a mechanism involving TGFβ. *Eur. J. Immunol.*, **34**, 1021–1030.

195. Pockaj, B.A., Basu, G.D., Pathangey, L.B., Gray, R.J., Hernandez, J.L., Gendler, S.J., and Mukherjee, P. (2004) Reduced T-cell and dendritic cell function is related to cyclooxygenase-2 overexpression and prostaglandin E2 secretion in patients with breast cancer. *Ann. Surg. Oncol.*, **11**, 328–339.

196. Puig-Kröger, A., Pello, O.M., Selgas, R., Criado, G., Bajo, M.A., Sánchez-Tomero, J.A., Alvarez, V. *et al.* (2003) Peritoneal dialysis solutions inhibit the differentiation and maturation of human monocyte-derived dendritic cells: effect of lactate and glucose-degradation products. *J. Leukoc. Biol.*, **73**, 482–492.

197. Konno, S., Eckman, J.A., Plunkett, B., Li, X., Berman, J.S., Schroeder, J., and Huang, S.K. (2006) Interleukin-10 and Th2 cytokines differentially regulate osteopontin expression in human monocytes and dendritic cells. *J. Interferon Cytokine Res.*, **26**, 562–567.

198. Grunewald, M., Avraham, I., Dor, Y., Bachar-Lustig, E., Itin, A., Jung, S., Chimenti, S. *et al.* (2006) VEGF-induced adult neovascularization: recruitment, retention, and role of accessory cells. *Cell*, **124**, 175–189.

199. Takakura, N., Watanabe, T., Suenobu, S., Yamada, Y., Noda, T., Ito, Y., Satake, M., and Suda, T. (2000) A role for hematopoietic stem cells in promoting angiogenesis. *Cell*, **102**, 199–209.

200. Janowska-Wieczorek, A., Marquez, L.A., Nabholtz, J.M., Cabuhat, M.L., Montano, J., Chang, H., Rozmus, J. *et al.* (1999) Growth factors and cytokines upregulate gelatinase expression in bone marrow CD34(+) cells and their transmigration through reconstituted basement membrane. *Blood*, **93**, 3379–3390.

201. Song, S., Ewald, A.J., Stallcup, W., Werb, Z., and Bergers, G. (2005) PDGFRbeta perivascular progenitor cells in tumours regulate pericyte differentiation and vascular survival. *Nat. Cell Biol.*, **7**, 870–879.

202. Bababeygy, S.R., Cheshier, S.H., Hou, L.C., Higgins, D.M., Weissman, I.L., and Tse, V.C. (2008) Hematopoietic stem cell-derived pericytic cells in brain tumor angio-architecture. *Stem Cells Dev.*, **17**, 11–18.

203. Aharinejad, S., Abraham, D., Paulus, P., Abri, H., Hofmann, M., Grossschmidt, K., Schäfer, R. *et al.* (2002) Colony-stimulating factor-1 antisense treatment suppresses growth of human tumor xenografts in mice. *Cancer Res.*, **62**, 5317–5324.

204. Loberg, R.D., Ying, C., Craig, M., Day, L.L., Sargent, E., Neeley, C., Wojno, K. *et al.* (2007) Targeting CCL2 with systemic delivery of neutralizing antibodies induces prostate cancer tumor regression in vivo. *Cancer Res.*, **67**, 9417–9424.

205. Egami, K., Murohara, T., Shimada, T., Sasaki, K., Shintani, S., Sugaya, T., Ishii, M. *et al.* (2003) Role of host angiotensin II type 1 receptor in tumor angiogenesis and growth. *J. Clin. Invest.*, **112**, 67–75.

206. Uemura, H., Nakaigawa, N., Ishiguro, H., and Kubota, Y. (2005) Antiproliferative efficacy of angiotensin II receptor blockers in prostate cancer. *Curr. Cancer Drug Targets*, **5**, 307–323.

207. Orimo, A., Gupta, P.B., Sgroi, D.C., Arenzana-Seisdedos, F., Delaunay, T., Naeem, R., Carey, V.J. *et al.* (2005) Stromal fibroblasts present in invasive human breast carcinomas promote tumor growth and angiogenesis through elevated SDF-1/CXCL12 secretion. *Cell*, **121**, 335–348.

208. Dong, J., Grunstein, J., Tejada, M., Peale, F., Frantz, G., Liang, W.C. *et al.* (2004) VEGF-null cells require PDGFR alpha signaling-mediated stromal fibroblast recruitment for tumorigenesis. *EMBO J.*, **23**, 2800–2810.

209. Fukumura, D., Xavier, R., Sugiura, T., Chen, Y., Park, E.C., Lu, N., Selig, M. *et al.* (1998) Tumor induction of VEGF promoter activity in stromal cells. *Cell*, **94**, 715–725.

210. Crawford, Y., Kasman, I., Yu, L., Zhong, C., Wu, X., Modrusan, Z., Kaminker, J., and Ferrara, N. (2009) PDGF-C mediates the angiogenic and tumorigenic properties of fibroblasts associated with tumors refractory to anti-VEGF treatment. *Cancer Cell*, **15**, 21–34.

211. Kalluri, R. and Zeisberg, M. (2006) Fibroblasts in cancer. *Nat. Rev. Cancer*, **6**, 392–401.

5
Comparison between Developmental and Tumor Angiogenesis

Andreas Bikfalvi

5.1
Introduction

The formation of new blood vessels (angiogenesis) is essential for embryonic development, postnatal growth, and wound healing. It also significantly contributes to pathologic conditions. Insufficient angiogenesis leads to tissue ischemia (ischemic heart disease, stroke), whereas excessive vascular growth promotes cancer, chronic inflammatory disorders (arthritis, psoriasis), or ocular neovascular disease [1]. All these disorders constitute major causes of morbidity and mortality in our Western societies. Algire in 1945 and Folkman in the 1970s suggested that blocking the blood supply to a tumor may actually destroy it [2]. During the following decades, researchers have gained essential insight into the biological mechanisms of angiogenic vessel growth. In this context, significant progress has been made to determine how new, proliferating endothelial cells differ from their nonproliferating "quiescent" counterparts and how their growth is controlled.

For many years, two vasoformative processes that include vasculogenesis and sprouting angiogenesis were recognized. However, in recent years, additional mechanisms have been recognized. These include angioblasts recruitment, cooption, mosaic vessels, or vasculogenic mimicry, with the latter still being controversial. These different mechanisms may exist concomitantly in the same tissue or may be selectively involved in a specific tissue type or host environment.

In this chapter, we will highlight some of the processes occurring in developmental and tumor angiogenesis and compare their significance in these two situations (Figure 5.1 and Table 5.1).

5.2
Vascularization by Sprouting Angiogenesis

The first mechanism that accounts for tumor vascularization was named *sprouting angiogenesis* and involves the proliferation and migration of endothelial cells from preexisting blood vessels and the organization of tubular vascular structures. The

Tumor Angiogenesis – From Molecular Mechanisms to Targeted Therapy. Edited by Francis S. Markland, Stephen Swenson, and Radu Minea
Copyright © 2010 WILEY-VCH Verlag GmbH & Co. KGaA, Weinheim
ISBN: 978-3-527-32091-2

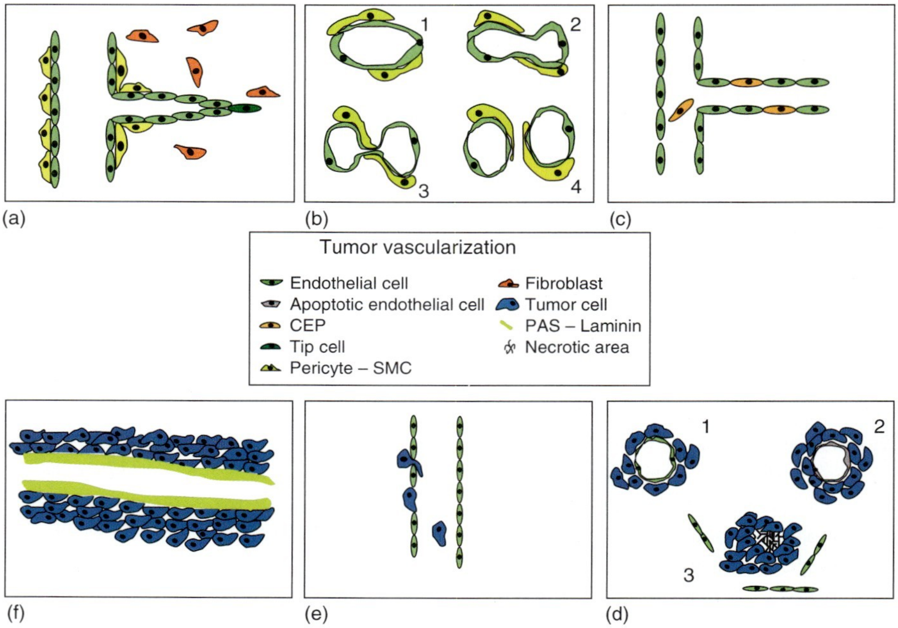

Figure 5.1 Different modes of vessel formation. The different mechanisms of tumor vascularization are depicted in the figure. These include: (a) sprouting angiogenesis; (b) intussusception; (c) recruitment of circulating endothelial precursors; (d) cooption; (e) mosaic vessels; and (f) vascular mimicry. SMC, smooth muscle cell. (From Auguste, P., Lemiere, S., Larrieu-Lahargue, F., and Bikfalvi, A. (2005) *Critical reviews in oncology/hematology* **54**(1), 53–61.)

Table 5.1 Differences between developmental and tumor angiogenesis.

	Developmental angiogenesis	Tumor angiogenesis
Sprouting angiogenesis	+	+
Defect in remodeling/pruning	–	+
Vasculogenesis	+	+/–
Cooption	+/–	+
Intussusception	+	+/–
Mosaic vessels	–	+
Mimicry	–	+

major players that control this process are VEGF family members, in particular VEGF-A [3]. VEGF-A induces vasodilatation of preexisting capillaries and increases permeability. Plasma proteins, particularly fibrinogen, extravasate and form a provisional matrix later used for endothelial cell migration and tube formation. Other players include angiopoietin-1 (Ang1), which has clearly sprouting effects on blood vessels and acts on the cytoskelatal machinery through Tie2 receptors [3, 4]. Furthermore, it is involved in endothelial cell–pericyte interactions and stabilizes vessels. On the contrary, Ang2 is stored in Weibel–Palade bodies, released by endothelial cells and antagonizes Ang1 activity on the Tie2 receptor. This inhibits sprouting and pericyte–endothelial cell interaction and favors an immature vascular network. Ang1 and Ang2 intervene in both developmental and tumor angiogenesis. In the tumor context, their role is less clear and conflicting results have been reported (Christofori *et al.* personal communication). For example, in some tumor types, Ang1 promotes invasion and metastasis while in others it does not.

Sprouting angiogenesis further involves metalloproteases and plasminogen activators that induce promigratory signals and lead to the degradation of the extracellular matrix permissive for endothelial cell migration [3].

Another important characteristic of sprouting angiogenesis is vessel guidance. Vessel guidance mechanisms have been identified in the retina and the hindbrain. It has been postulated that one important guidance cue is the VEGF/VEGFR system. Indeed, tip cells that are located at the invading front of the blood vessels, exhibit VEGFR2 receptor clusters in filipodia and seem to follow a VEGF gradient [5]. In addition, repulsive cues are also similarly involved as for axon guidance [6]. The repulsive netrin receptor UNC5B is expressed by endothelial tip cells and its disruption in mice leads to aberrant extension of endothelial tip cell filopodia and abnormal vessel navigation [6]. Whether guidance mechanisms also operate during tumor angiogenesis is not clearly established. Observations using the chicken chorioallantoic membrane (CAM) indicate that this might be so [7]. When Netrin-1 transfected cells are used and implanted into the CAM, repulsion of vessels is seen. It is possible that the guidance of blood vessels is aberrant to some extent in tumor tissue leading to a highly disorganized vasculature. Netrin-1 also seems to have survival effects on endothelial cells and the concept of dependency receptors through Unc5 receptors has been proposed [8]. In the absence of ligand cells survival is increased whereas in the presence of Netrin-1 no survival effect is seen. However, there are problems to reconcile the survival effect with the antirepulsion effect. Furthermore, it is less clear as to how the dependency concept functions in a tumor environment where vessels are permanently exposed to positive regulatory factors. Thus, the dependency concept is difficult to apply to a tumor angiogenesis context.

A very important molecular regulator of angiogenesis is the Delta4/Notch1 system [9]. It has been shown that mice heterozygous for the Delta4 allele exhibit increase in tip cell formation in the retina. This indicates that Delta4 is a negative regulator of tip cell formation. However, vessels in the Delta4 hemizygous mouse are nonfunctional. On the contrary, Jagged1, which is also a ligand for Notch1, and has opposite effects because it increases tip cells and sprouting. This differential effect seems to be due to posttranslational modifications of Notch-1 because glycosylation through Fring proteins plays a crucial role [10]. The Delta4/Notch system may also play a role in tumors [11]. Indeed, glioblastoma cells expressing Delta4 induce an increase of sprouting but with defective vessels. This defective angiogenesis, in turn, slows tumor growth. As a consequence, tumor growth is inhibited.

Another important event occurring during angiogenesis is pericyte recruitment. Besides the Ang system, another factor, PDGF-B, is also involved in pericyte recruitment and coverage and functions as a chemoattractant for pericytes [12, 13]. Furthermore, TGF-β and sphingosine-1-phosphate (S-1-P) also participate in vessel stabilization through their interaction with endoglin and the S-1-P receptor (Edg-1) [13]. These mechanisms are well tuned during development but not in tumors leading to a loosening of endothelial cell–pericyte interactions. This can explain why tumor capillaries are leaky and have abnormally high diameters. The molecular repertoire that tumor cells use to regulate angiogenic tube formation is diverse and is altered for a given tissue type or host environment. In tumors, the use of a specific repertoire at a given stage of vascular development follows a "Darwinian" selection process. It has been recognized that when the selective pressure on tumor cells is increased by, for instance, blocking a specific growth factor pathway, adaptation occurs and alternative growth factors and signaling molecules may be used [14]. Furthermore, highly invasive tumor cells may be also selected that exhibit up-regulation of signaling pathways involved in migration and invasion [15].

An interplay between classes of regulatory molecules during angiogenesis has been recognized. For example, fibroblast growth factor family members (FGFs) can directly bind and activate its type 1 receptor (FGFR1) present at the endothelial cell surface and induce protease production, migration, and proliferation [16]. It has been recently recognized that FGF signaling might be important in maintaining the cohesion of vascular endothelial cells by acting on the adherent junctional protein VE-cadherin [17]. Another explanation of FGF effects may be due to autocrine VEGF-A production induced by FGF-2 in endothelial cells; but, this is still controversial and may depend on the endothelial cell type. On the other end, autocrine FGFs loops have been demonstrated in tumor cells, such as glioblastoma and may increase VEGF-A production [18]. Besides its role in pericyte–vessel interactions and in vessel stabilization, PDGF-B, was also shown to be an angiogenic growth factor by acting on endothelial cells [19].

Hypoxia is a driving force during developmental and tumor sprouting angiogenesis [20, 21]. This has been extensively demonstrated for the development of retinal vessels where hypoxic astrocytes are found at the leading front that produce VEGF for vessel sprouting. The hypoxia-regulated system comprises among its principal components HIF-1α, HIF-2α, and proline hydroxylase 1, 2, and 3 (PhD 1, 2, and 3). Recent results indicate that the HIF system has distinct opposing effects in the endothelial cell compartment and in tumor cells [22]. In endothelial cells, HIF-2α is stabilized during hypoxia and will, in turn, activate the expression of soluble VEGFR1 that will neutralize VEGF activity. In tumor cells, HIF-1α is stabilized and this, in turn, induces VEGF expression. It is not clear whether these differences are also seen during developmental angiogenesis. In any case, hypoxia is a driving force for both developmental and tumor angiogenesis.

A very intriguing finding has been reported more recently. Intracellular non-secreted VEGF may regulate, specifically, endothelial cell survival [23]. When

VEGF is specifically silenced in endothelial cells using Tie2 Cre strategy in mice, endothelial cells undergo apoptosis *in vivo*. Thus it seems that intracellular VEGF has antiapoptotic effects *in vivo*. This may stabilize the endothelial cell tubes after sprouting has been achieved. This mechanism should be active in both developmental and tumor angiogenesis.

5.3 Intussusceptive Vascular Growth

Another variant of angiogenesis, different from sprouting, is called *intussusceptive vascular growth* [24]. During this process, an interstitial tissue column is inserted into the lumen of preexisting vessels inducing partition of the vessel lumen. This mechanism has been first characterized in lung development, but is now found in nearly all the organs studied and also occurs in tissue repair and tumor angiogenesis. It has been particularly well investigated in the chicken CAM [25].

During intussusception, two opposite endothelial cell membranes make contact ("kissing" contact) and interendothelial junctions develop at their edge. In the middle of the kissing contact, membranes are thinned and pressure induced by the cytoplasm opens them and separate the two vessels. Most of the time, a pericyte is at the origin of the kissing contact through its contractile activity. In the chicken CAM or in tumor models, only one membrane may make contact with the opposite membrane.

The gap between the two newly formed vessels is invaded by mesenchymal cells composed of fibroblast and pericytes to form a pillar or post (diameter < 2.5 m) or an interstitial or intervascular tissue structure (diameter > 2.5 m). Accumulation of extracellular matrix proteins, such as collagen or fibrin takes place within the pillar. Intussusception generates vessels more rapidly in a metabolically more economical manner than sprouting. No endothelial cell proliferation is needed. Endothelial cells only increase their volume and become thinner. Intussusception has been implicated in tumor angiogenesis and can explain a rapid remodeling of the vasculature [26].

The molecular mechanism of intussusception is not fully understood but shear stress plays a major role, as well as increasing blood flow rate. Shear stress can be sensed by endothelial cells and transduced inside the cell by platelet endothelial cell adhesion molecule-1 (PECAM-1) resulting in increased expression of angiogenic factors (TGF-1), eNOS, and adhesion molecules. Most of the factors implicated in endothelial–endothelial or endothelial–pericyte interactions are implicated in intussusception. In the chicken CAM, PDGF-B increases intussusception. This effect is blocked by neutralizing antibodies to PDGF-B. Additional factors such as erythropoietin are able to induce proliferation and migration of endothelial cells and *in vivo* angiogenesis in the CAM mainly by intussusception [27]. A better understanding of the molecular mechanism of intussusception may help to inhibit tumor angiogenesis.

5.4 Vasculogenesis and Angioblast Recruitment

During embryonic development, a common precursor of the hematopoietic and the endothelial cell lineages, the hemangioblast, is found in different areas, such as in the yolk sac, para-aortic splanchnopleura, and aorta-gonad-mesonephros regions (for a review for this and the whole section see [28, 29]). This precursor is VEGFR2 positive and acquires CD34, CD133, and VE-cadherin to give rise to a progenitor cell, the angioblast, which is able to differentiate into endothelial cells. In the adult organism, angioblasts are derived from multipotent adult progenitor cells (MAPCs) positive for VEGFR1, VEGFR2, AC133, Oct-4, and telomerase in the bone marrow. When stimulated by VEGF-A, MAPCs differentiate into endothelial cells *in vitro*. MAPCs can also generate other progenitor cells, such as mesenchymal stem cells that are at the origin of mural cells, adipocytes, myoblasts, and bone cartilage cells. Circulating endothelial precursors (CEPs) are recruited in the circulation by VEGF-A stimulation and/or Ang1 with two different kinetics, with Ang1 generating a lower but more sustained stimulation.

CEPs' recruitment to the circulation is also dependant on MMP-9-induced release of soluble kit ligand, which in turn increases cell proliferation and motility and CEPs' mobilization to the circulation. Furthermore, stromal-derived factor-1 (SDF-1) seems to play a critical role in endothelial progenitor cell (EPC) mobilization. The importance of CEP mobilization in tumor vascularization has been demonstrated in a mouse mutant for Id proteins. Id proteins belong to a family of proteins interacting with helix-loop-helix transcription factors implicated in early fetal development. Double mutants Id1+/−Id3−/− display vascular defects associated with impaired VEGF-A-induced mobilization and proliferation of CEPs and inhibition of tumor development. Tumors xenografted in lethally irradiated Id1+/−Id3−/− mice and reconstituted with wild type CEPs, grow as well as in control mice. Most of the endothelial cells are also Id3+. Moreover, in wild type mice, CEP recruitment to the tumor vasculature is completely inhibited by injection of a neutralizing antibody against, VEGFR2 but not against VEGFR1. Nevertheless, VEGFR1-positive hematopoietic stem and progenitor cells are recruited to the tumor vasculature and may participate at CEP incorporation into the endothelium. The extent of CEP recruitment into the tumor vasculature may depend on the tumor type [30], varying from 90% in lymphoma (B6RV2) to 5% in neuroblastoma implanted subcutaneously into mice. CEP recruitment is necessary for tumor implantation. This is suggested by the fact that in lymphoma most of the tumor vessels are derived from CEPs at day 2 but only 50% at day 14. CEPs may be a very powerful tool to deliver angiogenic inhibitors to the tumor vasculature.

CEPs in culture were successfully transduced with a retroviral vector expressing both soluble VEGFR2 and green fluorescent protein (GFP), and reintroduced into mice. In these mice, neuroblastoma growth was severely reduced. This demonstrates that CEP cells may be useful for antiangiogenic therapy. Alternatively, embryonic stem cells derived from the inner cell mass of mouse blastocysts and able to generate the endothelial cell lineage, can be used instead of CEPs to deliver antiangiogenic molecules.

The participation of endothelial cell precursor recruitment in tumors has been challenged by recent observations where only very few EPCs were accumulated into the tumor in a variety of mouse tumor models [31]. This has shed doubts on the validity of targeting EPCs for therapeutic antiangiogenesis treatment. The differences between this and former studies needs to be clarified.

5.5 Cooption

Many tumors develop in highly vascularized tissues, such as brain and lung. In these tissues, tumors can use another mechanism for vascularization, named cooption by Holash *et al.* [32] (for review see [33–35]). At the early stage of development, brain tumors are highly vascularized and vessels have a phenotype similar to normal brain. These vessels are surrounded, coopted by tumor cells and no sprouts are observed. This can explain why it is difficult to detect gliomas at early stages by MRI analysis. Coopted endothelial cells synthesize Ang2 and its receptor Tie2. Ang2 binds to its receptor present at the endothelial cell surface and induces dissociation of the mural cells from endothelial cells, inhibition of Ang1 activity, and increase in apoptosis. Ang2 activity results in a large decrease in tumor vessel number and an increase in vessel diameter. Consequently, the lack of vessels leads to hypoxia which up-regulates VEGF-A expression in tumor cells. As a consequence, strong angiogenesis develops mainly at the tumor periphery. In the tumor center, cells are organized in cuffs of pseudopalisades around the few surviving vessels. No other growth factors (FGFs) have been implicated in this mechanism.

Cooption depends on the site of tumor development. For example, rat mammary carcinoma is vascularized by cooption only if cells are injected inside the brain. Lewis lung carcinoma and melanoma cells metastasized respectively into lung or brain and are partially vascularized by cooption. Some melanoma tumors can coopt vessels but without evidence of vessel ingrowth. On the other hand, others have described a different mechanism of brain tumor vascularization, distinct from the cooption mechanism outlined above. Tumor cells or spheroids injected into the brain develop vascularization immediately by angiogenic sprouting with loss of the blood–brain barrier. Ang2, as well as VEGFR2 is increased in a subset of endothelial cells but no vessel regression is observed. Tumor cells are organized in cuffs of pseudopalisading cells around VEGFR2-positive vessels and use them to invade other brain areas.

Migration is facilitated by binding of tumor cell to a basement membrane rich in laminin, collagen IV, and tenascin. Tumor cells use vessels for oxygen and nutriment supply, and form pseudopalisading structures composed of proliferating tumor cells. These cuffs of tumor cells are difficult to kill by most of the classical antiangiogenic molecules. Only a very strong antiangiogenic molecules, such as high doses of VEGF-Trap able to induce complete vessel regression, may possibly kill these tumor cell cuffs [36].

5.6 Other Mechanisms of Vasoformation

There are two other mechanisms proposed that include vasculogenic mimicry and mosaic vessels [37, 38]. These mechanisms are operating exclusively in tumor angiogenesis. The concept of vasculogenic mimicry was proposed in 1999 by Maniotis *et al.* [37]. Aggressive uveal melanoma presents angiogenic vessels only at the border of the tumor. Inside the tumor, vessels delineated by endothelial cells are absent and curiously the tumor is not necrotic. A periodic acid stain (PAS) of the tumor showed a network of channels interconnected by loops. Despite a significant body of literature, vascular mimicry is not yet a universally accepted mechanism, as the concept has been challenged by several investigators. There are also some theoretical considerations that are difficult to reconcile with the model. For example, the endothelium constitutes a nonthrombogenic surface and massive thrombosis would follow in its absence, which would lead to tumor necrosis. However, it is possible that thrombosis is prevented by an increase in synthesis of tissue factor pathway inhibitor 1 (TFPI-1) in tumor cells. The fact that tumor cells are able to form tubular structures in matrigel or collagen gels is not only a characteristic of vascular cells since some epithelial cells, Madin-Darby canine kidney (MDCK), are also able to do so.

It has been postulated than tumor cells may localize into the wall of tumor blood vessels and form mosaic vessels [39]. These cells may be used by activated natural killer (NK) cells to penetrate inside the tumor. Chang *et al.* have described this mechanism in more detail and have named it *mosaic blood vessels* [39]. By using GFP-transfected tumor cells they showed that in colon carcinoma xenografts in ectopic (ovarian pedicle) or orthotopic (cecum) implantation, 15% of vessels are mosaic vessels, representing 4% of the total vascular surface. These mosaic vessels are functional because they can be perfused with a fluorescent lectin. Tumor cells in mosaic vessels are undergoing intravasation into the lumen and stay temporarily in the capillary vessel wall. Angiogenic growth factors, such as FGF-2 or VEGF-A activating matrix metalloproteinase-2 (MMP-2) activity, increase vessel basement membrane degradation and intravasation of tumor cells, thus increasing mosaic vessel number.

5.7 Developmental Angiogenesis and Tumor Angiogenesis: Similar and Different

As discussed above, developmental and tumor angiogenesis both undergo several modes of vascularization. Sprouting angiogenesis is operating in both situations. During development, vessels follow a strict vascular growth factor gradient, encounter positive and negative guidance cues, are then pruned and remodeled, and finally stabilized through pericyte recruitment. In contrast, in tumors there are defects in growth factor gradients, remodeling, and vessel stabilization. The driving forces are, in both situations, a similar set of regulatory molecules and cues. However, during developmental angiogenesis but not during tumor angiogenesis, the action of these factors is well tuned and they intervene in a coordinated fashion in both time and space. Furthermore, during developmental angiogenesis but not during tumor angiogenesis, the action of these factors is silenced when vascular development is completed. Thus, in tumors there is a lack of tuning, coordination, and silencing. In this case, the vasculature is made of restless endothelium where vessel turnover is permanently activated.

Other modes of vasoformation include intussusception, vasculogenesis and endothelial cell progenitor recruitment, cooption, vasculogenic mimicry, and mosaic vessels. Intussusception has been observed in developmental angiogenesis and is likely to also occur to some extent in tumors. Vasculogenesis is mainly active during development. The recruitment of endothelial cell progenitors has been reported in several solid tumors but the importance of this concept with regard to tumor angiogenesis has been recently challenged. Vessel cooption is observed in both, in development and in tumors, and has been particularly well studied in brain tumors. Lastly, vasculogenic mimicry is still a highly controversial concept, and mosaic vessels are exclusively observed in tumors. Taken together, tumor angiogenesis can be viewed as imperfect developmental angiogenesis that has acquired some specific additional mechanisms.

Acknowledgments

The work from our laboratory described in this review was supported by grants from the "Association de la Recherche sur le Cancer" and the "Ligue contre le Cancer." Parts of this chapter are derived from our review published in Critical reviews in Oncology/Hematology (Auguste, P., Lemiere, S., Larrieu-Lahargue, F., and Bikfalvi, A. (2005) *Critical reviews in oncology/hematology* **54**(1), 53–61).

References

1. Carmeliet, P. (2005) Angiogenesis in life, disease and medicine. *Nature*, **438**, 932–936.
2. Folkman, J. (1971) Tumor angiogenesis: therapeutic implications. *N. Engl. J. Med.*, **285**, 1182–1186.
3. Adams, R.H. and Alitalo, K. (2007) Molecular regulation of angiogenesis and lymphangiogenesis. *Nat. Rev. Mol. Cell Biol.*, **8**, 464–478.
4. Augustin, H.G., Koh, G.Y., Thurston, G., and Alitalo, K. (2009) Control of vascular morphogenesis and homeostasis through the angiopoietin-Tie system. *Nat. Rev. Mol. Cell Biol.*, **10**, 165–177.
5. Hellstrom, M., Phng, L.K., and Gerhardt, H. (2007) VEGF and Notch signaling: the yin and yang of angiogenic sprouting. *Cell Adhes. Migr.*, **1**, 133–136.
6. Larrivee, B., Freitas, C., Suchting, S., Brunet, I., and Eichmann, A. (2009) Guidance of vascular development:

lessons from the nervous system. *Circ. Res.*, **104**, 428–441.

7. Bouvree, K., Larrivee, B., Lv, X., Yuan, L., DeLafarge, B., Freitas, C., Mathivet, T., Breant, C., Tessier-Lavigne, M., Bikfalvi, A., Eichmann, A., and Pardanaud, L. (2008) Netrin-1 inhibits sprouting angiogenesis in developing avian embryos. *Dev. Biol.*, **318**, 172–183.
8. Castets, M., Coissieux, M.M., Delloye-Bourgeois, C., Bernard, L., Delcros, J.G., Bernet, A., Laudet, V., and Mehlen, P. (2009) Inhibition of endothelial cell apoptosis by netrin-1 during angiogenesis. *Dev. Cell*, **16**, 614–620.
9. Thurston, G. and Kitajewski, J. (2008) VEGF and Delta-Notch: interacting signalling pathways in tumour angiogenesis. *Br. J. Cancer*, **99**, 1204–1209.
10. Benedito, R., Roca, C., Sorensen, I., Adams, S., Gossler, A., Fruttiger, M., and Adams, R.H. (2009) The notch ligands Dll4 and Jagged1 have opposing effects on angiogenesis. *Cell*, **137**, 1124–1135.
11. Li, J.L. and Harris, A.L. (2009) Crosstalk of VEGF and Notch pathways in tumour angiogenesis: therapeutic implications. *Front Biosci.*, **14**, 3094–3110.
12. Andrae, J., Gallini, R., and Betsholtz, C. (2008) Role of platelet-derived growth factors in physiology and medicine. *Genes Dev.*, **22**, 1276–1312.
13. Gaengel, K., Genove, G., Armulik, A., and Betsholtz, C. (2009) Endothelial-mural cell signaling in vascular development and angiogenesis. *Arterioscler. Thromb. Vasc. Biol.*, **29**, 630–638.
14. Casanovas, O., Hicklin, D.J., Bergers, G., and Hanahan, D. (2005) Drug resistance by evasion of antiangiogenic targeting of VEGF signaling in late-stage pancreatic islet tumors. *Cancer Cell*, **8**, 299–309.
15. Paez-Ribes, M., Allen, E., Hudock, J., Takeda, T., Okuyama, H., Vinals, F., Inoue, M., Bergers, G., Hanahan, D., and Casanovas, O. (2009) Antiangiogenic therapy elicits malignant progression of tumors to increased local invasion and distant metastasis. *Cancer Cell*, **15**, 220–231.
16. Javerzat, S., Auguste, P., and Bikfalvi, A. (2002) The role of fibroblast growth factors in vascular development. *Trends Mol. Med.*, **8**, 483–489.
17. Murakami, M., Nguyen, L.T., Zhuang, Z.W., Moodie, K.L., Carmeliet, P., Stan, R.V., and Simons, M. (2008) The FGF system has a key role in regulating vascular integrity. *J. Clin. Invest.*, **118**, 3355–3366.
18. Auguste, P., Gursel, D.B., Lemiere, S., Reimers, D., Cuevas, P., Carceller, F., Di Santo, J.P., and Bikfalvi, A. (2001) Inhibition of fibroblast growth factor/fibroblast growth factor receptor activity in glioma cells impedes tumor growth by both angiogenesis-dependent and -independent mechanisms. *Cancer Res.*, **61**, 1717–1726.
19. Zhang, J., Cao, R., Zhang, Y., Jia, T., Cao, Y., and Wahlberg, E. (2009) Differential roles of PDGFR-alpha and PDGFR-beta in angiogenesis and vessel stability. *FASEB J.*, **23**, 153–163.
20. Semenza, G. (2009) Regulation of cancer cell metabolism by hypoxia-inducible factor 1. *Semin. Cancer Biol.*, **19**, 12–16.
21. Semenza, G.L. (2009) Regulation of oxygen homeostasis by hypoxia-inducible factor 1. *Physiology (Bethesda)*, **24**, 97–106.
22. Mazzone, M., Dettori, D., Leite de Oliveira, R., Loges, S., Schmidt, T., Jonckx, B., Tian, Y.M., Lanahan, A.A., Pollard, P., Ruiz de Almodovar, C., De Smet, F., Vinckier, S., Aragones, J., Debackere, K., Luttun, A., Wyns, S., Jordan, B., Pisacane, A., Gallez, B., Lampugnani, M.G., Dejana, E., Simons, M., Ratcliffe, P., Maxwell, P., and Carmeliet, P. (2009) Heterozygous deficiency of PHD2 restores tumor oxygenation and inhibits metastasis via endothelial normalization. *Cell*, **136**, 839–851.
23. Lee, S., Chen, T.T., Barber, C.L., Jordan, M.C., Murdock, J., Desai, S., Ferrara, N., Nagy, A., Roos, K.P., and Iruela-Arispe, M.L. (2007) Autocrine VEGF signaling is required for vascular homeostasis. *Cell*, **130**, 691–703.
24. Makanya, A.N., Hlushchuk, R., and Djonov, V.G. (2009) Intussusceptive angiogenesis and its role in vascular

morphogenesis, patterning, and remodeling. *Angiogenesis*, **12** (2), 113–123.

25. Makanya, A.N., Hlushchuk, R., Baum, O., Velinov, N., Ochs, M., and Djonov, V. (2007) Microvascular endowment in the developing chicken embryo lung. *Am. J. Physiol. Lung Cell Mol. Physiol.*, **292**, L1136–L1146.
26. Hlushchuk, R., Riesterer, O., Baum, O., Wood, J., Gruber, G., Pruschy, M., and Djonov, V. (2008) Tumor recovery by angiogenic switch from sprouting to intussusceptive angiogenesis after treatment with PTK787/ZK222584 or ionizing radiation. *Am. J. Pathol.*, **173**, 1173–1185.
27. Crivellato, E., Nico, B., Vacca, A., Djonov, V., Presta, M., and Ribatti, D. (2004) Recombinant human erythropoietin induces intussusceptive microvascular growth in vivo. *Leukemia*, **18**, 331–336.
28. Jin, S.W. and Patterson, C. (2009) The opening act: vasculogenesis and the origins of circulation. *Arterioscler. Thromb. Vasc. Biol.*, **29**, 623–629.
29. Kovacic, J.C., Moore, J., Herbert, A., Ma, D., Boehm, M., and Graham, R.M. (2008) Endothelial progenitor cells, angioblasts, and angiogenesis--old terms reconsidered from a current perspective. *Trends Cardiovasc. Med.*, **18**, 45–51.
30. Kerbel, R.S., Benezra, R., Lyden, D.C., Hattori, K., Heissig, B., Nolan, D.J., Mittal, V., Shaked, Y., Dias, S., Bertolini, F., and Rafii, S. (2008) Endothelial progenitor cells are cellular hubs essential for neoangiogenesis of certain aggressive adenocarcinomas and metastatic transition but not adenomas. *Proc. Natl. Acad. Sci. U.S.A.*, **105**, E54; author reply E55.
31. Purhonen, S., Palm, J., Rossi, D., Kaskenpaa, N., Rajantie, I., Yla-Herttuala, S., Alitalo, K., Weissman, I.L., and Salven, P. (2008) Bone marrow-derived circulating endothelial precursors do not contribute to vascular endothelium and are not needed for tumor growth. *Proc. Natl. Acad. Sci. U.S.A.*, **105**, 6620–6625.
32. Holash, J., Maisonpierre, P.C., Compton, D., Boland, P., Alexander, C.R., Zagzag, D., Yancopoulos, G.D., and Wiegand, S.J. (1999) Vessel cooption, regression, and growth in tumors mediated by angiopoietins and VEGF. *Science*, **284**, 1994–1998.
33. Auguste, P., Lemiere, S., Larrieu-Lahargue, F., and Bikfalvi, A. (2005) Molecular mechanisms of tumor vascularization. *Crit. Rev. Oncol. Hematol.*, **54**, 53–61.
34. Fischer, I., Gagner, J.P., Law, M., Newcomb, E.W., and Zagzag, D. (2005) Angiogenesis in gliomas: biology and molecular pathophysiology. *Brain Pathol.*, **15**, 297–310.
35. Monzani, E. and La Porta, C.A. (2008) Targeting cancer stem cells to modulate alternative vascularization mechanisms. *Stem Cell Rev.*, **4**, 51–56.
36. Kim, E.S., Serur, A., Huang, J., Manley, C.A., McCrudden, K.W., Frischer, J.S., Soffer, S.Z., Ring, L., New, T., Zabski, S., Rudge, J.S., Holash, J., Yancopoulos, G.D., Kandel, J.J., and Yamashiro, D.J. (2002) Potent VEGF blockade causes regression of coopted vessels in a model of neuroblastoma. *Proc. Natl. Acad. Sci. U.S.A.*, **99**, 11399–11404.
37. Folberg, R. and Maniotis, A.J. (2004) Vasculogenic mimicry. *Acta Pathol. Microbiol. Immunol. Scand.*, **112**, 508–525.
38. Zhang, S., Zhang, D., and Sun, B. (2007) Vasculogenic mimicry: current status and future prospects. *Cancer Lett.*, **254**, 157–164.
39. Chang, Y.S., di Tomaso, E., McDonald, D.M., Jones, R., Jain, R.K., and Munn, L.L. (2000) Mosaic blood vessels in tumors: frequency of cancer cells in contact with flowing blood. *Proc. Natl. Acad. Sci. U.S.A.*, **97**, 14608–14613.

6
Tumor Lymphangiogenesis

Swapnika Ramu and Young-Kwon Hong

6.1
Development of Lymphatic System

The human body contains two major circulatory systems, the blood vascular and the lymphatic system. In contrast to the intensively studied and well-characterized blood vascular system, relatively little was known about the origin and development of the lymphatic vessels. Within the last decade, however, the identification of reliable molecular markers of the lymphatic vasculature has led to a renewed interest in the study of basic lymphatic biology and has paved the way for many important studies of lymphatic origin and function.

The lymphatic vasculature is a linear, open-ended system, unlike the closed circulatory loop of the blood vasculature (Figure 6.1). It is composed of a network of thin-walled capillaries that collect the tissue fluid (lymph) that drains from the interstitial tissue spaces. This lymph is then carried by lymphatic capillaries to the larger collecting lymphatics and finally to the thoracic duct, from where the lymph enters the inferior vena cava for recirculation. The lymphatic vessels are covered by a single continuous layer of overlapping lymphatic endothelial cells (LECs) but, unlike blood vessels, are not associated with a continuous basement membrane; therefore, lymphatic capillaries are highly permeable (Figure 6.2). Lymphoid organs such as the lymph nodes, tonsils, Peyer's patches, spleen, and thymus are also part of the lymphatic system. The lymphatic vascular system plays key roles in the maintenance of tissue fluid homeostasis, uptake of intestinal lipids, and trafficking of immune cells to the sites of inflammation and infection. Recent studies have shown that in addition to these physiological functions, tumors can also utilize lymphatic vessels to migrate to distant sites and establish metastases (reviewed in [1]).

From a historic perspective, many important discoveries regarding the structure and origin of the lymphatic system have been followed by long periods of scientific obscurity. For example, although lymph had been described as "white blood" by Hippocrates, it was not until 1627 that lymphatic vessels were first identified by the Italian anatomist Aselli [2] as "milky veins." In contrast, the first detailed description of the blood vascular system was published as early as 1628, by William Harvey [3]. Despite the subsequent identification of other lymphatic vessels and

Tumor Angiogenesis – From Molecular Mechanisms to Targeted Therapy. Edited by Francis S. Markland, Stephen Swenson, and Radu Minea
Copyright © 2010 WILEY-VCH Verlag GmbH & Co. KGaA, Weinheim
ISBN: 978-3-527-32091-2

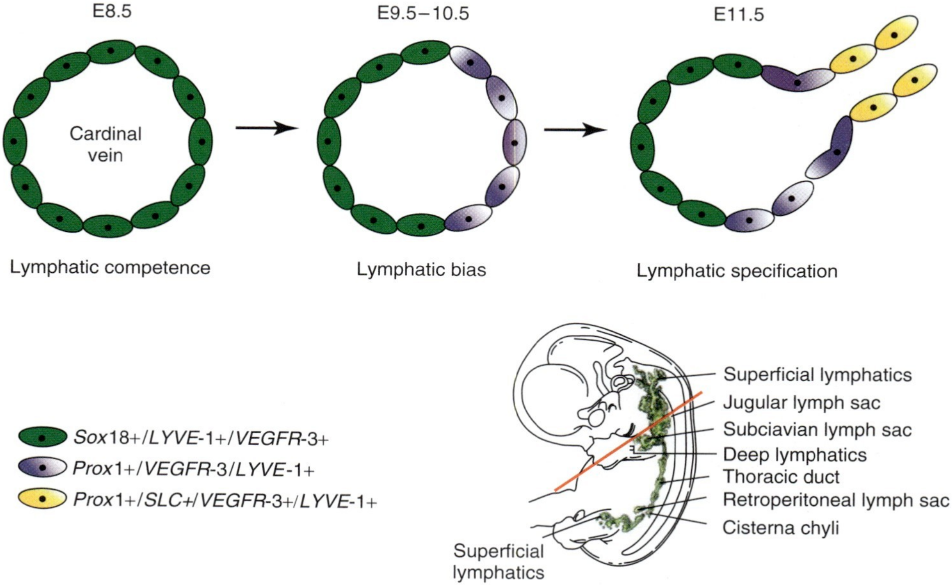

Figure 6.1 Current understanding of lymphatic development. At embryonic day 8.5, all cells of the cardinal vein, the first primitive vascular structure, express the lymphatic markers Sox18, lymphatic vessel endothelial hyaluronan receptor-1 (LYVE-1), and VEGFR-3 and exhibit lymphatic competence. Between embryonic days 9.5 and 10.5, a currently unidentified signal induces a subset of venous endothelial cells to upregulate the expression of Prox1, causing them to adopt a lymphatic bias. These cells then migrate and bud out of the cardinal vein starting at embryonic day 11.5, giving rise to the jugular lymph sac, the first primitive lymphatic vascular structure. These initial lymphatic endothelial cells also upregulate the expression of other lymphatic-specific markers such as secondary lymphatic chemokine (SLC).

their role in fluid drainage, it was not until 1902 that the first theory was proposed regarding the embryonic origin of lymphatic vessels. Based on the result from her ink-injection studies, Florence Sabin hypothesized that initial lymph sacs are derived by budding from embryonic veins, and that these primitive lymphatics then spread throughout the body to form mature lymphatics [4, 5]. A contrasting hypothesis put forth by Huntington and McClure proposed that lymphatics arose independently in the mesenchyme and became connected to the venous network later during development [6]. This issue was resolved only within the last decade, due to the identification of several lymphatic-specific molecular markers, spurring renewed interest in studies of lymphatic development and function [1, 7]. Molecular evidence has lent support to Sabin's hypothesis and has demonstrated the venous origin of lymphatics during embryonic development [7]. Figure 6.3 summarizes the current understanding of the molecular regulation of lymphatic development.

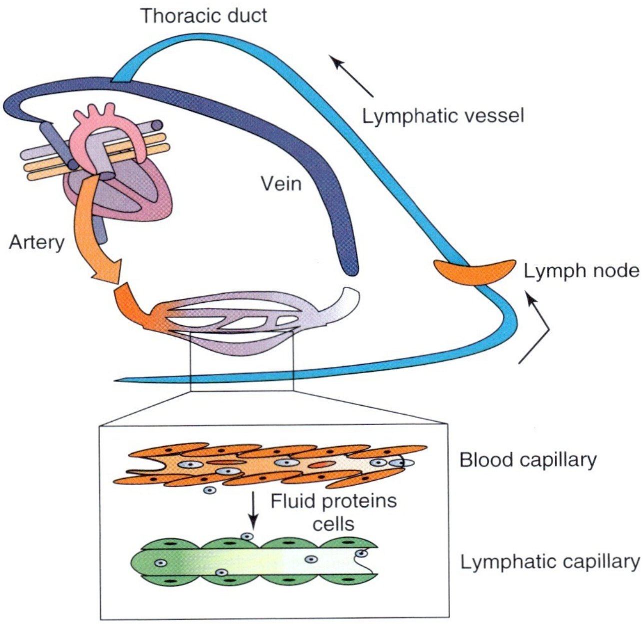

Figure 6.2 Macroscopic view of the blood and lymphatic systems. In contrast to the closed circular blood vascular system, the lymphatic vasculature is linear and open ended. Interstitial fluid drainage from tissues, carrying proteins and cells, is collected by lymphatic capillaries and enters lymphatic circulation. It passes through the lymph nodes and is eventually returned to the blood vascular network via the thoracic duct. (Modified from [158].)

6.2
Tumor Lymphangiogenesis

Given that the entry of tumor cells into lymphatic vessels is an early event in tumor progression, lymphatic involvement in tumor metastasis has considerable clinical implications. Lymph node metastasis is a commonly used clinical parameter for the staging of tumors, and is strongly correlated with patient survival as well as influencing treatment options such as chemotherapy, surgery, and radiation therapy [8].

While it was initially thought that metastasis via lymphatic vessels was a passive process, a growing body of evidence suggests that tumors can actively promote the growth of lymphatic vessels in order to disseminate to distant sites [9, 10] (Figure 6.4). However, the mechanisms by which such tumor-associated lymphangiogenesis may take place are currently unknown. For example, it is known that bone marrow–derived vascular endothelial progenitor cells are capable of incorporating into existing blood vessels to promote neovessel growth during cancer in multiple tumor types, including a mouse mammary tumor system [11, 12]. In comparison, the existence and relative contribution of lymphatic

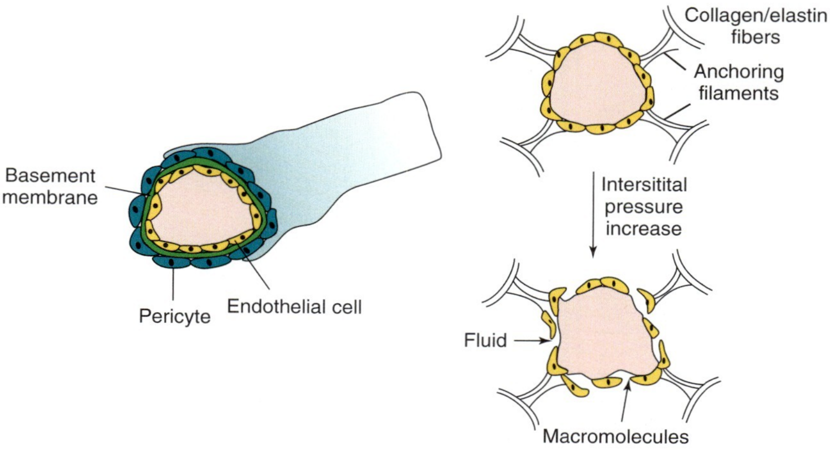

Figure 6.3 Comparison of blood and lymphatic vessel structure. Blood vessels (left) consist of a layer of endothelial cells with tightly associated pericytes (smooth muscle cells) and a continuous basement membrane layer and exhibit a regular lumen. In contrast, lymphatic vessels (right) consist of a single layer of endothelial cells with an irregular lumen and no associated basement membrane. Lymphatic endothelial cells are connected to the surrounding extracellular matrix via anchoring filaments. When there is an increase in interstitial fluid pressure, these anchoring filaments exert greater force on the lymphatic endothelial cells, creating gaps between the cells that allow for entry of fluid and macromolecules into lymphatic circulation.

endothelial progenitor cells during lymphangiogenesis remain unclear [13–15]. Interestingly, recent work has shown that bone marrow–derived vascular endothelial progenitor cells may also function to establish "premetastatic niches" that allow for the establishment and proliferation of tumor cells to form metastases [16, 17], suggesting that targeting of these precursor cells may be of therapeutic relevance in the prevention of tumor metastasis.

Tumor lymphangiogenesis may also be mediated through the transdifferentiation of other cell types into LECs. Using a mouse cornea model of inflammation-associated lymphangiogenesis, it has been shown that CD11b+ macrophages are able to transdifferentiate into LECs and are incorporated into the newly forming lymphatic vasculature [19, 20]. Intriguingly, tumor-associated macrophages have also been shown to express high levels of the prolymphangiogenic factors VEGF-C and VEGF-D, suggesting that they may be capable of stimulating peritumoral lymphangiogenesis [21].

It is known that lymphatic vessels are often present at the periphery of solid tumors (peritumoral lymphatics), and that invasion of tumor cells into these vessels is associated with a poor clinical outcome for various cancer types. However, it is still unclear whether functional lymphatics can be found within solid tumors (intratumoral lymphatics) and what, if any, role such vessels may play in lymphatic dissemination of tumor cells. Given that certain tumor types are able

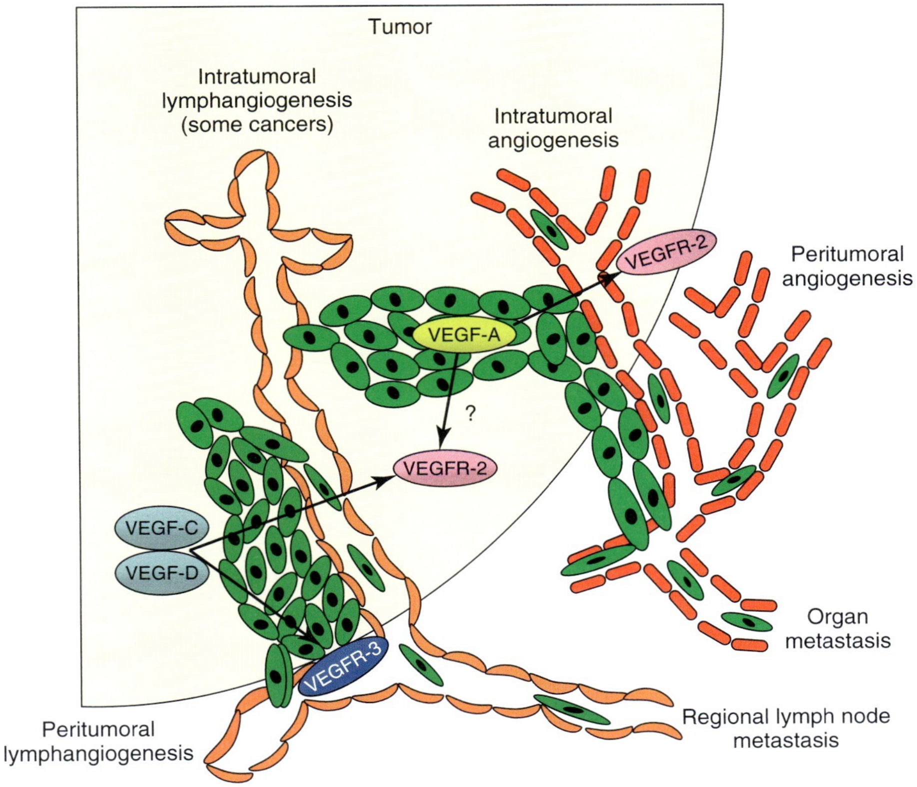

Figure 6.4 Current understanding of tumor angiogenesis and lymphangiogenesis. Tumor lymphangiogenesis is now thought to be an active process, with intratumoral lymphangiogenesis being observed in certain tumor types, promoting metastasis via lymphatics in addition to angiogenesis-associated metastasis. In addition, there is significant cross talk between the blood and lymphatic vessel systems, which may further promote tumor metastasis via both vascular systems. (Modified from [18].)

to metastasize to the lymph nodes in the absence of any intratumoral lymphatics, whereas other tumors exhibit an essential requirement for such lymphatics, the relative contribution of peri- versus intratumoral lymphatic vessels in lymphatic spread remains an open question [22].

However, metastasis via the lymphatic route offers many advantages for a tumor cell. The key role of lymphatic vessels in immune cell trafficking demonstrates that their structure and physiology is well suited for the entry and transport of cells [23]. In addition, lymphatic capillaries are much larger than blood capillaries and flow velocities are much lower [24]. Lymph fluid is also very similar to interstitial fluid and does not cause cell toxicity. By contrast, the environment within blood capillaries causes cell toxicity and mechanical deformation due to the high flow rate and resultant shear stress [25, 26].

6.3 Role of the VEGF-C/VEGF-D/VEGFR-3 Signaling Axis in Tumor Lymphangiogenesis

The best known signaling pathway implicated in tumor lymphangiogenesis involves two members of the vascular endothelial growth factor family, VEGF-C and VEGF-D, and their cognate cell surface receptor, VEGFR-3. The secreted glycoprotein VEGF-C was the first molecule to be identified as a lymphangiogenic growth factor [27]. VEGF-C exerts its lymphangiogenic effects by binding to the receptor tyrosine kinase VEGFR-3 [28, 29], which is expressed on the surface of LECs [30]. VEGF-C plays an essential role during normal lymphatic development by regulating the sprouting of the first lymphatic vessels from the cardinal vein [31]. The second molecule to be identified as a lymphangiogenic growth factor, VEGF-D, shares 61% homology with VEGF-C at the level of amino acid sequence and is also capable of binding to VEGFR-3 [32]. Similar to VEGF-C, overexpression of VEGF-D in transgenic mouse skin also induces lymphangiogenesis [33]. However, VEGF-D knockout mice did not display any lymphatic abnormalities, suggesting that either VEGF-D is dispensable for normal lymphatic development or VEGF-C or another unknown VEGFR-3 ligand may compensate for the lack of VEGF-D, thus preventing the occurrence of any observable lymphatic phenotype [34].

VEGFR-3 or Flt4, a member of the fms-like receptor tyrosine kinase family, was the first lymphatic-specific gene to be identified [30]. VEGFR-3, normally expressed on the surface of adult LECs, serves as a receptor for both VEGF-C and VEGF-D. Activation of VEGFR-3 induces LEC proliferation and migration as well as protection from apoptosis *in vitro* [35] and is sufficient to induce lymphangiogenesis *in vivo* [33]. VEGFR-3 knockout mice exhibit embryonic lethality at embryonic day 9.5 due to deficient vessel remodeling and disorganization of larger vessels, causing fluid accumulation and ultimately resulting in cardiovascular failure [36]. Functional blocking of VEGFR-3 by overexpression of a soluble form of VEGFR-3 in transgenic mouse skin causes lymphatic vessel regression and lymphedema, without affecting the blood vasculature [37]. Activation of VEGFR-3 by ligand binding results in receptor dimerization and induction of various signaling pathways, including protein kinase-C-dependent activation of the p42/p44 mitogen-activated protein (MAP) kinase signaling cascade and induction of Akt phosphorylation, thus contributing to the proliferative and antiapoptotic effects of VEGFR-3 [35].

The importance of VEGF-C in tumor lymphangiogenesis and metastasis has been established by studies using a variety of animal models. For example, injection of mice with human breast cancer cells engineered to overexpress VEGF-C resulted in increased tumor lymphangiogenesis and metastasis to lungs and regional lymph nodes [38]. Similarly, overexpression studies using mouse models of melanoma, fibrosarcoma [39], breast cancer [40], and gastric cancer [41] have all demonstrated the role of VEGF-C in promoting tumor lymphangiogenesis. Studies using the Rip1-Tag2 mouse model of pancreatic cancer showed that overexpression of VEGF-C in this system promoted the formation of extensive peritumoral lymphatic networks as well as metastasis to draining lymph nodes [42].

Alternative approaches to studying the role of VEGF-C in tumor lymphangiogenesis have employed soluble forms of VEGFR-3 that sequester and thereby inactivate VEGF-C. A lung cancer xenograft model in mice where the tumor cells expressed soluble VEGFR-3 showed the presence of fewer intratumoral lymphatics and decreased metastasis to draining lymph nodes [43]. An alternative approach using an adenovirus expression system confirmed these findings. Soluble VEGFR-3 exerted similar inhibitory effects on lymphangiogenesis and metastasis in a rodent model of breast cancer [44]. A similar approach employed in a mouse melanoma model showed inhibition of lymph node metastasis [45]. siRNA-mediated gene silencing of VEGF-C in a mouse mammary tumor model resulted in decreased lymphangiogenesis, lymph node metastasis, and spontaneous lung metastasis [46]. Most recently, both soluble forms of VEGFR-3 and VEGFR-3 blocking antibodies have been shown to reduce tumor lymphangiogenesis and metastasis without affecting tumor growth in mouse models of androgen-dependent and androgen-independent prostate cancer [47].

While the exact molecular mechanism by which VEGF-C promotes lymphangiogenesis is still under investigation, imaging studies using intravital microscopy of draining lymphatics in the mouse ear have shown that VEGF-C aids tumor metastasis by inducing hyperplasia of peritumoral lymphatics and increasing flow rate in these vessels. As a result, delivery of tumor cells to lymphatics and lymph nodes is enhanced, causing a significant increase in tumor cell accumulation within the lymph node and subsequent lymph node metastasis [48].

Using a mouse xenograft tumor model, it has been shown that VEGF-D can induce the formation of intratumoral lymphatics and can promote tumor spread to lymph nodes [49]. These effects could be blocked by using a VEGF-D-specific antibody. VEGF-D overexpression was correlated with high lymphatic vascularization in human pancreatic cancer specimens. Pancreatic xenograft tumors derived from VEGF-D overexpressing cells displayed increased intra- and peritumoral lymphatics, increased lymphatic vessel invasion, and a higher number of lymphatic metastases [50]. Similar results were observed using the Rip1-Tag2 mouse model of pancreatic cancer, wherein VEGF-D overexpression promoted lymphangiogenesis and formation of lung and lymph node metastases [51]. VEGF-D has also been correlated with tumor stage and lymph node metastasis in hepatocellular carcinoma. Subcutaneous tumors derived from VEGF-D overexpressing pancreatic carcinoma cell lines showed increase in volume, increased number of intra- and peritumoral lymphatics, and higher frequency of metastasis to lymph nodes. Higher lymph node metastasis was also seen in an orthotopic tumor model. Finally, the effects of VEGF-D overexpression could be inhibited by coexpression of a soluble form of VEGFR-3 [52].

A correlation between the expression of prolymphangiogenic factors and lymphatic metastasis has been established for several human tumors (Table 6.1, adapted from [9]). Alterations in VEGF-C and VEGF-D levels have been associated with a number of different tumor types. However, the question of whether tumors can also overexpress VEGFR-3 remains controversial. An initial report suggested that VEGFR-3 was highly overexpressed in various tumors, and that the

Table 6.1 Association of lymphangiogenesis and cancer.

Tumor type	Involvement of lymphatic metastasis	References
Breast cancer	Expression of VEGF-C showed significant correlation with lymphatic vessel invasion; five-year disease-free survival rate of VEGF-C positive group was significantly poorer as compared to negative group	[56]
Cervical cancer	VEGF-C mRNA was the sole independent factor influencing pelvic lymph node metastasis and its expression was correlated with poor prognosis	[57]
Colorectal cancer	Expression of VEGF-A, but not VEGF-C or VEGF-D, was correlated with lymphatic spread	[58]
	Significant correlation between VEGF-C mRNA expression and lymph node metastasis, lymphatic involvement, and depth of invasion	[59]
	Expression of VEGF-D showed a positive correlation with the extent of lymphatic involvement	[60]
Endometrial carcinoma	VEGF-C expression correlated with lymphatic vessel invasion, depth of invasion, and lymph node metastasis	[61]
Esophageal cancer	VEGF-C expression correlated with depth of tumor invasion, tumor stage, venous invasion, lymphatic invasion, and lymph node metastasis	[62]
Gastric cancer	Strong correlation was found between VEGF-C expression and lymph node status, lymphatic invasion, venous invasion, and tumor infiltrating patterns	[63]
	VEGF-C expression was associated with increased lymphatic and venous expression and poor prognosis	[64]
	Positive expression of VEGF-C in lymphatic invasion-positive early gastric cancer (EGC) was significantly higher than in lymphatic invasion-negative EGC	[65]
Head and neck squamous cell carcinoma	Increased expression of VEGF and VEGF-C in tumors; increased VEGF-C expression levels and decreased VEGF-D levels had predictive value for cervical node metastases	[66]
Lung adenocarcinoma	High levels of VEGF-C correlated with lymph node metastasis; low VEGF-D : VEGF-C ratio was associated with lymphatic invasion	[67]
Non–small cell lung cancer	Expression of VEGF-A and VEGF-C was associated with nodal microdissemination; VEGF-C expression was significantly associated with lymph node metastasis, lymphatic invasion, and poor prognosis	[68, 69]
Ovarian carcinoma	VEGF-C expression was correlated with increased angiogenesis, lymphangiogenesis, invasive phenotype, and poor prognosis	[70]
Pancreatic cancer	VEGF-C expression was associated with increased lymphatic invasion and lymph node metastasis, but not decreased patient survival	[71]
Prostate cancer	Increased expression of VEGF-C was observed in lymph node positive tumors	[72]
Thyroid cancer	Increased expression of VEGF-C correlated with lymph node invasion	[73]

VEGFR-3–VEGF-C axis might contribute to the invasive capacity of these tumors [53]. However, subsequent reports argued that VEGFR-3 expression was solely restricted to blood and lymphatic vessels in solid tumors [54], and at present the discrepancy in these observations has not been resolved [55].

In addition to overexpression of specific lymphangiogenic factors, lymphatic density has also been used as a prognostic factor for various tumor types. For example, it has been shown that the lymphatic density associated with primary melanomas is a more sensitive prognostic marker for sentinel lymph node lymphangiogenesis, as compared to measurements of tumor thickness [74]. Lymphatic vascular density can also be used as a prognostic marker in esophageal adenocarcinoma [75], non–small cell lung cancer [76], oral squamous cell carcinoma [77], renal cell carcinoma [78], colorectal cancer [79], and several other tumor types.

6.4 Cross Talk between Angiogenesis and Lymphangiogenesis

Several recent findings have highlighted the existence of extensive cross talk between the VEGF-C/VEGF-D/VEGFR-3 and VEGF-A/VEGFR-2 signaling networks, suggesting that these signaling cascades can mediate interactions between the lymphatic and blood vascular endothelial systems. These findings are of great importance for the design of therapeutic agents to block tumor-associated angiogenesis and lymphangiogenesis.

The first evidence for cross talk between these two vascular systems came with the findings that proteolytically processed forms of VEGF-C and VEGF-D are capable of binding to the VEGFR-2, activation of which had been classically associated with angiogenesis [28, 80]. It has now been shown that both VEGF-C and VEGF-D are able to induce angiogenesis *in vivo* [81, 82]. In addition, the expression of VEGFR-3 has been detected on tumor-associated blood vessels [83].

VEGFR-3 expression has also been detected in a small subset of CD34+ positive cells that coexpress CD133. Although present in very small numbers in the circulation of healthy adults, it is suggested that these cells may form a unique subpopulation of lymphatic/vascular progenitor cells that are capable of differentiating into lineage and may play a role in vessel development during postnatal angiogenesis and lymphangiogenesis [84].

Inhibition of VEGFR-3–ligand interactions by means of blocking antibodies was able to inhibit the growth of several human tumor xenografts in immunocompromised mice. The blood vessel density of these tumors was significantly decreased and increased areas of hypoxia and necrosis were observed. As expected, lymphangiogenesis was also significantly reduced in these tumors, suggesting a dual role for VEGFR-3 in promoting both tumor-associated angiogenesis and lymphangiogenesis [85].

The argument that VEGFR-3 may contribute to tumor-associated angiogenesis has been further strengthened by the recent findings that VEGFR-3 plays a key role in sprouting angiogenesis of endothelial tip cells during early embryonic development, postnatal development, and in tumors [86]. Further, combined blockade of both VEGFR-2 and VEGFR-3 showed additive antitumor and antiangiogenic effects. Similarly, combination antibody treatment was also able to more potently reduce lymph node and lung metastases than each antibody alone [87]. This suggests that combined blockade of both signaling axes should be considered in the development of new antiangiogenic agents.

6.5
Role of Other Factors

6.5.1
VEGF-A

While VEGF-A is best known for its role as a proangiogenic growth factor, there is some evidence to suggest that it may play a role in lymphangiogenesis as well. Expression of VEGF-A has been detected on the surface of some LECs [88], and adenovirally expressed and recombinant VEGF-A was found to promote lymphangiogenesis in the mouse ear and cornea, respectively [89, 90]. Using a chemically induced skin carcinogenesis model, it has been shown that VEGF-A expression leads to increased lymphangiogenesis and metastasis to sentinel and distant lymph nodes [91]. Similar findings were observed in a mouse xenograft model using human fibrosarcoma cells [92]. A VEGF-A neutralizing antibody was able to reduce lymphatic vessel density and lymph node metastasis in an orthotopic breast cancer model [93]. However, VEGF-A was not able to induce lymphangiogenesis or lymph node metastasis in a tumor xenograft model using human embryonic kidney 293 (HEK293) cells [49], or in a mouse model of pancreatic cancer [94]. Therefore, it appears that VEGF-A is able to exert its lymphangiogenic effects only in certain specific contexts. Furthermore, since the molecular mechanism underlying these findings remains to be elucidated, it is possible that VEGF-A functions in an indirect rather than a direct manner, potentially through cross talk with the VEGF-C/VEGF-D/VEGFR-3 signaling cascade.

6.5.2
PDGF-BB

It has been shown that expression of platelet-derived growth factor (PDGF)-BB in murine fibrosarcoma cells can induce the formation of intratumoral lymphatics in syngeneic mice, thus promoting lymph node metastasis [95]. In human esophageal squamous cell carcinomas, the expression of PDGF-BB was found to correlate with increased lymphatic vessel density, lymphatic invasion, and lymph node metastasis, suggesting that PDGF-BB can also function as a lymphangiogenic factor in some human cancers [96]. Similar to VEGF-A, it is not presently clear whether PDGF-BB can function in a direct or indirect manner to induce lymphangiogenesis.

6.5.3
Hepatocyte Growth Factor

Hepatocyte growth factor (HGF) and its cognate tyrosine kinase receptor, HGF receptor (also called *c-Met*) are known to play an important role in tumor metastasis, and are considered important targets for therapeutic intervention [97]. c-Met is expressed on the surface of LECs, and HGF treatment of primary LECs *in vitro* can induce cell proliferation, migration, and tube formation, as well as lymphangiogenesis *in vivo* [98]. Since a soluble form of VEGFR-3 was capable of partially blocking HGF-induced lymphangiogenesis, this growth factor may function in an indirect manner. In a mouse model of mammary tumors, expression of HGF induced the formation of peritumoral lymphatics. Recent findings also suggest that HGF expression levels may serve as a prognostic indicator of lymph node metastasis in prostate cancer [99].

6.5.4
Insulin-Like Growth Factors

Considerable evidence suggests that members of the insulin-like growth factor (IGF) family play an important role in tumor progression [100]. Expression of the growth factors IGF-1 and IGF-2 has been correlated with malignant progression and poor prognosis [101]. IGF-1 and -2 can also induce proliferation and migration of primary LECs *in vitro*. Using a mouse cornea model, it was found that these two factors can also induce lymphangiogenesis *in vitro*. In contrast to HGF-mediated lymphangiogenesis, IGF-mediated lymphangiogenesis was not blocked by a soluble form of VEGFR-3, suggesting that these factors may be functioning directly to induce lymphatic vessel growth in this system [102]. However, it has also been shown that the IGF-1 receptor (IGF-1R) is capable of promoting VEGF-C expression and subsequent lymph node metastasis in a Lewis lung carcinoma model. This pathway may therefore serve as an alternate mechanism to promote lymphatic metastasis [103].

6.5.5
Angiopoietins

Angiopoietins (Angs), ligands for the endothelial cell surface tyrosine kinase receptor Tie2, are known to be proangiogenic growth factors. However, Ang-2 exhibits a dual function in both postnatal angiogenesis and developmental lymphangiogenesis [104]. Ang-2-deficient mice exhibit defects in lymphatic vasculature remodeling and maturation [105]; it has been previously shown that defects in Ang-2 mice can be rescued by Ang-1, suggesting that both these ligands may have a function in promoting lymphangiogenesis [106]. Furthermore, Ang-1 is also capable of promoting lymphangiogenesis in a mouse corneal model and in the skin of transgenic mice [107, 108]. However, at present there is no evidence regarding the involvement of

Ang-1 or Ang-2 in tumor lymphangiogenesis and lymph node metastasis in animal models or in human cancers.

6.5.6
Fibroblast Growth Factor-2

Low doses of fibroblast growth factor-2 (FGF-2) have been shown to stimulate lymphangiogenesis in a mouse cornea model, while at higher levels FGF can induce angiogenesis. FGF-2 also upregulates the expression of VEGF-C and VEGF-D, and its lymphangiogenic effects can be blocked by a soluble form of VEGFR-3, suggesting that the interaction of FGF-2 with the VEGF-C/VEGF-D/VEGFR-3 signaling axis is responsible for its ability to induce lymphatic vessel growth [109, 110]. The role of FGF-2 in promoting tumor lymphangiogenesis and lymph node metastasis remains to be explored. The importance of FGF-mediated signaling in LEC proliferation and lymphatic development has been further underscored with the finding that the fibroblast growth factor receptor-3 (FGFR3) is directly regulated by *Prox1*, the master control gene of lymphatic development, and that FGF can also promote LEC proliferation, migration, and survival independent of VEGFR-3.

6.5.7
Chemokines

Chemokines are a family of chemoattractant cytokines that bind transmembrane G protein-coupled receptors on their target cells, thereby allowing these cells to undergo directed migration along a chemokine concentration gradient [111]. While their role in controlling leukocyte migration has been well established, they may also serve to direct tumor cell migration into lymphatics. For example, both the chemokine receptor CCR7 and its cognate ligand, CCL21, have been implicated in lymphatic metastasis. A mouse model of melanoma using the murine cell line B16 showed that overexpression of CCR7 promoted the migration of tumor cells to draining lymph nodes, and neutralizing antibodies to the ligand CCL21 were able to block this migration [112]. Using an *in vivo* melanoma model, it was observed that melanoma cells grew toward LECs due to secretion of CCL21 by LECs, thus suggesting that lymphatics may actively secrete chemotactic agents that promote tumor cell migration and lymphatic invasion [113]. Multiple molecular mechanisms may account for the lymphangiogenic effect of this chemokine. For example, the CCR7/CCL21 signaling axis may promote lymphangiogenesis via synergistic interaction with VEGF-C [114]. Alternatively, the upregulation of CCR7 by cyclooxygenase-2 has been shown to enhance lymphatic invasion of breast cancer cells [115], a finding which may account for the association of cyclooxygenase-2 with lymph node metastasis in human breast cancer [116].

Other chemokines implicated in tumor lymphangiogenesis include CXCR3, which was able to promote lymph node metastasis in mouse models of melanoma and colon cancer [117, 118], and CXCR4, which has been shown to promote lymph node metastasis in an orthotopic mouse model of oral squamous cell carcinoma [119, 120]. Blocking the interaction of CXCR4 with its ligand, CXCL12, significantly impaired lymph node metastasis of human breast cancer cells in a mouse model [121]. Finally, interleukin-7 (IL-7) and its cognate receptor IL-7R have been shown to promote lymphangiogenesis in lung cancer via the c-Fos/c-Jun signaling pathway [122].

These chemokines have also been implicated in promoting lymphatic metastasis of various human cancers, including gastric carcinoma [123], colorectal carcinoma [124], breast cancer [125, 126], pancreatic cancer [127], hepatocellular carcinoma [128], oral squamous cell carcinoma [129], lung adenocarcinoma [130], and papillary thyroid carcinoma [131].

6.5.8 Other Factors

Recent work has shown that the growth hormone receptor (GHR) is expressed on the surface of LECs and treatment of LECs with the cognate ligand, growth hormone (GH), is capable of inducing proliferation, migration, sprouting, and tube formation *in vitro* independent of VEGFR-2 or VEGFR-3 activation. Intradermal implantation of GH-containing Matrigels promoted the enlargement and proliferation of cutaneous lymphatic vessels, suggesting that GH can function as a lymphangiogenic factor both *in vitro* and *in vivo* [132].

Integrins are a large family of heterodimeric transmembrane glycoprotein receptors that mediate cell–cell interactions as well as cellular interactions with the extracellular matrix. Their role in promoting tumor angiogenesis has been intensively studied; in addition, they may also function to induce tumor lymphangiogenesis [133]. The role of integrin $\alpha 9$ in normal lymphatic development has been established [134]. It is also known to function as a receptor for both VEGF-C and VEGF-D and may use this pathway to promote lymphangiogenesis [135]. The integrin $\alpha 4\beta 1$ is expressed on the surface of tumor-associated LECs and antagonists of this integrin can block tumor metastasis via lymphatics [136]. Integrins $\alpha 1\beta 1$ and $\alpha 2\beta 1$ are expressed on LECs at sites of wound healing, and blocking of these integrins inhibits lymphangiogenesis associated with wound healing [137]. In a mouse corneal inflammation model, a small population of lymphatic vessels expressed the integrin $\alpha 5\beta 1$, and blockade of this integrin by small molecule antagonists significantly blocked the outgrowth of new lymphatic vessels during inflammation [138].

Connexins are a family of transmembrane proteins that assemble to form vertebrate gap junction channels that are required for intercellular communication and play an essential role in many physiological processes [139]. Recent evidence suggests that some members of the connexin protein family may be associated with lymph node metastasis in human tumors. For example, connexin 26 was found to be associated with lymph vessel invasion in papillary and follicular thyroid cancers, as well as in breast cancer [140, 141].

6.6 Lymph Node Lymphangiogenesis

A new concept that has emerged in the field of tumor lymphangiogenesis is that of lymph node lymphangiogenesis, whereby tumors are able to induce proliferation and expansion within the sentinel lymph node before any tumor cells have metastasized to this site (reviewed in [142]). For example, using a transgenic mouse model wherein VEGF-A was overexpressed in the skin, it was shown that induction of skin carcinogenesis by VEGF-A was accompanied by proliferation of tumor-associated lymphatics. Unexpectedly, primary tumors were also able to induce lymphatic proliferation within sentinel lymph nodes prior to tumor metastasis [91]. In contrast, VEGF-C overexpression in mouse skin had no effect on tumor-associated lymphatics, but was capable of mediating sentinel lymph node lymphangiogenesis. In addition, VEGF-C overexpression was also positively correlated with additional metastasis of tumor cells from lymph nodes to other distant sites such as distal lymph nodes and lungs [143]. Similar observations have been made in various other tumor models, including nasopharyngeal carcinoma [144] and melanoma [145]. Interestingly, recent work *in vitro* has shown that human breast cancer cell lines are capable of inducing the expression of adhesion molecules on the surface of primary LECs, possibly through the secretion of ATP [146]. This is in keeping with the concept of a "premetastatic niche," by which primary tumors produce alterations in the local microenvironment in sites of distant metastasis, allowing for subsequent homing of tumor cells to these sites [16].

6.7 Therapeutic Implications

Within the last decade, the identification of several key molecular determinants of lymphatic cell identity and function has spurred renewed interest in the field of lymphatic biology. The growing clinical importance of tumor-associated lymphatics makes them a relevant target for therapeutic considerations (reviewed in [147]). The three major strategies adopted to target tumor lymphangiogenesis include (i) the use of blocking antibodies against either VEGF-C/VEGF-D or VEGFR-3 to effectively prevent receptor-ligand interaction; (ii) the use of soluble forms of VEGFR-3 or VEGFR-2 to sequester and inactivate ligands; and (iii) the use of nonspecific small molecule inhibitors that can also inhibit VEGFR-2 and VEGFR-3. Table 6.2 summarizes the current status of several antilymphangiogenic approaches.

Additionally, a recent clinically relevant finding is the observation that LECs isolated from tumor lymphatic vasculature show a significantly altered pattern of gene expression as compared to their normal counterparts, and that similar changes can also be observed in the lymphatic vasculature of some human cancers [157]. These differences between tumor and normal lymphatics provide insight into the molecular mechanisms of tumor cell invasion into lymphatics, and identify molecules that may be attractive therapeutic targets for inhibiting this invasion.

Table 6.2 Clinical applications of antilymphangiogenic therapy.

Inhibitor	Type of compound	Target	Mode of action	Clinical trial status	References
Soluble VEGFR-3	Ig fusion protein	VEGF-C, VEGF-D, other VEGFR-3 ligands	Prevents activation of surface VEGFR-3 by ligand	None	[40, 43, 44]
VEGF-D antibody	Mouse monoclonal antibody; humanized version under development	VEGF-D	Blocks binding to VEGFR-2 and VEGFR-3	None	[49, 148]
VEGF-C antibody	Monoclonal antibody	VEGF-C	Blocks VEGF-C binding to receptor	None	Unpublished
VEGFR-3 antibody	Monoclonal antibody	VEGFR-3 extracellular domain	Blocks VEGF-C and VEGF-D binding to receptor	None	[85, 87, 149]
BAY 43-9006 (sorafenib)	Bi-aryl urea	VEGFR-3, VEGFR-2, Raf-1, PGDFRβ, c-Kit, CD135/flt3, FGFR-1, BRAF tyrosine kinases	Inhibition of tyrosine kinase activity	Phase III for renal cell cancer; phase II for CML and prostate, ovarian, pancreatic, head and neck, breast, thyroid, and lung cancers; phase I for glioma	[150]

(*continued overleaf*)

Table 6.2 *(continued)*

Inhibitor	Type of compound	Target	Mode of action	Clinical trial status	References
CEP-7055	*N,N*-Dimethyl glycine ester	VEGFR-3, VEGFR-2, VEGFR-1, CD135/Flt3, Mlk1, Mlk 3 tyrosine kinases	Inhibition of tyrosine kinase activity	Phase I–II for various malignancies	[151]
MAE87, MAE106, MAZ51	Indolinone	VEGFR-3, VEGFR-2	Inhibition of tyrosine kinase activity	None	[152]
PTK787/ZK222584 (vatalanib)	Chloroanilino-pyridylmethyl phthalazine succinate	VEGFR-3, VEGFR-2. VEGFR-1, c-Kit, c-Fms, PGDFRβ	Inhibition of tyrosine kinase activity	Phase III for colorectal cancer, phase II for mesothelioma and myelodysplastic syndromes, phase I and II for chronic myelogenous leukemia (CML), acute myeloid leukemia (AML), and agnogenic myeloid metaplasia (AMM)	[153, 154]
Avastin	Monoclonal antibody	VEGF-A	Blocks receptor binding	Marketed/phase III–IV	[155]
Veglin	Antisense oligonucleotide	VEGF-A, VEGF-C, VEGF-D	Reduction in protein levels	Phase I	[156]

6.8
Summary

The importance of tumor lymphangiogenesis in mediating tumor metastasis has now been well established with extensive experimental data and clinicopathological evidence. Several key molecular regulators of tumor lymphangiogenesis have also been identified. However, many questions still remain unanswered. For example, the relative contribution and importance of peri- versus intratumoral lymphatics remains to be conclusively established. Similarly, it is still unclear whether tumors can stimulate lymphangiogenesis by promoting the growth of existing lymphatics or by promoting the recruitment of progenitor cells. Understanding the molecular interactions between the tumor and surrounding stroma within the tumor microenvironment will help to further clarify this issue. Recent evidence has shown that tumor-associated LECs show a distinct gene expression profile, with an increase in the expression of certain adhesion molecules. These findings may help to clarify the interactions between tumor cells and LECs and may provide mechanistic insight into tumor invasion of lymphatic vessels, as well as suggesting new possibilities for therapeutic intervention. While the VEGF-C/VEGF-D/VEGFR-3 axis has been the focus of intense interest, the relative contributions of other molecules, such as members of the PDGF family, angiopoietins, and chemokines, remain to be assessed. Finally, it is now increasingly clear that the most clinically relevant results will be obtained only by simultaneous targeting of the VEGFR-2 and VEGFR-3 signaling pathways, possibly in combination with other chemotherapeutic agents that directly target tumor cells. Depending on the extent of lymphatic involvement, it may be possible to design tailored regimens of combination chemotherapy for specific tumor types in order to ensure maximal efficiency. Therefore, understanding the mechanisms underlying lymphatic metastasis, clarifying the extent of lymphatic dependence in metastasis of various tumors, and finally the design and clinical testing of novel therapeutic agents to block this process will be of great importance in the future.

References

1. Oliver, G. and Alitalo, K. (2005) The lymphatic vasculature: recent progress and paradigms. *Annu. Rev. Cell Dev. Biol.*, **21**, 457–483.
2. Asellius, G. (1627) *De Lactibus Sive Lacteis Venis*, Mediolani, Milan.
3. Harvey, W. (1628) *An Anatomical Study of the Motion of the Heart and Blood in Animals*, Richard Lownde, London.
4. Sabin, F.R. (1902) On the origin of the lymphatic system from the veins and the development of the lymph hearts and thoracic duct in the pig. *Am. J. Anat.*, **1**, 367–391.
5. Sabin, F.R. (1904) On the development of the superficial lymphatics in the skin of the pig. *Am. J. Anat.*, **3**, 183–195.
6. Huntington, G.S. and McClure, C.F.W. (1910) The anatomy and development of the jugular lymph sac in the domestic cat (Felis domestica). *Am. J. Anat.*, **10**, 177–311.
7. Oliver, G. and Srinivasan, R.S. (2008) Lymphatic vasculature development: current concepts. *Ann. N. Y. Acad. Sci.*, **1131**, 75–81.

8. Das, S. and Skobe, M. (2008) Lymphatic vessel activation in cancer. *Ann. N. Y. Acad. Sci.*, **1131**, 235–241.
9. Stacker, S.A., Achen, M.G., Jussila, L., Baldwin, M.E., and Alitalo, K. (2002) Lymphangiogenesis and cancer metastasis. *Nat. Rev. Cancer*, **2** (8), 573–583.
10. Wong, S.Y. and Hynes, R.O. (2006) Lymphatic or hematogenous dissemination: how does a metastatic tumor cell decide? *Cell Cycle*, **5** (8), 812–817.
11. Lyden, D., Hattori, K., Dias, S., Costa, C., Blaikie, P., Butros, L., Chadburn, A., Heissig, B., Marks, W., Witte, L., Wu, Y., Hicklin, D., Zhu, Z., Hackett, N.R., Crystal, R.G., Moore, M.A., Hajjar, K.A., Manova, K., Benezra, R., and Rafii, S. (2001) Impaired recruitment of bone-marrow-derived endothelial and hematopoietic precursor cells blocks tumor angiogenesis and growth. *Nat. Med.*, **7** (11), 1194–1201.
12. Nolan, D.J., Ciarrocchi, A., Mellick, A.S., Jaggi, J.S., Bambino, K., Gupta, S., Heikamp, E., McDevitt, M.R., Scheinberg, D.A., Benezra, R., and Mittal, V. (2007) Bone marrow-derived endothelial progenitor cells are a major determinant of nascent tumor neovascularization. *Genes Dev.*, **21** (12), 1546–1558.
13. He, Y., Rajantie, I., Ilmonen, M., Makinen, T., Karkkainen, M.J., Haiko, P., Salven, P., and Alitalo, K. (2004) Preexisting lymphatic endothelium but not endothelial progenitor cells are essential for tumor lymphangiogenesis and lymphatic metastasis. *Cancer Res.*, **64** (11), 3737–3740.
14. Religa, P., Cao, R., Bjorndahl, M., Zhou, Z., Zhu, Z., and Cao, Y. (2005) Presence of bone marrow-derived circulating progenitor endothelial cells in the newly formed lymphatic vessels. *Blood*, **106** (13), 4184–4190.
15. Kerjaschki, D., Huttary, N., Raab, I., Regele, H., Bojarski-Nagy, K., Bartel, G., Krober, S.M., Greinix, H., Rosenmaier, A., Karlhofer, F., Wick, N., and Mazal, P.R. (2006) Lymphatic endothelial progenitor cells contribute to de novo lymphangiogenesis in human renal transplants. *Nat. Med.*, **12** (2), 230–234.
16. Kaplan, R.N., Rafii, S., and Lyden, D. (2006) Preparing the "soil": the premetastatic niche. *Cancer Res.*, **66** (23), 11089–11093.
17. Gao, D., Nolan, D.J., Mellick, A.S., Bambino, K., McDonnell, K., and Mittal, V. (2008) Endothelial progenitor cells control the angiogenic switch in mouse lung metastasis. *Science*, **319** (5860), 195–198.
18. Detmar, M. and Hirakawa, S. (2002) The formation of lymphatic vessels and its importance in the setting of malignancy. *J. Exp. Med.*, **196** (6), 713–718.
19. Maruyama, K., Ii, M., Cursiefen, C., Jackson, D.G., Keino, H., Tomita, M., Van Rooijen, N., Takenaka, H., D'Amore, P.A., Stein-Streilein, J., Losordo, D.W., and Streilein, J.W. (2005) Inflammation-induced lymphangiogenesis in the cornea arises from CD11b-positive macrophages. *J. Clin. Invest.*, **115** (9), 2363–2372.
20. Kerjaschki, D. (2005) The crucial role of macrophages in lymphangiogenesis. *J. Clin. Invest.*, **115** (9), 2316–2319.
21. Schoppmann, S.F., Birner, P., Stockl, J., Kalt, R., Ullrich, R., Caucig, C., Kriehuber, E., Nagy, K., Alitalo, K., and Kerjaschki, D. (2002) Tumor-associated macrophages express lymphatic endothelial growth factors and are related to peritumoral lymphangiogenesis. *Am. J. Pathol.*, **161** (3), 947–956.
22. Jain, R.K. and Fenton, B.T. (2002) Intratumoral lymphatic vessels: a case of mistaken identity or malfunction? *J. Natl. Cancer Inst.*, **94** (6), 417–421.
23. Randolph, G.J., Angeli, V., and Swartz, M.A. (2005) Dendritic-cell trafficking to lymph nodes through lymphatic vessels. *Nat. Rev. Immunol.*, **5** (8), 617–628.
24. Swartz, M.A., Hubbell, J.A., and Reddy, S.T. (2008) Lymphatic drainage function and its immunological implications: from dendritic cell homing to vaccine design. *Semin. Immunol.*, **20** (2), 147–156.

25. Liotta, L.A., Steeg, P.S., and Stetler-Stevenson, W.G. (1991) Cancer metastasis and angiogenesis: an imbalance of positive and negative regulation. *Cell*, **64** (2), 327–336.
26. Weiss, L. and Schmid-Schonbein, G.W. (1989) Biomechanical interactions of cancer cells with the microvasculature during metastasis. *Cell Biophys.*, **14** (2), 187–215.
27. Jeltsch, M., Kaipainen, A., Joukov, V., Meng, X., Lakso, M., Rauvala, H., Swartz, M., Fukumura, D., Jain, R.K., and Alitalo, K. (1997) Hyperplasia of lymphatic vessels in VEGF-C transgenic mice. *Science*, **276** (5317), 1423–1425.
28. Joukov, V., Pajusola, K., Kaipainen, A., Chilov, D., Lahtinen, I., Kukk, E., Saksela, O., Kalkkinen, N., and Alitalo, K. (1996) A novel vascular endothelial growth factor, VEGF-C, is a ligand for the Flt4 (VEGFR-3) and KDR (VEGFR-2) receptor tyrosine kinases. *EMBO J.*, **15** (2), 290–298.
29. Lee, J., Gray, A., Yuan, J., Luoh, S.M., Avraham, H., and Wood, W.I. (1996) Vascular endothelial growth factor-related protein: a ligand and specific activator of the tyrosine kinase receptor Flt4. *Proc. Natl. Acad. Sci. U.S.A.*, **93** (5), 1988–1992.
30. Kaipainen, A., Korhonen, J., Mustonen, T., van Hinsbergh, V.W., Fang, G.H., Dumont, D., Breitman, M., and Alitalo, K. (1995) Expression of the fms-like tyrosine kinase 4 gene becomes restricted to lymphatic endothelium during development. *Proc. Natl. Acad. Sci. U.S.A.*, **92** (8), 3566–3570.
31. Karkkainen, M.J., Haiko, P., Sainio, K., Partanen, J., Taipale, J., Petrova, T.V., Jeltsch, M., Jackson, D.G., Talikka, M., Rauvala, H., Betsholtz, C., and Alitalo, K. (2004) Vascular endothelial growth factor C is required for sprouting of the first lymphatic vessels from embryonic veins. *Nat. Immunol.*, **5** (1), 74–80.
32. Achen, M.G., Jeltsch, M., Kukk, E., Makinen, T., Vitali, A., Wilks, A.F., Alitalo, K., and Stacker, S.A. (1998) Vascular endothelial growth factor D (VEGF-D) is a ligand for the tyrosine kinases VEGF receptor 2 (Flk1) and VEGF receptor 3 (Flt4). *Proc. Natl. Acad. Sci. U.S.A.*, **95** (2), 548–553.
33. Veikkola, T., Jussila, L., Makinen, T., Karpanen, T., Jeltsch, M., Petrova, T.V., Kubo, H., Thurston, G., McDonald, D.M., Achen, M.G., Stacker, S.A., and Alitalo, K. (2001) Signalling via vascular endothelial growth factor receptor-3 is sufficient for lymphangiogenesis in transgenic mice. *EMBO J.*, **20** (6), 1223–1231.
34. Baldwin, M.E., Halford, M.M., Roufail, S., Williams, R.A., Hibbs, M.L., Grail, D., Kubo, H., Stacker, S.A., and Achen, M.G. (2005) Vascular endothelial growth factor D is dispensable for development of the lymphatic system. *Mol. Cell Biol.*, **25** (6), 2441–2449.
35. Makinen, T., Veikkola, T., Mustjoki, S., Karpanen, T., Catimel, B., Nice, E.C., Wise, L., Mercer, A., Kowalski, H., Kerjaschki, D., Stacker, S.A., Achen, M.G., and Alitalo, K. (2001) Isolated lymphatic endothelial cells transduce growth, survival and migratory signals via the VEGF-C/D receptor VEGFR-3. *EMBO J.*, **20** (17), 4762–4773.
36. Dumont, D., Jussila, L., Taipale, J., Lymboussaki, A., Mustonen, T., Pajusola, K., Breitman, M., and Alitalo, K. (1998) Cardiovascular failure in mouse embryos deficient in VEGF receptor-3. *Science*, **282** (5390), 946–949.
37. Makinen, T., Jussila, L., Veikkola, T., Karpanen, T., Kettunen, M.I., Pulkkanen, K.J., Kauppinen, R., Jackson, D.G., Kubo, H., Nishikawa, S., Yla-Herttuala, S., and Alitalo, K. (2001) Inhibition of lymphangiogenesis with resulting lymphedema in transgenic mice expressing soluble VEGF receptor-3. *Nat. Med.*, **7** (2), 199–205.
38. Skobe, M., Hawighorst, T., Jackson, D.G., Prevo, R., Janes, L., Velasco, P., Riccardi, L., Alitalo, K., Claffey, K., and Detmar, M. (2001) Induction of tumor lymphangiogenesis by VEGF-C promotes breast cancer metastasis. *Nat. Med.*, **7** (2), 192–198.
39. Padera, T.P., Kadambi, A., di Tomaso, E., Carreira, C.M., Brown, E.B., Boucher, Y., Choi, N.C., Mathisen, D.,

Wain, J., Mark, E.J., Munn, L.L., and Jain, R.K. (2002) Lymphatic metastasis in the absence of functional intratumor lymphatics. *Science*, **296** (5574), 1883–1886.

40. Karpanen, T., Egeblad, M., Karkkainen, M.J., Kubo, H., Yla-Herttuala, S., Jaattela, M., and Alitalo, K. (2001) Vascular endothelial growth factor C promotes tumor lymphangiogenesis and intralymphatic tumor growth. *Cancer Res.*, **61** (5), 1786–1790.
41. Yanai, Y., Furuhata, T., Kimura, Y., Yamaguchi, K., Yasoshima, T., Mitaka, T., Mochizuki, Y., and Hirata, K. (2001) Vascular endothelial growth factor C promotes human gastric carcinoma lymph node metastasis in mice. *J. Exp. Clin. Cancer Res.*, **20** (3), 419–428.
42. Mandriota, S.J., Jussila, L., Jeltsch, M., Compagni, A., Baetens, D., Prevo, R., Banerji, S., Huarte, J., Montesano, R., Jackson, D.G., Orci, L., Alitalo, K., Christofori, G., and Pepper, M.S. (2001) Vascular endothelial growth factor-C-mediated lymphangiogenesis promotes tumour metastasis. *EMBO J.*, **20** (4), 672–682.
43. He, Y., Kozaki, K., Karpanen, T., Koshikawa, K., Yla-Herttuala, S., Takahashi, T., and Alitalo, K. (2002) Suppression of tumor lymphangiogenesis and lymph node metastasis by blocking vascular endothelial growth factor receptor 3 signaling. *J. Natl. Cancer Inst.*, **94** (11), 819–825.
44. Krishnan, J., Kirkin, V., Steffen, A., Hegen, M., Weih, D., Tomarev, S., Wilting, J., and Sleeman, J.P. (2003) Differential in vivo and in vitro expression of vascular endothelial growth factor (VEGF)-C and VEGF-D in tumors and its relationship to lymphatic metastasis in immunocompetent rats. *Cancer Res.*, **63** (3), 713–722.
45. Lin, J., Lalani, A.S., Harding, T.C., Gonzalez, M., Wu, W.W., Luan, B., Tu, G.H., Koprivnikar, K., VanRoey, M.J., He, Y., Alitalo, K., and Jooss, K. (2005) Inhibition of lymphogenous metastasis using adeno-associated virus-mediated gene transfer of a soluble VEGFR-3 decoy receptor. *Cancer Res.*, **65** (15), 6901–6909.
46. Chen, Z., Varney, M.L., Backora, M.W., Cowan, K., Solheim, J.C., Talmadge, J.E., and Singh, R.K. (2005) Down-regulation of vascular endothelial cell growth factor-C expression using small interfering RNA vectors in mammary tumors inhibits tumor lymphangiogenesis and spontaneous metastasis and enhances survival. *Cancer Res.*, **65** (19), 9004–9011.
47. Burton, J.B., Priceman, S.J., Sung, J.L., Brakenhielm, E., An, D.S., Pytowski, B., Alitalo, K., and Wu, L. (2008) Suppression of prostate cancer nodal and systemic metastasis by blockade of the lymphangiogenic axis. *Cancer Res.*, **68** (19), 7828–7837.
48. Hoshida, T., Isaka, N., Hagendoorn, J., di Tomaso, E., Chen, Y.L., Pytowski, B., Fukumura, D., Padera, T.P., and Jain, R.K. (2006) Imaging steps of lymphatic metastasis reveals that vascular endothelial growth factor-C increases metastasis by increasing delivery of cancer cells to lymph nodes: therapeutic implications. *Cancer Res.*, **66** (16), 8065–8075.
49. Stacker, S.A., Caesar, C., Baldwin, M.E., Thornton, G.E., Williams, R.A., Prevo, R., Jackson, D.G., Nishikawa, S., Kubo, H., and Achen, M.G. (2001) VEGF-D promotes the metastatic spread of tumor cells via the lymphatics. *Nat. Med.*, **7** (2), 186–191.
50. Von Marschall, Z., Scholz, A., Stacker, S.A., Achen, M.G., Jackson, D.G., Alves, F., Schirner, M., Haberey, M., Thierauch, K.H., Wiedenmann, B., and Rosewicz, S. (2005) Vascular endothelial growth factor-D induces lymphangiogenesis and lymphatic metastasis in models of ductal pancreatic cancer. *Int. J. Oncol.*, **27** (3), 669–679.
51. Kopfstein, L., Veikkola, T., Djonov, V.G., Baeriswyl, V., Schomber, T., Strittmatter, K., Stacker, S.A., Achen, M.G., Alitalo, K., and Christofori, G. (2007) Distinct roles of vascular endothelial growth factor-D in lymphangiogenesis and metastasis. *Am. J. Pathol.*, **170** (4), 1348–1361.
52. Thelen, A., Scholz, A., Benckert, C., von Marschall, Z., Schroder, M.,

Wiedenmann, B., Neuhaus, P., Rosewicz, S., and Jonas, S. (2008) VEGF-D promotes tumor growth and lymphatic spread in a mouse model of hepatocellular carcinoma. *Int. J. Cancer*, **122** (11), 2471–2481.

53. Su, J.L., Yang, P.C., Shih, J.Y., Yang, C.Y., Wei, L.H., Hsieh, C.Y., Chou, C.H., Jeng, Y.M., Wang, M.Y., Chang, K.J., Hung, M.C., and Kuo, M.L. (2006) The VEGF-C/Flt-4 axis promotes invasion and metastasis of cancer cells. *Cancer Cell*, **9** (3), 209–223.
54. Petrova, T.V., Bono, P., Holnthoner, W., Chesnes, J., Pytowski, B., Sihto, H., Laakkonen, P., Heikkila, P., Joensuu, H., and Alitalo, K. (2008) VEGFR-3 expression is restricted to blood and lymphatic vessels in solid tumors. *Cancer Cell*, **13** (6), 554–556.
55. Su, J.L., Chen, P.S., Chien, M.H., Chen, P.B., Chen, Y.H., Lai, C.C., Hung, M.C., and Kuo, M.L. (2008) Further evidence for expression and function of the VEGF-C/VEGFR-3 axis in cancer cells. *Cancer Cell*, **13** (6), 557–560.
56. Kinoshita, J., Kitamura, K., Kabashima, A., Saeki, H., Tanaka, S., and Sugimachi, K. (2001) Clinical significance of vascular endothelial growth factor-C (VEGF-C) in breast cancer. *Breast Cancer Res. Treat.*, **66** (2), 159–164.
57. Hashimoto, I., Kodama, J., Seki, N., Hongo, A., Yoshinouchi, M., Okuda, H., and Kudo, T. (2001) Vascular endothelial growth factor-C expression and its relationship to pelvic lymph node status in invasive cervical cancer. *Br. J. Cancer*, **85** (1), 93–97.
58. George, M.L., Tutton, M.G., Janssen, F., Arnaout, A., Abulafi, A.M., Eccles, S.A., and Swift, R.I. (2001) VEGF-A, VEGF-C, and VEGF-D in colorectal cancer progression. *Neoplasia*, **3** (5), 420–427.
59. Akagi, K., Ikeda, Y., Miyazak, i.M., Abe, T., Kinoshita, J., Maehara, Y., and Sugimachi, K. (2000) Vascular endothelial growth factor-C (VEGF-C) expression in human colorectal cancer tissues. *Br. J. Cancer*, **83** (7), 887–891.
60. White, J.D., Hewett, P.W., Kosuge, D., McCulloch, T., Enholm, B.C., Carmichael, J., and Murray, J.C. (2002) Vascular endothelial growth factor-D expression is an independent prognostic marker for survival in colorectal carcinoma. *Cancer Res.*, **62** (6), 1669–1675.
61. Hirai, M., Nakagawara, A., Oosaki, T., Hayashi, Y., Hirono, M., and Yoshihara, T. (2001) Expression of vascular endothelial growth factors (VEGF-A/VEGF-1 and VEGF-C/VEGF-2) in postmenopausal uterine endometrial carcinoma. *Gynecol. Oncol.*, **80** (2), 181–188.
62. Kitadai, Y., Amioka, T., Haruma, K., Tanaka, S., Yoshihara, M., Sumii, K., Matsutani, N., Yasui, W., and Chayama, K. (2001) Clinicopathological significance of vascular endothelial growth factor (VEGF)-C in human esophageal squamous cell carcinomas. *Int. J. Cancer*, **93** (5), 662–666.
63. Yonemura, Y., Endo, Y., Fujita, H., Fushida, S., Ninomiya, I., Bandou, E., Taniguchi, K., Miwa, K., Ohoyama, S., Sugiyama, K., and Sasaki, T. (1999) Role of vascular endothelial growth factor C expression in the development of lymph node metastasis in gastric cancer. *Clin. Cancer Res.*, **5** (7), 1823–1829.
64. Ichikura, T., Tomimatsu, S., Ohkura, E., and Mochizuki, H. (2001) Prognostic significance of the expression of vascular endothelial growth factor (VEGF) and VEGF-C in gastric carcinoma. *J. Surg. Oncol.*, **78** (2), 132–137.
65. Kabashima, A., Maehara, Y., Kakeji, Y., and Sugimachi, K. (2001) Overexpression of vascular endothelial growth factor C is related to lymphogenous metastasis in early gastric carcinoma. *Oncology*, **60** (2), 146–150.
66. O-charoenrat, P., Rhys-Evans, P., and Eccles, S.A. (2001) Expression of vascular endothelial growth factor family members in head and neck squamous cell carcinoma correlates with lymph node metastasis. *Cancer*, **92** (3), 556–568.

67. Niki, T., Iba, S., Tokunou, M., Yamada, T., Matsuno, Y., and Hirohashi, S. (2000) Expression of vascular endothelial growth factors A, B, C, and D and their relationships to lymph node status in lung adenocarcinoma. *Clin. Cancer Res.*, **6** (6), 2431–2439.
68. Ohta, Y., Shridhar, V., Bright, R.K., Kalemkerian, G.P., Du, W., Carbone, M., Watanabe, Y., and Pass, H.I. (1999) VEGF and VEGF type C play an important role in angiogenesis and lymphangiogenesis in human malignant mesothelioma tumours. *Br. J. Cancer*, **81** (1), 54–61.
69. Kajita, T., Ohta, Y., Kimura, K., Tamura, M., Tanaka, Y., Tsunezuka, Y., Oda, M., Sasaki, T., and Watanabe, G. (2001) The expression of vascular endothelial growth factor C and its receptors in non-small cell lung cancer. *Br. J. Cancer*, **85** (2), 255–260.
70. Ueda, M., Hung, Y.C., Terai, Y., Kanda, K., Kanemura, M., Futakuchi, H., Yamaguchi, H., Akise, D., Yasuda, M., and Ueki, M. (2005) Vascular endothelial growth factor-C expression and invasive phenotype in ovarian carcinomas. *Clin. Cancer Res.*, **11** (9), 3225–3232.
71. Tang, R.F., Itakura, J., Aikawa, T., Matsuda, K., Fujii, H., Korc, M., and Matsumoto, Y. (2001) Overexpression of lymphangiogenic growth factor VEGF-C in human pancreatic cancer. *Pancreas*, **22** (3), 285–292.
72. Tsurusaki, T., Kanda, S., Sakai, H., Kanetake, H., Saito, Y., Alitalo, K., and Koji, T. (1999) Vascular endothelial growth factor-C expression in human prostatic carcinoma and its relationship to lymph node metastasis. *Br. J. Cancer*, **80** (1-2), 309–313.
73. Bunone, G., Vigneri, P., Mariani, L., Buto, S., Collini, P., Pilotti, S., Pierotti, M.A., and Bongarzone, I. (1999) Expression of angiogenesis stimulators and inhibitors in human thyroid tumors and correlation with clinical pathological features. *Am. J. Pathol.*, **155** (6), 1967–1976.
74. Dadras, S.S., Lange-Asschenfeldt, B., Velasco, P., Nguyen, L., Vora, A., Muzikansky, A., Jahnke, K., Hauschild, A., Hirakawa, S., Mihm, M.C., and Detmar, M. (2005) Tumor lymphangiogenesis predicts melanoma metastasis to sentinel lymph nodes. *Mod. Pathol.*, **18** (9), 1232–1242.
75. Saad, R.S., Lindner, J.L., Liu, Y., and Silverman, J.F. (2009) Lymphatic vessel density as prognostic marker in esophageal adenocarcinoma. *Am. J. Clin. Pathol.*, **131** (1), 92–98.
76. Kadota, K., Huang, C.L., Liu, D., Ueno, M., Kushida, Y., Haba, R., and Yokomise, H. (2008) The clinical significance of lymphangiogenesis and angiogenesis in non-small cell lung cancer patients. *Eur. J. Cancer*, **44** (7), 1057–1067.
77. Miyahara, M., Tanuma, J., Sugihara, K., and Semba, I. (2007) Tumor lymphangiogenesis correlates with lymph node metastasis and clinicopathologic parameters in oral squamous cell carcinoma. *Cancer*, **110** (6), 1287–1294.
78. Ishikawa, Y., Aida, S., Tamai, S., Akasaka, Y., Kiguchi, H., Akishima-Fukasawa, Y., Hayakawa, M., Soh, S., Ito, K., Kimura-Matsumoto, M., Ishiguro, S., Nishimura, C., Kamata, I., Shimokawa, R., and Ishii, T. (2007) Significance of lymphatic invasion and proliferation on regional lymph node metastasis in renal cell carcinoma. *Am. J. Clin. Pathol.*, **128** (2), 198–207.
79. Saad, R.S., Kordunsky, L., Liu, Y.L., Denning, K.L., Kandil, H.A., and Silverman, J.F. (2006) Lymphatic microvessel density as prognostic marker in colorectal cancer. *Mod. Pathol.*, **19** (10), 1317–1323.
80. Stacker, S.A., Stenvers, K., Caesar, C., Vitali, A., Domagala, T., Nice, E., Roufail, S., Simpson, R.J., Moritz, R., Karpanen, T., Alitalo, K., and Achen, M.G. (1999) Biosynthesis of vascular endothelial growth factor-D involves proteolytic processing which generates non-covalent homodimers. *J. Biol. Chem.*, **274** (45), 32127–32136.
81. Cao, Y., Linden, P., Farnebo, J., Cao, R., Eriksson, A., Kumar, V., Qi, J.H., Claesson-Welsh, L., and Alitalo, K. (1998) Vascular endothelial growth factor C induces angiogenesis in vivo.

Proc. Natl. Acad. Sci. U.S.A., **95** (24), 14389–14394.

82. Rissanen, T.T., Markkanen, J.E., Gruchala, M., Heikura, T., Puranen, A., Kettunen, M.I., Kholova, I., Kauppinen, R.A., Achen, M.G., Stacker, S.A., Alitalo, K., and Yla-Herttuala, S. (2003) VEGF-D is the strongest angiogenic and lymphangiogenic effector among VEGFs delivered into skeletal muscle via adenoviruses. *Circ. Res.*, **92** (10), 1098–1106.

83. Valtola, R., Salven, P., Heikkila, P., Taipale, J., Joensuu, H., Rehn, M., Pihlajaniemi, T., Weich, H., de Waal, R., and Alitalo, K. (1999) VEGFR-3 and its ligand VEGF-C are associated with angiogenesis in breast cancer. *Am. J. Pathol.*, **154** (5), 1381–1390.

84. Salven, P., Mustjoki, S., Alitalo, R., Alitalo, K., and Rafii, S. (2003) VEGFR-3 and CD133 identify a population of CD34+ lymphatic/vascular endothelial precursor cells. *Blood*, **101** (1), 168–172.

85. Laakkonen, P., Waltari, M., Holopainen, T., Takahashi, T., Pytowski, B., Steiner, P., Hicklin, D., Persaud, K., Tonra, J.R., Witte, L., and Alitalo, K. (2007) Vascular endothelial growth factor receptor 3 is involved in tumor angiogenesis and growth. *Cancer Res.*, **67** (2), 593–599.

86. Tammela, T., Zarkada, G., Wallgard, E., Murtomaki, A., Suchting, S., Wirzenius, M., Waltari, M., Hellstrom, M., Schomber, T., Peltonen, R., Freitas, C., Duarte, A., Isoniemi, H., Laakkonen, P., Christofori, G., Yla-Herttuala, S., Shibuya, M., Pytowski, B., Eichmann, A., Betsholtz, C., and Alitalo, K. (2008) Blocking VEGFR-3 suppresses angiogenic sprouting and vascular network formation. *Nature*, **454** (7204), 656–660.

87. Roberts, N., Kloos, B., Cassella, M., Podgrabinska, S., Persaud, K., Wu, Y., Pytowski, B., and Skobe, M. (2006) Inhibition of VEGFR-3 activation with the antagonistic antibody more potently suppresses lymph node and distant metastases than inactivation of VEGFR-2. *Cancer Res.*, **66** (5), 2650–2657.

88. Saaristo, A., Veikkola, T., Enholm, B., Hytonen, M., Arola, J., Pajusola, K., Turunen, P., Jeltsch, M., Karkkainen, M.J., Kerjaschki, D., Bueler, H., Yla-Herttuala, S., and Alitalo, K. (2002) Adenoviral VEGF-C overexpression induces blood vessel enlargement, tortuosity, and leakiness but no sprouting angiogenesis in the skin or mucous membranes. *FASEB J.*, **16** (9), 1041–1049.

89. Nagy, J.A., Vasile, E., Feng, D., Sundberg, C., Brown, L.F., Detmar, M.J., Lawitts, J.A., Benjamin, L., Tan, X., Manseau, E.J., Dvorak, A.M., and Dvorak, H.F. (2002) Vascular permeability factor/vascular endothelial growth factor induces lymphangiogenesis as well as angiogenesis. *J. Exp. Med.*, **196** (11), 1497–1506.

90. Cursiefen, C., Chen, L., Borges, L.P., Jackson, D., Cao, J., Radziejewski, C., D'Amore, P.A., Dana, M.R., Wiegand, S.J., and JW., S. (2004) VEGF-A stimulates lymphangiogenesis and hemangiogenesis in inflammatory neovascularization via macrophage recruitment. *J. Clin. Invest.*, **113** (7), 1040–1050.

91. Hirakawa, S., Kodama, S., Kunstfeld, R., Kajiya, K., Brown, L.F., and Detmar, M. (2005) VEGF-A induces tumor and sentinel lymph node lymphangiogenesis and promotes lymphatic metastasis. *J. Exp. Med.*, **201** (7), 1089–1099.

92. Björndahl, M.A., Cao, R., Burton, J.B., Brakenhielm, E., Religa, P., Galter, D., Wu, L., and Cao, Y. (2005) Vascular endothelial growth factor-a promotes peritumoral lymphangiogenesis and lymphatic metastasis. *Cancer Res.*, **65** (20), 9261–9268.

93. Whitehurst, B., Flister, M.J., Bagaitkar, J., Volk, L., Bivens, C.M., Pickett, B., Castro-Rivera, E., Brekken, R.A., Gerard, R.D., and Ran, S. (2007) Anti-VEGF-A therapy reduces lymphatic vessel density and expression of VEGFR-3 in an orthotopic breast tumor model. *Int. J. Cancer*, **121** (10), 2181–2191.

94. Gannon, G., Mandriota, S.J., Cui, L., Baetens, D., Pepper, M.S., and

Christofori, G. (2002) Overexpression of vascular endothelial growth factor-A165 enhances tumor angiogenesis but not metastasis during beta-cell carcinogenesis. *Cancer Res.*, **62** (2), 603–608.

95. Cao, R., Björndahl, M.A., Religa, P., Clasper, S., Garvin, S., Galter, D., Meister, B., Ikomi, F., Tritsaris, K., Dissing, S., Ohhashi, T., Jackson, D.G., and Cao, Y. (2004) PDGF-BB induces intratumoral lymphangiogenesis and promotes lymphatic metastasis. *Cancer Cell*, **6** (4), 333–345.

96. Matsumoto, S., Yamada, Y., Narikiyo, M., Ueno, M., Tamaki, H., Miki, K., Wakatsuki, K., Enomoto, K., Yokotani, T., and Nakajima, Y. (2007) Prognostic significance of platelet-derived growth factor-BB expression in human esophageal squamous cell carcinomas. *Anticancer Res.*, **27** (4B), 2409–2414.

97. Comoglio, P.M., Giordano, S., and Trusolino, L. (2008) Drug development of MET inhibitors: targeting oncogene addiction and expedience. *Nat. Rev. Drug Discov.*, **7** (6), 504–516.

98. Kajiya, K., Hirakawa, S., Ma, B., Drinnenberg, I., and Detmar, M. (2005) Hepatocyte growth factor promotes lymphatic vessel formation and function. *EMBO J.*, **24** (16), 2885–2895.

99. Gupta, A., Karakiewicz, P.I., Roehrborn, C.G., Lotan, Y., Zlotta, A.R., and Shariat, S.F. (2008) Predictive value of plasma hepatocyte growth factor/scatter factor levels in patients with clinically localized prostate cancer. *Clin. Cancer Res.*, **14** (22), 7385–7390.

100. Pollak, M. (2008) Insulin and insulin-like growth factor signalling in neoplasia. *Nat. Rev. Cancer*, **8** (12), 915–928.

101. Mita, K., Nakahara, M., and Usui, T. (2000) Expression of the insulin-like growth factor system and cancer progression in hormone-treated prostate cancer patients. *Int. J. Urol.*, **7** (9), 321–329.

102. Björndahl, M., Cao, R., Nissen, L.J., Clasper, S., Johnson, L.A., Xue, Y., Zhou, Z., Jackson, D., Hansen, A.J., and Cao, Y. (2005) Insulin-like growth factors 1 and 2 induce lymphangiogenesis in vivo. *Proc. Natl. Acad. Sci. U.S.A.*, **102** (43), 15593–15598.

103. Tang, Y., Zhang, D., Fallavollita, L., and Brodt, P. (2003) Vascular endothelial growth factor C expression and lymph node metastasis are regulated by the type I insulin-like growth factor receptor. *Cancer Res.*, **63** (6), 1166–1171.

104. Thurston, G. (2003) Role of Angiopoietins and Tie receptor tyrosine kinases in angiogenesis and lymphangiogenesis. *Cell Tissue Res.*, **314** (1), 61–68.

105. Dellinger, M., Hunter, R., Bernas, M., Gale, N., Yancopoulos, G., Erickson, R., and Witte, M. (2008) Defective remodeling and maturation of the lymphatic vasculature in Angiopoietin-2 deficient mice. *Dev. Biol.*, **319** (2), 309–320.

106. Gale, N.W., Thurston, G., Hackett, S.F., Renard, R., Wang, Q., McClain, J., Martin, C., Witte, C., Witte, M.H., Jackson, D., Suri, C., Campochiaro, P.A., Wiegand, S.J., and Yancopoulos, G.D. (2002) Angiopoietin-2 is required for postnatal angiogenesis and lymphatic patterning, and only the latter role is rescued by Angiopoietin-1. *Dev. Cell*, **3** (3), 411–423.

107. Morisada, T., Oike, Y., Yamada, Y., Urano, T., Akao, M., Kubota, Y., Maekawa, H., Kimura, Y., Ohmura, M., Miyamoto, T., Nozawa, S., Koh, G.Y., Alitalo, K., and Suda, T. (2005) Angiopoietin-1 promotes LYVE-1-positive lymphatic vessel formation. *Blood*, **105** (12), 4649–4656.

108. Tammela, T., Saaristo, A., Lohela, M., Morisada, T., Tornberg, J., Norrmén, C., Oike, Y., Pajusola, K., Thurston, G., Suda, T., Yla-Herttuala, S., and Alitalo, K. (2005) Angiopoietin-1 promotes lymphatic sprouting and hyperplasia. *Blood*, **105** (12), 4642–4648.

109. Kubo, H., Cao, R., Brakenhielm, E., Makinen, T., Cao, Y., and Alitalo, K. (2002) Blockade of vascular endothelial growth factor receptor-3 signaling inhibits fibroblast growth factor-2-induced lymphangiogenesis in mouse cornea. *Proc. Natl. Acad. Sci. U.S.A.*, **99** (13), 8868–8873.

110. Chang, L.K., Garcia-Cardena, G., Farnebo, F., Fannon, M., Chen, E.J., Butterfield, C., Moses, M.A., Mulligan, R.C., Folkman, J., and Kaipainen, A. (2004) Dose-dependent response of FGF-2 for lymphangiogenesis. *Proc. Natl. Acad. Sci. U.S.A.*, **101** (32), 11658–11663.
111. Mackay, C.R. (2001) Chemokines: immunology's high impact factors. *Nat. Immunol.*, **2** (2), 95–101.
112. Wiley, H.E., Gonzalez, E.B., Maki, W., Wu, M.T., and Hwang, S.T. (2001) Expression of CC chemokine receptor-7 and regional lymph node metastasis of B16 murine melanoma. *J. Natl. Cancer Inst.*, **93** (21), 1638–1643.
113. Shields, J.D., Emmett, M.S., Dunn, D.B., Joory, K.D., Sage, L.M., Rigby, H., Mortimer, P.S., Orlando, A., Levick, J.R., and Bates, D.O. (2007) Chemokine-mediated migration of melanoma cells towards lymphatics–a mechanism contributing to metastasis. *Oncogene*, **26** (21), 2997–3005.
114. Issa, A., Le, T.X., Shoushtari, A.N., Shields, J.D., and Swartz, M.A. (2009) Vascular endothelial growth factor-C and C-C chemokine receptor 7 in tumor cell-lymphatic cross-talk promote invasive phenotype. *Cancer Res.*, **69** (1), 349–357.
115. Pan, M.R., Hou, M.F., Chang, H.C., and Hung, W.C. (2008) Cyclooxygenase-2 up-regulates CCR7 via EP2/EP4 receptor signaling pathways to enhance lymphatic invasion of breast cancer cells. *J. Biol. Chem.*, **283** (17), 11155–11163.
116. Costa, C., Soares, R., Reis-Filho, J.S., Leitão, D., Amendoeira, I., and Schmitt, F.C. (2002) Cyclo-oxygenase 2 expression is associated with angiogenesis and lymph node metastasis in human breast cancer. *J. Clin. Pathol.*, **55** (6), 429–434.
117. Kawada, K., Sonoshita, M., Sakashita, H., Takabayashi, A., Yamaoka, Y., Manabe, T., Inaba, K., Minato, N., Oshima, M., and Taketo, M.M. (2003) Pivotal role of CXCR3 in melanoma cell metastasis to lymph nodes. *Cancer Res.*, **64** (11), 4010–4017.
118. Kawada, K., Hosogi, H., Sonoshita, M., Sakashita, H., Manabe, T., Shimahara, Y., Sakai, Y., Takabayashi, A., Oshima, M., and Taketo, M.M. (2007) Chemokine receptor CXCR3 promotes colon cancer metastasis to lymph nodes. *Oncogene*, **26** (32), 4679–4688.
119. Uchida, D., Begum, N.M., Tomizuka, Y., Bando, T., Almofti, A., Yoshida, H., and Sato, M. (2004) Acquisition of lymph node, but not distant metastatic potentials, by the overexpression of CXCR4 in human oral squamous cell carcinoma. *Lab. Invest.*, **84** (12), 1538–1546.
120. Uchida, D., Onoue, T., Tomizuka, Y., Begum, N.M., Miwa, Y., Yoshida, H., and Sato, M. (2007) Involvement of an autocrine stromal cell derived factor-1/CXCR4 system on the distant metastasis of human oral squamous cell carcinoma. *Mol. Cancer Res.*, **5** (5), 685–694.
121. Müller, A., Homey, B., Soto, H., Ge, N., Catron, D., Buchanan, M.E., McClanahan, T., Murphy, E., Yuan, W., Wagner, S.N., Barrera, J.L., Mohar, A., Verástegui, E., and Zlotnik, A. (2001) Involvement of chemokine receptors in breast cancer metastasis. *Nature*, **410** (6824), 50–56.
122. Ming, J., Zhang, Q., Qiu, X., and Wang, E. (2009) Interleukin 7/interleukin 7 receptor induce c-Fos/c-Jun-dependent vascular endothelial growth factor-D up-regulation: a mechanism of lymphangiogenesis in lung cancer. *Eur. J. Cancer* **45** (5), 866–873.
123. Mashino, K., Sadanaga, N., Yamaguchi, H., Tanaka, F., Ohta, M., Shibuta, K., Inoue, H., and Mori, M. (2002) Expression of chemokine receptor CCR7 is associated with lymph node metastasis of gastric carcinoma. *Cancer Res.*, **62** (10), 2937–2941.
124. Günther, K., Leier, J., Henning, G., Dimmler, A., Weissbach, R., Hohenberger, W., and Förster, R. (2005) Prediction of lymph node metastasis in colorectal carcinoma by expression of chemokine receptor CCR7. *Int. J. Cancer*, **116** (5), 726–733.

125. Cabioglu, N., Yazici, M., Arun, B., Broglio, K.R., Hortobagyi, G.N., Price, J.E., and Sahin, A. (2005) CCR7 and CXCR4 as novel biomarkers predicting axillary lymph node metastasis in T1 breast cancer. *Clin. Cancer Res.*, **11** (16), 5686–5693.
126. Salvucci, O., Bouchard, A., Baccarelli, A., Deschênes, J., Sauter, G., Simon, R., Bianchi, R., and Basik, M. (2006) The role of CXCR4 receptor expression in breast cancer: a large tissue microarray study. *Breast Cancer Res. Treat.*, **97** (3), 275–283.
127. Nakata, B., Fukunagaa, S., Noda, E., Amano, R., Yamada, N., and Hirakawa, K. (2008) Chemokine receptor CCR7 expression correlates with lymph node metastasis in pancreatic cancer. *Oncology*, **74** (1-2), 69–75.
128. Schimanski, C.C., Bahre, R., Gockel, I., Müller, A., Frerichs, K., Hörner, V., Teufel, A., Simiantonaki, N., Biesterfeld, S., Wehler, T., Schuler, M., Achenbach, T., Junginger, T., Galle, P.R., and Moehler, M. (2006) Dissemination of hepatocellular carcinoma is mediated via chemokine receptor CXCR4. *Br. J. Cancer*, **95** (2), 210–217.
129. Ishikawa, T., Nakashiro, K., Hara, S., Klosek, S.K., Li, C., Shintani, S., and Hamakawa, H. (2006) CXCR4 expression is associated with lymph-node metastasis of oral squamous cell carcinoma. *Int. J. Oncol.*, **28** (1), 61–66.
130. Maekawa, S., Iwasaki, A., Shirakusa, T., Kawakami, T., Yanagisawa, J., Tanaka, T., Shibaguchi, H., Kinugasa, T., Kuroki, M., and Kuroki, M. (2008) Association between the expression of chemokine receptors CCR7 and CXCR3, and lymph node metastatic potential in lung adenocarcinoma. *Oncol. Rep.*, **19** (6), 1461–1468.
131. Wagner, P.L., Moo, T., Arora, N., Liu, Y.F., Zarnegar, R., Scognamiglio, T., and Fahey, T.J.III (2008) The chemokine receptors CXCR4 and CCR7 are associated with tumor size and pathologic indicators of tumor aggressiveness in papillary thyroid carcinoma. *Ann. Surg. Oncol.*, **15** (10), 2833–2841.
132. Banziger-Tobler, N.E., Halin, C., Kajiya, K., and Detmar, M. (2008) Growth hormone promotes lymphangiogenesis. *Am. J. Pathol.*, **173** (2), 586–597.
133. Avraamides, C.J., Garmy-Susini, B., and Varner, J.A. (2008) Integrins in angiogenesis and lymphangiogenesis. *Nat. Rev. Cancer*, **8** (8), 604–617.
134. Huang, X.Z., Wu, J.F., Ferrando, R., Lee, J.H., Wang, Y.L., Farese, R.V.J., and Sheppard, D. (2000) Fatal bilateral chylothorax in mice lacking the integrin alpha9beta1. *Mol. Cell Biol.*, **20** (14), 5208–5215.
135. Vlahakis, N.E., Young, B.A., Atakilit, A., and Sheppard, D. (2005) The lymphangiogenic vascular endothelial growth factors VEGF-C and -D are ligands for the integrin alpha9beta1. *J. Biol. Chem.*, **280** (6), 4544–4552.
136. Garmy-Susini, B., Makale, M., Fuster, M., and Varner, J.A. (2007) Methods to study lymphatic vessel integrins. *Methods Enzymol.*, **426**, 415–438.
137. Hong, Y.K., Lange-Asschenfeldt, B., Velasco, P., Hirakawa, S., Kunstfeld, R., Brown, L.F., Bohlen, P., Senger, D.R., and Detmar, M. (2004) VEGF-A promotes tissue repair-associated lymphatic vessel formation via VEGFR-2 and the alpha1beta1 and alpha2beta1 integrins. *FASEB J.*, **18** (10), 1111–1113.
138. Dietrich, T., Onderka, J., Bock, F., Kruse, F.E., Vossmeyer, D., Stragies, R., Zahn, G., and Cursiefen, C. (2007) Inhibition of inflammatory lymphangiogenesis by integrin alpha5 blockade. *Am. J. Pathol.*, **171** (1), 361–372.
139. Wei, C.J., Xu, X., and Lo, C.W. (2004) Connexins and cell signaling in development and disease. *Annu. Rev. Cell Dev. Biol.*, **20**, 811–838.
140. Naoi, Y., Miyoshi, Y., Taguchi, T., Kim, S.J., Arai, T., Maruyama, N., Tamaki, Y., and Noguchi, S. (2008) Connexin26 expression is associated with aggressive phenotype in human papillary and follicular thyroid cancers. *Cancer Lett.*, **262** (2), 248–256.
141. Naoi, Y., Miyoshi, Y., Taguchi, T., Kim, S.J., Arai, T., Tamaki, Y., and Noguchi, S. (2007) Connexin26 expression is associated with lymphatic vessel invasion

and poor prognosis in human breast cancer. *Breast Cancer Res. Treat.*, **106** (1), 11–17.

142. Ji, R.C. (2009) Lymph node lymphangiogenesis: a new concept for modulating tumor metastasis and inflammatory process. *Histol. Histopathol.*, **24** (3), 377–384.

143. Hirakawa, S., Brown, L.F., Kodama, S., Paavonen, K., Alitalo, K., and Detmar, M. (2007) VEGF-C-induced lymphangiogenesis in sentinel lymph nodes promotes tumor metastasis to distant sites. *Blood*, **109** (3), 1010–1017.

144. Qian, C.N., Berghuis, B., Tsarfaty, G., Bruch, M., Kort, E.J., Ditlev, J., Tsarfaty, I., Hudson, E., Jackson, D.G., Petillo, D., Chen, J., Resau, J.H., and Teh, B.T. (2006) Preparing the "soil": the primary tumor induces vasculature reorganization in the sentinel lymph node before the arrival of metastatic cancer cells. *Cancer Res.*, **66** (21), 10365–10376.

145. Harrell, M.I., Iritani, B.M., and Ruddell, A. (2007) Tumor-induced sentinel lymph node lymphangiogenesis and increased lymph flow precede melanoma metastasis. *Am. J. Pathol.*, **170** (2), 774–786.

146. Kawai, Y., Kaidoh, M., and Ohhashi, T. (2008) MDA-MB-231 produces ATP-mediated ICAM-1-dependent facilitation of the attachment of carcinoma cells to human lymphatic endothelial cells. *Am. J. Physiol. Cell Physiol.*, **295** (5), C1123–C1132.

147. Stacker, S.A. and Achen, M.G. (2008) From anti-angiogenesis to anti-lymphangiogenesis: emerging trends in cancer therapy. *Lymphat. Res. Biol.*, **6** (2-4), 165–172.

148. Achen, M.G., Roufail, S., Domagala, T., Catimel, B., Nice, E.C., Geleick, D.M., Murphy, R., Scott, A.M., Caesar, C., Makinen, T., Alitalo, K., and Stacker, S.A. (2000) Monoclonal antibodies to vascular endothelial growth factor-D block its interactions with both VEGF receptor-2 and VEGF receptor-3. *Eur. J. Biochem.*, **267** (9), 2505–2515.

149. Persaud, K., Tille, J.C., Liu, M., Zhu, Z., Jimenez, X., Pereira, D.S., Miao, H.Q., Brennan, L.A., Witte, L., Pepper, M.S., and Pytowski, B. (2004) Involvement of the VEGF receptor 3 in tubular morphogenesis demonstrated with a human anti-human VEGFR-3 monoclonal antibody that antagonizes receptor activation by VEGF-C. *J. Cell Sci.*, **117** (Pt 13), 2745–2756.

150. Wilhelm, S.M., Adnane, L., Newell, P., Villanueva, A., Llovet, J.M., and Lynch, M. (2008) Preclinical overview of sorafenib, a multikinase inhibitor that targets both Raf and VEGF and PDGF receptor tyrosine kinase signaling. *Mol. Cancer Ther.*, **7** (10), 3129–3140.

151. Ruggeri, B., Singh, J., Gingrich, D., Angeles, T., Albom, M., Yang, S., Chang, H., Robinson, C., Hunter, K., Dobrzanski, P., Jones-Bolin, S., Pritchard, S., Aimone, L., Klein-Szanto, A., Herbert, J.M., Bono, F., Schaeffer, P., Casellas, P., Bourie, B., Pili, R., Isaacs, J., Ator, M., Hudkins, R., Vaught, J., Mallamo, J., and Dionne, C. (2003) CEP-7055: a novel, orally active pan inhibitor of vascular endothelial growth factor receptor tyrosine kinases with potent antiangiogenic activity and antitumor efficacy in preclinical models. *Cancer Res.*, **63** (18), 5978–5991.

152. Kirkin, V., Thiele, W., Baumann, P., Mazitschek, R., Rohde, K., Fellbrich, G., Weich, H., Waltenberger, J., Giannis, A., and Sleeman, J.P. (2004) MAZ51, an indolinone that inhibits endothelial cell and tumor cell growth in vitro, suppresses tumor growth in vivo. *Int. J. Cancer*, **112** (6), 986–993.

153. Lin, B., Podar, K., Gupta, D., Tai, Y.T., Li, S., Weller, E., Hideshima, T., Lentzsch, S., Davies, F., Li, C., Weisberg, E., Schlossman, R.L., Richardson, P.G., Griffin, J.D., Wood, J., Munshi, N.C., and Anderson, K.C. (2002) The vascular endothelial growth factor receptor tyrosine kinase inhibitor PTK787/ZK222584 inhibits growth and migration of multiple myeloma cells in the bone marrow microenvironment. *Cancer Res.*, **62** (17), 5019–5026.

154. Schomber, T., Zumsteg, A., Strittmatter, K., Crnic, I., Antoniadis, H., Littlewood-Evans, A., Wood, J., and Christofori, G. (2009) Differential effects of the vascular endothelial growth factor receptor inhibitor PTK787/ZK222584 on tumor angiogenesis and tumor lymphangiogenesis. *Mol. Cancer Ther.*, **8** (1), 55–63.

155. Ferrara, N. and Kerbel, R.S. (2005) Angiogenesis as a therapeutic target. *Nature*, **438** (7070), 967–974.

156. Jain, R.K., Duda, D.G., Clark, J.W., and Loeffler, J.S. (2006) Lessons from phase III clinical trials on anti-VEGF therapy for cancer. *Nat. Clin. Pract. Oncol.*, **3** (1), 24–40.

157. Clasper, S., Royston, D., Baban, D., Cao, Y., Ewers, S., Butz, S., Vestweber, D., and Jackson, D.G. (2008) A novel gene expression profile in lymphatics associated with tumor growth and nodal metastasis. *Cancer Res.*, **68** (18), 7293–7303.

158. Karkkainen, M.J., Makinen, T., and Alitalo, K., (2002) Lymphatic endothelium: a new frontier of metastasis research. *Nat. Cell. Biol.*, **4** (1), E2–E5.

Part III
Signal Transduction and Angiogenesis

Tumor Angiogenesis – From Molecular Mechanisms to Targeted Therapy. Edited by Francis S. Markland, Stephen Swenson, and Radu Minea
Copyright © 2010 WILEY-VCH Verlag GmbH & Co. KGaA, Weinheim
ISBN: 978-3-527-32091-2

7
Integrin Involvement in Angiogenesis

Abebe Akalu, Liangru Contois, and Peter C. Brooks

7.1
Introduction

A critical requirement for homeostasis and tissue expansion in organisms involves the efficient delivery of oxygen and nutrients and the removal of metabolic wastes [1]. Blood and lymphatic vessels largely facilitate these fundamental processes, although recent evidence also suggests that fluid-conducting channels lined by either tumor cells or a mosaic of tumor and endothelial cells may contribute to the perfusion of a subset of malignant tumors [2]. In general, vascular networks regulate physiological processes such as embryonic development, wound healing, adipogenesis, and reproduction. The vasculature also plays important roles in tumor growth, metastasis, and other pathologies such as diabetic retinopathy, arthritis, and psoriasis [3, 4]. The formation of functional blood vessels can occur by a number of mechanisms. For example, vasculogenesis is a process by which blood vessels form *de novo* from precursor cells [5]. Arteriogenesis is a process by which vessels form by a mechanism of activation, dilation, and remodeling of small, preexisting nonfunctional vessels [6]. Finally, *angiogenesis* can be defined as the development of new blood vessels from preexisting vessels [7].

Angiogenesis may occur by at least two different processes, which include sprouting angiogenesis and intussusceptive angiogenesis, both of which have been documented to occur during tumor growth [8, 9]. Sprouting angiogenesis involves growth factor–stimulated proliferation and invasion of individual cells from an existing vessel, resulting in a solid cord of cells that ultimately forms a lumen by a process of canalization or lumenogenesis. In contrast, intussusceptive angiogenesis involves the splitting of existing vessels by the formation of translumenal pillary bridges to form higher order branching patterns [8, 9]. Intussusceptive angiogenesis may occur more rapidly than sprouting angiogenesis and may not require the extensive proliferation, and extracellular matrix (ECM) remodeling typically associated with sprouting angiogenesis [10].

While many similarities are to likely exist between the cellular and molecular mechanisms that contribute to the formation of functional vessels, distinct differences might also exist. Studies utilizing peptide phage libraries and

Tumor Angiogenesis – From Molecular Mechanisms to Targeted Therapy. Edited by Francis S. Markland, Stephen Swenson, and Radu Minea
Copyright © 2010 WILEY-VCH Verlag GmbH & Co. KGaA, Weinheim
ISBN: 978-3-527-32091-2

differential microarrays have identified tissue-specific vascular markers [11, 12]. In another study, tumor endothelial cells exhibited altered cytoskeletal reorganization in response to mechanical tension as compared to endothelial cells isolated from normal tissues [13]. While the cellular and molecular structure of tumor blood vessels is known to differ from that of normal blood vessels, marked differences also exist between normal vessels from different tissue compartments such as the brain, kidney, and liver [14–16]. Blood vessels from different vascular beds may vary in expression of cell surface molecules including adhesion receptors and growth factor receptors [17–19]. In addition, the extent of mural cell coverage and the integrity and composition of the ECM may also vary considerably [17–19]. These molecular and structural differences probably reflect in part, the unique functions attributed to these organs.

7.2
Sprouting Angiogenesis

Given the complexity of the process by which blood vessels form, we will limit our discussion primarily to sprouting angiogenesis. Sprouting angiogenesis can be organized into at least three temporal phases including an initiation phase, an invasive phase, and a maturation phase (Figure 7.1). It is important to note that while angiogenesis can be organized into these three phases, the cellular and molecular processes that help characterize these stages exhibit extensive

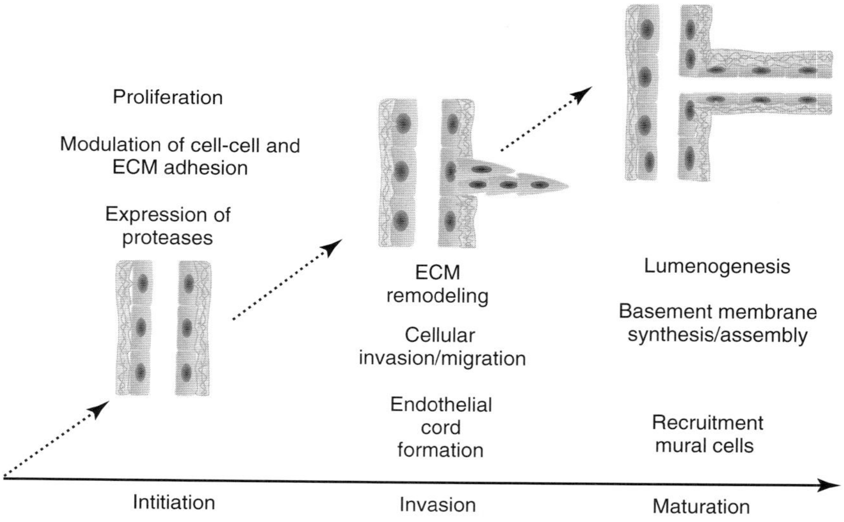

Figure 7.1 Three phases of sprouting angiogenesis. Sprouting angiogenesis involves an interconnected series of cellular, biochemical, and molecular events leading to the formation of new blood vessels from preexisting vessels. Schematic representation of the three general phases of sprouting angiogenesis, which includes the initiation, invasive, and maturation steps.

overlap, as the angiogenic cascade is an interconnected continuum of events rather than a set of isolated steps. During the initiation phase, angiogenic growth factors such as vascular endothelial growth factor (VEGF), fibroblast growth factor (FGF), insulin-like growth factor-1 (IGF-1), and hepatocyte growth factor (HGF) can bind to their respective cell surface receptors, thereby initiating signaling cascades that ultimately lead to proangiogenic patterns of gene expression. Growth factor stimulation induces proliferation, expression of enzymes and modulates the expression of cell–cell and cell–matrix adhesion molecules [20]. Following initiation, the local ECM can be remodeled to create a microenvironment permissive for a subset of activated endothelial cells to invade, migrate, and form endothelial cell cords [21]. In the maturation phase, a complex set of events occurs to form a functional lumen, a process that has been termed *lumenogenesis* [22]. Finally, the capillary sprouts become stabilized by the establishment of new cell–cell and cell–ECM interactions and the recruitment of supporting cells.

7.3 Cellular Regulators of Angiogenesis

Angiogenesis requires a coordinated balance of both stimulatory and inhibitory factors. Thus, it is not surprising that in addition to endothelial and smooth muscle cells, a wide array of cell types play active roles in angiogenesis. Given this diversity, we will limit our discussion to a few well-characterized examples to illustrate this cellular complexity (Figure 7.2). In the case of tumor angiogenesis,

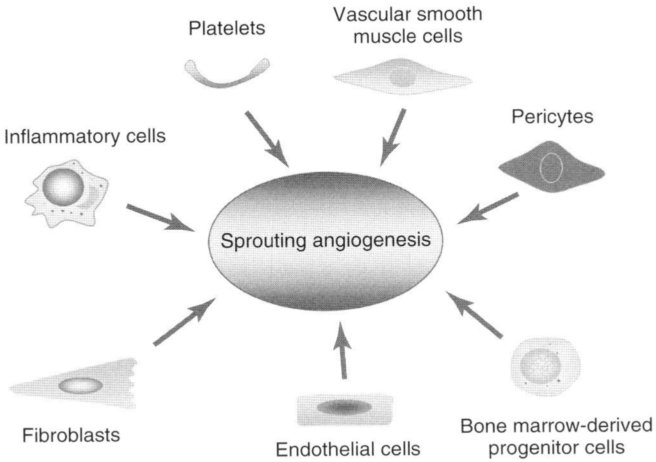

Figure 7.2 Cellular contribution to sprouting angiogenesis. Numerous cell types, in addition to endothelial cells, play active roles in sprouting angiogenesis. Schematic representation illustrating important examples of different cell types thought to contribute to sprouting angiogenesis.

gradients of chemoattractants such as VEGF, PlGF, and chemokines such as CXCL12 and CCL4 contribute to the recruitment of myeloid cells to hypoxic areas of neovascularization [23, 24]. Moreover, tumor-associated fibroblasts can also release regulatory proteins. Many of these cell types promote angiogenesis by releasing growth factors, inflammatory cytokines, and proteolytic enzymes that impact endothelial cells, smooth muscle cells, and pericytes. These infiltrating cells may also generate or release inhibitory factors and promote maturation of blood vessels [23, 24]. In addition, circulating endothelial progenitor cells may be incorporated into the growing vessel [25, 26]. The extent to which endothelial progenitor cells contribute to new vessel formation may depend on the particular vascular bed in question and whether the angiogenesis is associated with a specific process such as wound healing or tumor formation.

The recruitment of vascular smooth muscle cells and pericytes to the growing vessels is critical for their maturation and stability. The differential impact of growth factors on endothelial cells and pericytes illustrates an important example of the dynamic and coordinated balance of stimulators and inhibitors required for vessel development. While VEGF stimulates endothelial cell proliferation, it may also have opposing functions in mural cells. A recent study suggests that VEGF might actually inhibit pericyte coverage by a mechanism involving suppression of PDGF-Rβ signaling thereby inhibiting vessel maturation [27]. Thus, the coordinated temporal and spatial control of expression and activity of growth factors is critical for proper blood vessel development. Platelets have also been suggested to influence angiogenesis by the differential release of angiogenic regulatory factors at sites of neovascularization. Platelets may sequester and release both proangiogenic factors such as bFGF and VEGF as well as antiangiogenic molecules such as endostatin, angiostatin, and thrombospondin-1 via distinct α-granules [28, 29]. As can be appreciated from these examples, a great diversity of cell types regulate the balanced expression and activity of pro- and antiangiogenic factors needed for blood vessel formation.

7.4
Molecular Regulators of Angiogenesis

It would be beyond the scope of our review to discuss all the known molecular regulators of blood vessel formation, thus we have chosen a few examples to highlight some of the major families of molecules that play active roles in each of the three phases of sprouting angiogenesis. During the initiation phase, a number of proangiogenic factors can be secreted from many different cell types. The distribution and bioactivity of proangiogenic growth factors such as VEGF and FGF2 can be regulated by mechanisms including sequestration in the ECM, binding alternative cell surface receptors such as integrins, and modulation by proteolytic cleavage [30–32]. The expression and activity of these proangiogenic factors must be high enough to overcome the levels of endogenous angiogenesis inhibitors that function in keeping new vessel formation in check. These proangiogenic growth factors can bind their respective cell surface receptor kinases thereby initiating downstream signaling events. For example, VEGF-A binds and activates VEGFR2. Phosphorylation of VEGFR2 activates a number of signaling cascades including the MAP/Erk and PI3k/Akt pathways [33]. These major signaling cascades have been shown to regulate biochemical and molecular processes crucial to angiogenesis [1–4].

As mentioned above, growth factor stimulation can activate an integrated network of gene expression leading to elevated levels of molecules required for the invasive phase of angiogenesis. Growth factor stimulation is known to differentially regulate the expression and activity of cell–cell and cell–ECM adhesion receptors such as cadherins and integrins [34, 35]. Differential expression and activation of these receptors modulate the interactions between endothelial and other supporting cells as well as physical connections between cells and the underlying ECM. These important changes in adhesive capacity play crucial roles in facilitating endothelial cell migration away from the preexisting vessels. In coordination with the modified adhesive capacity, enzymes can be released from a number of sources. Enzymes such as serine proteases and matrix metalloproteinases (MMPs) function in a highly orchestrated manner with their respective inhibitors to modify the local microenvironment in a number of important ways. Proteolytic enzymes degrade restrictive physical barriers as well as expose functional cryptic regulatory elements within the ECM, thereby altering cellular proliferation, invasion, and survival [36–38]. Moreover, proteolytic enzymes have also been shown to play roles in activating growth factors, releasing stimulatory and inhibitor factors that are sequestered in the matrix and facilitating mobilization of progenitor and stem cells from functional niches that may contribute to angiogenesis [31, 39, 40].

Finally, in the maturation phase, differential expression and activity of molecules such as growth factor, proteolytic enzymes, and adhesion receptors contribute to lumenogenesis, and recruitment of accessory cells to stabilize vessels [22]. During lumenogenesis, endothelial cells undergo a number of processes that are not completely understood. Recent work has provided evidence that molecules such as integrin $\alpha 2\beta 1$, cytosolic signaling molecules such as CDC42, Rac-1 and proteolytic enzymes such as MT1-MMP play roles in lumenogenesis [22]. As the nascent blood vessels mature, pericytes and smooth muscle cells are recruited to stabilize the vessel and can secrete new ECM molecules such as collagen type-IV and laminin that are organized into a functional basement membrane. Integrin interactions with this new basement membrane as well as cell–cell interactions can initiate signaling pathways that promote differentiation [41].

7.5 Integrins and Angiogenesis

As can be appreciated from our brief introduction to sprouting angiogenesis, the stromal microenvironment including the cellular and extracellular components, play critical roles in new blood vessel formation. Integrins are a major family of cell surface molecules, which facilitate bidirectional communication between different cell types and the extracellular compartment. Given the importance of bidirectional information flow between cells and the acellular microenvironment, we will focus our remaining discussion on integrins and the multiple roles by which they may regulate angiogenesis.

Integrins are a multifunctional family of heterodimeric adhesion receptors composed of α and β chains derived from separate gene products. At least 24 distinct integrins have been described which are generated from at least 18 α and 8 β subunits [42]. Endothelial cells express a variety of integrin heterodimers including $\beta 1$, $\beta 3$, $\beta 4$, $\beta 5$, and $\beta 8$ containing integrins and many of these integrins have been shown to play functional roles in angiogenesis [43]. Importantly, integrins expressed in cells such as fibroblasts, pericytes, smooth muscle cells, inflammatory cells, and tumor cells may also contribute to new vessel formation. These transmembrane receptors can mediate both cell–cell and cell–ECM interactions. In addition to mediating physical connections between individual cells and between the cell and the ECM, they can also facilitate bidirectional signaling from inside the cell to the outside and from outside to inside. Integrins not only bind ECM molecules, but also regulatory molecules that play active roles in neovascularization including proteases and protease inhibitors, other cell adhesion molecules, growth factors and growth factor receptors, and cytoplasmic signaling molecules [44]. Given the diversity of these molecular interactions, it is not surprising that integrins play multiple roles in regulating angiogenesis. In this regard, evidence from both genetic models and pharmacological inhibitor studies has provided a wealth of support for integrin function in angiogenesis. For example, while $\alpha 5$ integrin null mice formed blood vessels, they were defective in maturation and stability suggesting that $\alpha 5\beta 1$ integrin may not be critical in early vasculogenesis, but may be more important for the later maturation stage of angiogenesis [45]. Moreover, αv null mice also exhibited defects in vessels within the central nervous system and gastrointestinal tract [46–48]. While $\beta 3$ integrin null mice exhibited few defects during embryonic vascular development, $\beta 3$ signaling-deficient knockin mice showed reduced postnatal growth factor–induced and pathological angiogenesis [49]. Recent studies also demonstrated alterations in maturation of coronary capillaries in male $\beta 3$ null mice [50]. Taken together, these studies suggest that integrins may play different roles during embryonic vascular development and postnatal angiogenesis. Pharmacological inhibition studies have also provided evidence that integrins play roles in angiogenesis. For example, antagonists of integrins such as $\alpha 1\beta 1$, $\alpha 2\beta 1$, $\alpha 4\beta 1$, $\alpha 5\beta 1$ αv$\beta 3$, and αv$\beta 5$ have all been shown to inhibit angiogenesis [51–55].

7.6 Integrin Structure

The capacity of integrins to regulate angiogenesis depends in large part on their structure. Important insight has been gained from NMR as well as crystallographic studies of the extracellular regions of integrins in complex with their ligands. Early studies on αv$\beta 3$ revealed molecular details into general integrin structure [56]. In this regard, integrins can be organized into an extracellular domain composed of α and β chains which form a ligand binding surface. A transmembrane domain and a short cytoplasmic tail follow this extracellular domain. The *N*-terminal domain of the α-chain is organized as a head region with a β-propeller type structure. Additional structural features of the α-chain include a leg-like region composed of an Ig-like thigh domain and two regions termed *calf domains* [57]. The α-chains also have a short transmembrane and cytoplasmic domain. The *N*-terminal portion of the β-chain is organized into a head-like region containing a βA domain followed by an Ig-like hybrid domain and leg-like structures containing PSI and EGF-like repeats [57]. The β-chains also have a short transmembrane region and a cytoplasmic tail that can bind a variety of cytoplasmic adaptor proteins [57].

Given their membrane spanning regions and flexible molecular structures, these receptors are well suited to function as mechanotransducers, thereby allowing the cells to sense and in turn, respond to their dynamic microenvironment. New studies have provided evidence that integrins exist within at least three general conformations, which provide an important mechanism to control bidirectional flow of information between the extracellular microenvironment and the cells. Studies suggest that integrins can exist in a low affinity state with a closed headpiece and a bent conformation in which the head region is in close proximity to the plasma membrane. An intermediate affinity state characterized by an extended leg region with a closed headpiece has also been described [58]. Finally, a fully extended conformation with an open headpiece characterizing the high affinity state has also been shown [58]. The equilibrium between these various conformations regulates integrin ligand binding and subsequent signal transduction.

7.7 Integrin Binding and Bidirectional Signaling

As mentioned above, integrin heterodimers have the capacity to bind a variety of ECM proteins, other transmembrane receptors, and soluble ligands (Figure 7.3). These molecular interactions are largely dependant on integrin conformation. At least two major models have been proposed to help explain the mechanisms by which integrin conformation facilitates ligand binding [58]. These models were largely developed based on studies focusing on inside-out signaling. Experiments have uncovered a critical role for the cytosolic protein talin as a central control point in integrin function. For example, exposure of a cryptic integrin-binding site within talin promotes interactions with the β integrin cytoplasmic tail [59]. This talin-β–integrin interaction is thought to facilitate physical separation of the α and β tails and may play a key role in integrin activation. For example, the switchblade model involves a physical separation of the cytoplasmic tails of the α- and β-chains which propagate structural changes through the transmembrane region that ultimately allow extension of the leg-like regions and movement of the headpiece away from the membrane, allowing enhanced ligand access [58]. An alternative mechanism has also been proposed which involves a piston-like movement of the α- and β-chains within the plane of the membrane which again, results in integrin leg extension and separation of the headpiece away from the plasma membrane

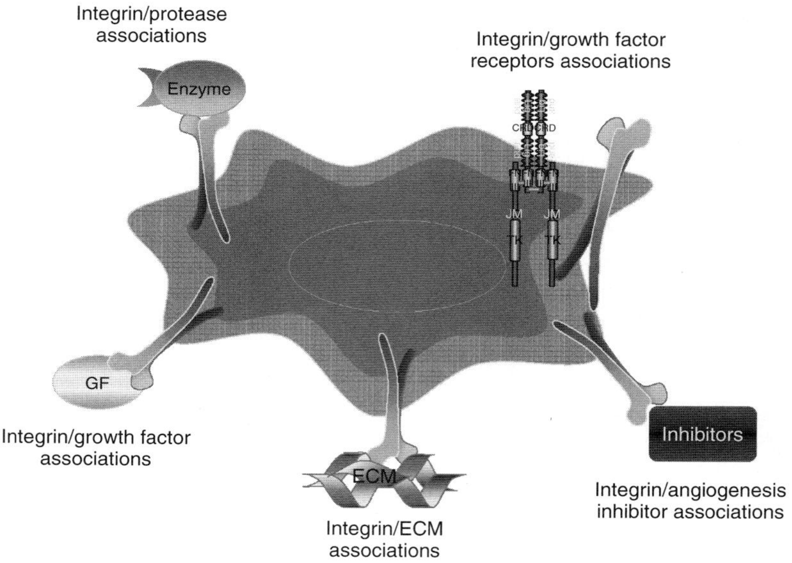

Figure 7.3 Integrin associations with molecular regulators of sprouting angiogenesis. Sprouting angiogenesis involves a wide array of distinct families of regulatory molecules expressed in numerous cell types. Integrin receptors associate with a variety of molecules in addition to ECM proteins. Schematic representation illustrating important examples of molecular regulators of sprouting angiogenesis thought to associate with integrins.

[58]. Whether these structural changes are associated with physical separation of the transmembrane domains is still not completely understood, but studies have suggested that dimerization of the transmembrane domains may contribute to integrin clustering, which in turn may promote signaling [60].

While a detailed discussion of the molecular mechanisms that regulate integrin ligand recognition would be well beyond the scope of this review, an important concept in this respect involves the movement of the integrin headpiece away from the plasma membrane. This physical separation may allow access to amino acids within the cation-regulated I and I-like domains of the α- and β-chains [58]. Metal binding motifs such as the MIDAS (metal ion dependent adhesive site) and AMIDAS (adjacent metal ion dependent adhesive site) sites bind cations such as calcium, magnesium, and manganese, all of which regulate integrin-binding activity [61]. Exposure of residues within the binding surface of the integrin allows coordination with carboxylate groups of acidic residues such as aspartic acid (D) of the well-known integrin-binding motif RGD [61].

In addition to the inside-out signaling that can enhance binding to extracellular ligands, these receptors also participate in outside-in signaling. For example integrin binding to ECM proteins, other cell surface receptors, and soluble factors has been shown to modulate signaling cascades including MAPK/Erk, PI3k/Akt, Rho/MLCK, and the JNk pathways [62, 63]. In particular, integrin binding has been associated with enhanced phosphorylation of the cytoplasmic tails, which can in turn associate with a variety of adaptor molecules as well as non–receptor kinases such as FAK (focal adhesion kinase), Src, and ILK [64]. Interestingly, FAK has been shown to have numerous binding sites for additional signaling molecules, thereby promoting the formation of large signaling complexes. These signaling complexes are known to affect downstream pathways, which ultimately regulate gene expression, cell cycle, and cell survival [62–64].

7.8
Involvement of Integrins in the Initiation Phase of Angiogenesis

As mentioned above, integrin interactions with the local ECM can regulate growth factor signaling, thereby modulating crucial events during the initiation phase of angiogenesis including proliferation, cell adhesion, and expression of proteolytic enzymes. Angiogenesis can be initiated by proangiogenic factors released from many sources such as endothelial cells, fibroblasts, inflammatory infiltrates, tumor cells, mural cells, and platelets. Some growth factors may also be found immobilized within ECM [30]. These ECM-immobilized growth factors may be recognized in a matrix-bound state or released in a bioactive form to help stimulate proliferation and motility [30, 31]. Given the diversity of proangiogenic growth factors, cytokines, and chemokines as well as antiangiogenic factors, it is likely that a number of different mechanisms exist to coordinate and regulate the impact of these diverse factors during new vessel growth. To this end, integrin receptors have been shown to play fundamental roles in regulating the response of cells to external stimuli. For example, immediate early genes important for G0/G1 cell cycle transition such as *c-fos*, *c-myc*, and *c-jun* can be induced in endothelial cells by growth factors, but optimal expression of these molecules required integrin-mediated interactions with ECM components such as fibronectin [65]. The specific type of integrin interactions with a particular ECM protein can also alter the ability of endothelial cells to respond to growth factors. Studies suggest that FGF2 can stimulate endothelial cell cycle progression when plated on fibronectin, but not on laminin [66]. Alterations in integrin signaling following growth factor stimulation may also be modulated by the integrity of the ECM. Growth factor stimulation of vascular smooth muscle cells failed to induce S-phase entry when plated on polymerized collagen type-I, but readily entered S-phase when plated on denatured monomeric collagen type-I [67].

7.9
Integrin and Growth Factor Interactions in Angiogenesis

In recent years, great strides have been made in understanding the molecular mechanism by which integrins cooperate with growth factors and their respective receptors to control endothelial cellular behavior and angiogenesis. As discussed above, a number of secreted growth factors such as VEGF, FGF, IGF-1, and EGF can be immobilized and stored in the ECM [30, 31]. Provisional ECM proteins including fibronectin, fibrin, and vitronectin, which are often associated with tissues undergoing angiogenesis, are major ECM molecules known to bind these factors [30, 31]. Localization of these proangiogenic molecules within the ECM may either enhance or inhibit their activity by altering receptor recognition, thereby providing a mechanism to modulate angiogenesis. Integrins may associate with immobilized proangiogenic growth factors. For example, integrins such as $\alpha v\beta 3$, $\alpha 3\beta 1$, and $\alpha 9\beta 1$ may interact with specific isoforms of VEGF-A [31, 68]. Integrins $\alpha v\beta 3$ and $\alpha 3\beta 1$ interactions with immobilized VEGF enhanced endothelial cell adhesion, migration, and survival *in vitro* [31]. Moreover, blocking antibodies directed to $\alpha 9\beta 1$ was shown to inhibit VEGF but not FGF2-induced angiogenesis [68]. In another study, $\alpha 9\beta 1$ was shown to interact with thrombospondin-1 (TSP-1) and promote angiogenesis [69].

In addition to VEGF, a number of other factors thought to regulate angiogenesis have also been shown to bind integrins including FGF1, Angiopoietins (Ang1), connective tissue growth factor (CTGF), and cysteine-rich angiogenic protein 61 (Cyr61) [70–73]. Studies have shown that $\alpha v\beta 3$ may directly bind immobilized FGF1 and enhance cell adhesion, migration, and proliferation [70]. Moreover, studies have suggested that $\alpha 5\beta 1$ may associate with Ang-1 and that this interaction facilitates signaling leading to vascular cell migration that is independent from the Ang-1 receptor Tie-2 [71]. Finally, a number of antiangiogenic factors are also thought to bind integrins, thereby regulating the balance of pro- and anti-angiogenic signaling activity [74–76]. As mentioned previously, platelets can store and release antiangiogenic molecules such as TSP-1, which has been believed to bind integrin $\alpha 9\beta 1$. In addition, new evidence suggests that the antiangiogenic chemokine CXCL4 produced in platelets may bind $\alpha v\beta 3$ in endothelial cells and inhibit adhesion and migration [77]. Angiogenesis inhibitors such as such as canstatin and tumstatin may bind $\beta 1$ and $\beta 3$ integrins, leading to inhibition of endothelial cell proliferation and survival by altering expression of cell cycle inhibitors and apoptosis regulatory proteins [78, 79].

7.10
Integrin and Growth Factor Receptor Interactions in Angiogenesis

An expanding number of studies demonstrate associations between growth factor receptors and integrins. Among the most well-studied examples is the association of VEGFR2 with $\alpha v\beta 3$. VEGF-A binding to VEGFR2 recruits Src and facilitates phosphorylation of the $\beta 3$ cytoplasmic tail [80]. Phosphorylation of the $\beta 3$-tail may promote $\alpha v\beta 3$-VEGFR2 complex formation, which in turn, is thought to enhance downstream signaling leading to endothelial proliferation and survival [80]. In addition to interacting with the VEGFR2, $\alpha v\beta 3$ may associate with other growth factor receptors including FGF, IGF-1, PDGF, and HGF receptors [81–84]. While the molecular mechanisms governing these associations are not fully understood, these $\alpha v\beta 3$-growth factor receptor complexes are likely to enhance the activation of numerous signaling cascades such as the MAP/Erk and PI3k/Akt pathways.

While much attention has been focused on the association of $\alpha v\beta 3$ with a variety of binding partners, other integrins have also been shown to associate with growth factor receptors. For example, the $\beta 1$-EGFR complex may facilitate PI3kinase signaling [85]. Moreover, $\alpha 5\beta 1$ was shown to associate with Tie-2 receptor, which is known to bind Ang-1 thereby regulating its stimulatory activity [86]. The complex formation between Tie-2 and $\alpha 5\beta 1$ may function to reduce the threshold amount of Ang-1 needed to promote Tie-2 phosphorylation and subsequent regulation of endothelial cell behavior [86]. The HGF receptor MET, as well as FGFR3 and the semaphorin receptor plexinB1, may all interact with $\alpha 5\beta 1$ [87–89]. Finally, $\alpha 6$-containing integrins such as $\alpha 6\beta 1$ and $\alpha 6\beta 4$ have also been shown to associate with MET and ErB2 [90]. Importantly, interactions between integrins and ECM components can also influence growth factor signaling. For example, $\beta 1$ integrin interactions with collagen type-I inhibited VEGF2 activation by a mechanism involving recruitment or the tyrosine phosphatase Shp-2, thereby dephosphorylating VEGFR2 and inhibiting downstream signaling [91]. Intriguing new findings also suggest that the tissue inhibitor of metalloproteinase-2 (TIMP-2) may inhibit angiogenesis by an MMP-independent mechanism involving TIMP-2 binding to $\alpha 3\beta 1$, leading to Shp-1 dependent inactivation of FGF and VEGF receptors, and thereby inhibiting endothelial cell proliferation and migration [92].

7.11 Integrin-Mediated Regulation of Cell Adhesion

An important group of molecules that mediate cell–cell adhesion are the cadherins and a number of cadherins are expressed on vascular cells [93]. Alterations in adhesion junctions (AJs) and tight junctions (TJs) are thought to play vital roles in angiogenesis and embryonic vascular development, as mice deficient in VE-cadherin exhibited significant vascular abnormalities [94]. While the disruption of endothelial cell–cell interaction is likely to contribute to the enhanced proliferation and migration of endothelial tip cells, cadherins may also play important roles in later stages of vessel formation including pericyte and smooth muscle cell coverage. Experimental evidence has demonstrated that integrin interaction with ECM proteins such as fibronectin, can disrupt localization of E-cadherin [95]. Moreover while integrin interactions with collagen decreases, E-cadherin interactions with laminin have been shown to upregulate its expression [96]. In other studies, integrin binding to collagen was shown to enhance activation of the MMP-9, which led to shedding of E-cadherin from the surface of tumor cells [97]. It would be interesting to determine whether similar mechanisms operate in vascular cells.

Experimental support for the role of integrins in regulating cadherin-dependent endothelial cell behavior comes from work examining the regulation of VE-cadherin. Evidence suggests that $\alpha v\beta 3$ interactions with fibronectin can disrupt VE-cadherin distribution in endothelial cells and that this disruption was associated with the activation of Src but not Ras [95]. These findings are consistent with a possible role for $\alpha v\beta 3$-mediated signaling in disrupting VE-cadherin function leading to decreased endothelial cell–cell interactions, which might ultimately promote migration. Previous studies have also suggested that FGF signaling may play an important role in regulating cadherin expression in endothelial cells. FGF2 stimulation of endothelial cells *in vitro* was shown to decrease the surface expression of several cadherins including E, N, P, and VE-cadherins, while VEGF stimulation exhibited minimal effects [98]. The inhibition of cell surface expression appeared to depend at least in part on JNK signaling. Integrin-dependent signaling is known to modulate the JNK pathway. Integrins such as $\alpha v\beta 3$ and $\alpha 5\beta 1$ have been shown to associate with FGF and FGF receptors, which may contribute to altered cadherin expression. Thus, integrin cross-talk with other cell adhesion receptors may play roles in regulating cell–cell adhesion thereby providing endothelial tip cells with an enhanced capacity to migrate.

7.12 Integrin-Mediated Regulation of Protease Expression

The ability of vascular cells to remodel the local extracellular microenvironment is a critical step in angiogenesis and the expression of proteolytic enzymes such as MMPs and serine proteases, two major families of enzymes, are thought to contribute to ECM remodeling. The coordinated expression of enzymes and their respective inhibitors is probably required for vascular cells to enter the invasive phase of angiogenesis. Here again, integrin receptors play an important role. Many different cell types important for angiogenesis use integrin interactions with distinct ECM molecules to differentially regulate the expression of MMPs. In addition to endothelial cells, tumor cells, fibroblasts, macrophages, and many others contribute to the enhanced expression of proteases within the microenvironment in which angiogenesis is initiated. For example, $\alpha 5\beta 1$ interaction with the 120 kD cell-binding domain of fibronectin was shown to upregulate expression of MMP-3 and MMP-9 in fibroblasts [99]. In contrast, cellular interaction with the CS-1 domain of fibronectin suppressed expression of MMPs [100]. Integrin $\alpha 2\beta 1$ interactions with collagen were shown to upregulate expression of MMP-1 in endothelial cells, while $\alpha 5\beta 1$ expressed in macrophages contributed to increased expression of MMP-9 [101, 102]. Interestingly, a cryptic peptide in laminin-1 was shown to initiate a MAP kinase–dependent upregulation of both MMP-9 and the serine protease urokinase plasminogen activator (uPA) in macrophages [103]. Finally, studies have also shown that overexpression of $\alpha v\beta 6$ in squamous cell carcinoma resulted in enhanced expression of MMP-3 and MMP-9 [104, 105]. The molecular mechanisms by which integrins regulate proteases probably depends on the enzyme and cell type. In studies examining the regulation of MMP-1 in endothelial cells, the Ets-1 transcription factor was shown to play a role in this $\beta 1$-integrin-dependent upregulation of MMP-1 following collagen binding [106]. These studies are consistent with the possibility that Ets factors may play roles in integrin-mediated regulation of MMPs during the early stages of angiogenesis. In an interesting new study, evidence was provided that $\alpha 3\beta 1$ may play a functional role in modulating the stability of MMP-9 mRNA [107]. Taken together, these examples help highlight the complex roles that integrins may play during initiation of angiogenesis.

7.13 Involvement of Integrins in the Invasive Phase of Angiogenesis

As discussed above, angiogenesis is a highly integrated continuum of cellular and molecular events, which depend functionally on the successful completion of prior steps in the cascade. The invasive phase of angiogenesis can be characterized by remodeling of the local microenvironment. While experimental evidence demonstrates that integrins can regulate protease expression, many of these enzymes are secreted in a catalytically inactive form and thus, need to be activated as well as localized to specific functional sites to facilitate efficient cellular invasion. In this respect, integrins not only regulate expression, but also activation and functional localization of matrix-degrading proteases. This enhanced proteolytic activity may lead to structural alterations in the local ECM, providing less restrictive microenvironments for invasion and migration. In addition, proteolytic alteration in the structure of the ECM has been shown to expose unique matrix-immobilized cryptic extracellular matrix epitopes (MICEEs) that can regulate integrin-dependent endothelial cell adhesion, proliferation, and migration [36–38, 108–110].

7.14 Involvement of Integrins in Protease Activation and Cell Surface Localization

A growing set of studies demonstrates a role for integrin interactions with specific ECM proteins in regulating the activation of proteolytic enzymes. Several studies have suggested that $\beta 1$ integrin interactions with collagen enhanced the activation of MMPs including MMP-2, MMP-9, and MT1-MMP in a variety of cell types [111–113]. The increase in MMP-2 activation following collagen binding appeared to depend on integrin interactions with native collagen type-I, as binding to denatured collagen exhibited little if any effect on MMP-2 activation [114]. Moreover, this enhanced activation of MMP-2 was associated with elevated levels of MT1-MMP. In other studies, $\beta 1$-mediated activation of MMP-2 could be reduced by over expression of $\alpha v\beta 3$, suggesting that distinct integrin signaling events may have opposing effects in certain cell types [111]. Further evidence supporting a role for integrins in regulating protease activity comes from studies examining $\alpha v\beta 3$ in vascular smooth muscle cells, which suggested that $\alpha v\beta 3$ may play a role in activating the serine protease tPA [115].

In addition to playing roles in expression and activation of proteolytic enzymes, evidence now indicates that integrins might directly bind proteolytic enzymes at the cell surface thereby localizing activity to sites of active migration and invasion (Figure 7.4). In this regard, we previously made the novel observation that $\alpha v\beta 3$ co-localized with MMP-2 on angiogenic blood vessels and invasive tumor cells [116]. This protease–integrin interaction was associated with enhanced collagen degradation [116]. Blocking $\alpha v\beta 3$ interactions with MMP-2 by soluble hemopexin domain of MMP-2, or a small molecule inhibitor of this interaction reduced angiogenesis and tumor growth, demonstrating the biological importance of this association [117, 118]. Following these initial observations, an extensive set of studies has demonstrated many new examples of integrin-protease associations and their functional implications. For example, in addition to MMP-2- $\alpha v\beta 3$ interaction, $\alpha v\beta 3$ has been shown to associate with the serine protease thrombin, the serine protease receptor uPAR as well as the secreted lysosomal protease procathepsin X [119–121]. Interestingly, in endothelial cells, MT1-MMP was found associated predominately with $\alpha v\beta 3$ in endothelial tips cells during migration, while it was largely associated with $\beta 1$ integrin in non-migrating endothelial cells, suggesting that $\alpha v\beta 3$ association with MT1-MMP may facilitate endothelial tip cell invasion and migration [122].

In addition to MMP-2 and MT1-MMPs, MMP-9 has also been shown to associate with a number of integrins including $\alpha M\beta 2$, $\alpha 5\beta 1$, $\alpha 3\beta 1$, and $\alpha v\beta 5$ in diverse cell types including tumor cells, epithelial cells, and neutrophils [122–125]. In addition, the interstitial collagenase MMP-1 was shown to interact with collagen binding integrin $\alpha 2\beta 1$ [126]. Finally, a diverse set of disintegrin and metalloproteinase-like molecules termed ADAM (A-disintegrin and metalloproteinase), have been shown to associate with $\beta 1$, $\beta 3$, and $\beta 7$ containing integrins [127]. Thus, it is clear that the integrin–protease interactions are widespread among different tissue and cell

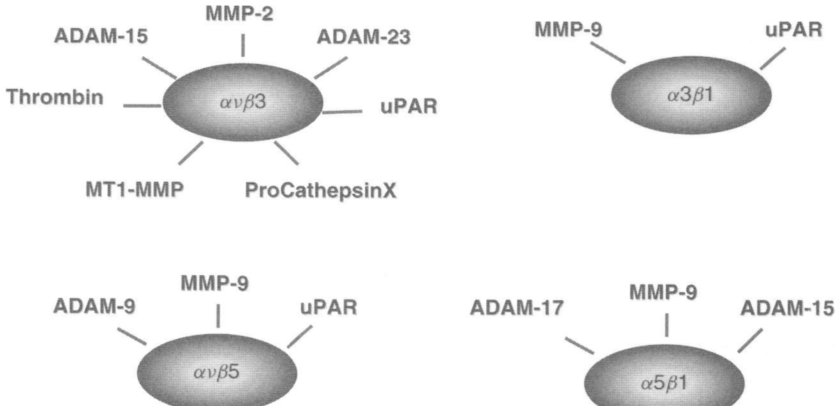

Figure 7.4 Integrin associations with proteolytic enzymes. Proteolytic enzymes from diverse classes are known to play functional roles in sprouting angiogenesis. Representative examples of proteolytic enzymes known to associate with integrins.

types. While many of these integrin–protease interactions have not been directly linked to angiogenesis, given the role of many of these cell types in angiogenesis, it would not be surprising that some of these interactions facilitate this invasive process. For example, studies suggest that proteolytic enzymes such as MMP-9 may release VEGF, which may enhance endothelial cell invasion and migration, while in other studies MMP-9 has been implicated in mobilization of progenitor cells that may contribute to blood vessel development [128, 129].

Interestingly, our recent evidence suggests that proteolytic remodeling of the ECM may not simply create a less restrictive pathway for vascular cell migration, but may also result in the exposure of cryptic regulatory elements (MICEE) buried in the three-dimensional structure of ECM proteins [36–38, 108–110]. We recently identified a number of MICEE including the HUIV26 and the HU177 cryptic sites within collagen and the STQ cryptic site within laminin, which were shown to be selectively expressed around angiogenic blood vessels [36–38]. Integrin $\alpha v\beta 3$ binds the HUIV26 cryptic collagen site and this interaction was shown to regulate endothelial and tumor cell adhesion and migration *in vitro* and angiogenesis, tumor growth, and metastasis *in vivo* [36–38, 108–110]. Moreover, cellular interaction with the HU177 cryptic site was shown not only to regulate endothelial cell adhesion and migration, but also to regulate the expression of $P27^{KIP1}$ and endothelial cell proliferation *in vitro* and angiogenesis *in vivo* [37]. These studies lead to the development of a humanized antibody (D93/TRC093) directed to the HUI77 site, which is currently being evaluated clinically for the treatment of solid tumors [110].

7.15
Involvement of Integrins in the Maturation Phase of Angiogenesis

During angiogenesis, endothelial cells move away from the preexisting vessel and begin to organize into endothelial cords and re-establish cell–cell interactions as part of their maturation. Interestingly, evidence suggests that bone marrow–derived progenitor cells may be mobilized and become incorporated into the newly forming vessel [23–25]. Experimental evidence suggests that $\alpha 4\beta 1$ integrin may play a functional role in the mobilization and recruitment of monocytes to the vasculature [130]. In fact, blocking $\alpha 4\beta 1$ integrin was shown to reduce the extent of monocyte-stimulated angiogenesis [130]. After endothelial cord formation, lumenogenesis is thought to occur. While relatively little is known about this important process, new molecular insight is beginning to emerge and integrins may play an important functional role. Lumenogenesis involves a process by which intracellular vacuoles form and merge [22]. Lumenogenesis can be organized into several steps which include cell wrapping, budding, cavitation, and cord hollowing that ultimately leads to the formation of a functional lumen [22]. Intriguing new studies have implicated molecules such as CDC42, Rac-1, and MT-1 MMP in lumenogenesis and integrins may impact the functional activity of all these regulator molecules [22]. Evidence suggests an important role for $\alpha 2\beta 1$ in lumen formation as antibodies directed to $\alpha 2\beta 1$ can inhibit endothelial lumen formation [22].

As the nascent vessel continues to mature, additional stabilization processes occur such as synthesis and assembly of new vascular basement membranes as well as the recruitment of mural cells such as pericytes and smooth muscle cells. Here again, integrins have the potential to play multiple roles in all these processes. Fibroblasts, pericytes, and vascular smooth muscle cells are all thought to contribute to the formation of basement membranes and their synthesis and proper assembly may be integrin dependent. Genetic evidence supporting this notion comes from studies demonstrating defects and duplications of basement membranes in $\alpha 3$ integrin–deficient mice [131]. Moreover, glomerial basement membranes in the kidney were shown to exhibit defects in podocyte-specific $\beta 1$ integrin null mice and defects in basement membranes were also detected in conditional $\beta 1$ integrin null mice [132]. Interestingly, Akt-1 deficient mice were shown to exhibit defects in the assembly of fibronectin matrix associated with blood vessels, while treatment of fibroblasts derived from these Akt-1 deficient mice with a stimulating $\beta 1$ antibody reversed this effect [133]. Additional studies have also shown that the synthesis and assembly of laminin and collagen type-IV molecules into functional basement membranes may depend on $\beta 1$ integrins [134, 135]. Collectively, these studies are consistent with a role for integrins in the formation of basement membranes that stabilize the maturing blood vessel.

Finally, a crucial process in promoting blood vessel stability involves pericyte and smooth muscle cell coverage [27, 52]. Several studies have suggested that recruitment of mural cells to blood vessels may depend in part on $\beta 1$ integrins. For example, vascular maturation and development was impaired in mutant mice in which $\beta 1$ integrin were selectively reduced in mural cells [27, 52, 136]. In other studies, blocking $\alpha 4\beta 1$ integrin interactions with VCAM-1 resulted in reduced pericyte coverage of blood vessels, increased apoptosis, and decreased angiogenesis [52]. As mentioned previously, studies have implicated myeloid cells in contributing to angiogenesis by several mechanisms including recruitment to site of tumor vascularization leading to enhanced release of proangiogenic factors. Other studies also indicate that myeloid cells may differentiate into mural cells such as pericytes, thereby contributing to vessel stability. Moreover, $\alpha 4$ integrin-deficient mice were shown to be associated with a reduction in pericyte and smooth muscle cell coverage in cranial vessels [137]. These examples highlight the roles of integrins during the maturation phase of angiogenesis.

7.16 Conclusions

As can be appreciated from our discussion, many exciting studies have provided more in depth and at times, surprising new insights into the molecular mechanisms that control angiogenesis and the roles that integrins play in these processes. These examples of how integrins may contribute to the control of angiogenesis represent only a few of the many interesting studies that implicate integrins in all the stages of new vessel development. It is rapidly becoming clear that integrins may be an important therapeutic target for the control of pathological angiogenesis. In fact, a number of clinical trials are underway to evaluate the effects of targeting this important class of multifunctional cell adhesion receptors in the treatment of malignant tumors. Thus, the exciting new molecular insight into the functions of integrins during angiogenesis will no doubt have a positive impact on the strategic use of integrin antagonists for therapeutic intervention.

Acknowledgments

This work was supported in part by grants 2ROICA91645 to PCB and a grant from Cancer Innovations Inc to PCB. Additional support was provided from grant P20RR15555 to Robert Friesel and subproject to PCB.

References

1. Folkman, J. (2007) Angiogenesis: an organizing principle for drug discovery? *Nat. Rev. Drug Discov.*, **6**, 273–286.
2. Zhang, S., Zhang, D., and Sun, B. (2007) Vasculogenic mimicry: current status and future prospects. *Cancer Lett.*, **254**, 157–164.
3. Carmeliet, P. (2005) Angiogenesis in life, disease and medicine. *Nature*, **438**, 932–936.
4. Akalu, A., Cretu, A., and Brooks, P.C. (2005) Targeting integrins for the control of tumor angiogenesis. *Expert Opin. Invest. Drugs*, **14**, 1475–1486.
5. Jakobsson, L., Kreuger, J., and Claesson-Welsh, L. (2007) Building blood vessels--stem cell models in vascular biology. *J. Cell Biol.*, **177**, 751–755.
6. Heil, M. and Schaper, W. (2004) Influence of mechanical, cellular, and molecular factors on collateral artery growth (arteriogenesis). *Circ. Res.*, **95**, 449.
7. Fischer, C., Schneider, M., and Carmeliet, P. (2006) Principles and therapeutic implications of angiogenesis, vasculogenesis and arteriogenesis. *Handb. Exp. Pharmacol.*, **176**, 157–212.
8. Gagne, P., Akalu, A., and Brooks, P.C. (2004) Challenges facing antiangiogenic therapy for cancer: impact of the tumor extracellular environment. *Expert Rev. Anticancer Ther.*, **4**, 129–140.
9. Burri, P.H., Hlushchuk, P.R., and Djonov, V. (2004) Intussusceptive angiogenesis: its emergence, its characteristics, and its significance. *Dev. Dyn.*, **231**, 474–488.
10. Hlushchuk, R., Riesterer, O., Baum, O., Wood, J., Gruber, G., Pruschy, M., and Djonov, V. (2008) Tumor recovery by angiogenic switch from sprouting to intussusceptive angiogenesis after treatment with PTK787/ZK222584 or ionizing radiation. *Am. J. Pathol.*, **173**, 1173–1185.
11. Ruoslahti, E. (2004) Vascular zip codes in angiogenesis and metastasis. *Biochem. Soc. Trans.*, **32**, 397–402.
12. Maria, L.B. and Claudio, N. (2007) Novel challenges in exploring peptide ligands and corresponding tissue-specific endothelial receptors. *Eur. J. Cancer.*, **43**, 1242–1250.
13. Ghosh, K., Thodeti, C.K., Dudley, A.C., Mammoto, A., Klagsbrun, M., and Ingber, D.E. (2008) Tumor-derived endothelial cells exhibit aberrant Rho-mediated mechanosensing and abnormal angiogenesis in vitro. *Proc. Natl. Acad. Sci. U.S.A.*, **105**, 11305–11310.
14. Lee, J.S., Semela, D., Iredale, J., and Shah, V.H. (2007) Sinusoidal remodeling and angiogenesis: a new function for the liver-specific pericyte? *Hepatology*, **45**, 817–8125.

15. Stenman, J.M., Rajagopal, J., Carroll, T.J., Ishibashi, M., McMahon, J., and McMahon, A.P. (2008) Canonical Wnt signaling regulates organ-specific assembly and differentiation of CNS vasculature. *Science*, **322**, 1247–1250.
16. Kawasaki, Y., Suzuki, J., Nozawa, R., Sakai, N., Tannji, M., Isome, M., Suzuki, H., and Nozawa, Y. (2004) FB21, a monoclonal antibody that reacts with a sialic-acid-dependent carbohydrate epitope is a marker for glomerular endothelial cell injury. *Am. J. Kidney Dis.*, **44**, 239–249.
17. Corti, A., Curnis, F., Arap, W., and Pasqualini, R. (2008) The neovasculature homing motif NGR: more than meets the eye. *Blood*, **112**, 2628–2635.
18. Kermani, P. and Hempstead, B. (2007) Brain-derived neurotrophic factor: a newly described mediator of angiogenesis. *Trends Cardiovasc. Med.*, **17**, 140–143.
19. Hallmann, R., Horn, N., Selg, M., Wendler, O., Pausch, F., and Sorokin, L.M. (2005) Expression and function of laminins in the embryonic and mature vasculature. *Physiol. Rev.*, **85**, 979–1000.
20. Comoglio, P.M., Boccaccio, C., and Trusolino, L. (2003) Interactions between growth factor receptors and adhesion molecules: breaking the rules. *Curr. Opin. Cell Biol.*, **15**, 565–571.
21. van Hinsbergh, V.W. and Koolwijk, P. (2008) Endothelial sprouting and angiogenesis: matrix metalloproteinases in the lead. *Cardiovasc. Res.*, **78**, 203–212.
22. Davis, D.E., Koh, A.W., and Stratman, N. (2007) Mechanisms controlling human endothelial lumen formation and tube assembly in three-dimensional extracellular matrices. *Birth Defects Res. Embryo Today*, **81**, 270–285.
23. Murdoch, C., Muthana, M., Coffelt, S.B., and Lewis, C.E. (2008) The role of myeloid cells in the promotion of tumour angiogenesis. *Nat. Rev. Cancer*, **8**, 618–631.
24. Colmone, A. and Sipkins, D.A. (2008) Beyond angiogenesis: the role of endothelium in the bone marrow vascular niche. *Transl. Res.*, **151**, 1–9.
25. Hristov, M., Erl, W., and Weber, P.C. (2003) Endothelial progenitor cells: mobilization, differentiation, and homing. *Trends Cardiovasc. Med.*, **3**, 201–216.
26. Gao, D., Nolan, D.J., Mellick, A.S., Bambino, K., McDonnell, K., and Mittal, V. (2008) Endothelial progenitor cells control the angiogenic switch in mouse lung metastasis. *Science*, **319**, 95–98.
27. Greenberg, J.I., Shields, D.J., Barillas, S.G., Acevedo, L.M., Murphy, E., Huang, J., Scheppke, L., Stockmann, C., Johnson, R.S., Angle, N., and Cheresh, D.A. (2008) A role for VEGF as a negative regulator of pericyte function and vessel maturation. *Nature*, **456**, 809–813.
28. Italiano, J.E., Richardson, J.L., Patel-Hett, S., Battinelli, E., Zaslavsky, A., Short, S., Ryeom, S., Folkman, J., and Klement, G.L. (2008) Angiogenesis is regulated by a novel mechanism: pro- and antiangiogenic proteins are organized into separate platelet alpha granules and differentially released. *Blood*, **111**, 1227–1233.
29. Klement, G.L., Yip, T.T., Cassiola, F., Kikuchi, L., Cervi, D., Podust, V., Italiano, J.E., Wheatley, E., Abou-Slaybi, A., Bender, E., Almog, N., Kieran, M., and Folkman, J. (2008) Platelets actively sequester angiogenesis regulators. *Blood*, **5**, 125–135.
30. Clark, R.A. (2008) Synergistic signaling from extracellular matrix-growth factor complexes. *J. Invest. Dermatol.*, **128**, 1354–1355.
31. Hutchings, H., Ortega, N., and Plouët, J. (2003) Extracellular matrix-bound vascular endothelial growth factor promotes endothelial cell adhesion, migration, and survival through integrin ligation. *FASEB J.*, **11**, 1520–1522.
32. Miyazawa, K., Shimomura, T., Naka, D., and Kitamura, N. (1994) Proteolytic activation of hepatocyte growth factor in response to tissue injury. *J. Biol. Chem.*, **269**, 8966–8970.
33. Matsumoto, T. and Claesson-Welsh, L. (2001) VEGF receptor signal transduction. *Sci. STKE*, **112**, RE21.
34. Nakamura, Y., Patrushev, N., Inomata, H., Mehta, D., Urao, N., Kim, H.W.,

Razvi, M., Kini, V., Mahadev, K., Goldstein, B.J., McKinney, R., Fukai, T., and Ushio-Fukai, M. (2008) Role of protein tyrosine phosphatase 1B in vascular endothelial growth factor signaling and cell-cell adhesions in endothelial cells. *Circ. Res.*, **102**, 1182–1191.

35. Sepp, N.T., Li, L.J., Lee, K.H., Brown, E.J., Caughman, S.W., Lawley, T.J., and Swerlick, R.A. (1994) Basic fibroblast growth factor increases expression of the alpha v beta 3 integrin complex on human microvascular endothelial cells. *J. Invest. Dermatol.*, **103**, 295–299.

36. Xu, J., Rodriguez, D., Petitclerc, E., Kim, J.J., Hangai, M., Moon, Y.S., Davis, G.E., and Brooks, P.C. (2001) Proteolytic exposure of a cryptic site within collagen type IV is required for angiogenesis and tumor growth in vivo. *J. Cell Biol.*, **154**, 1069–1079.

37. Akalu, A., Roth, J.M., Caunt, M., Policarpio, D., Liebes, L., and Brooks, P.C. (2007) Inhibition of angiogenesis and tumor metastasis by targeting a matrix immobilized cryptic extracellular matrix epitope in laminin. *Cancer Res.*, **67**, 4353–4363.

38. Cretu, A., Roth, J.M., Caunt, M., Akalu, A., Policarpio, D., Formenti, S., Gagne, P., Liebes, L., and Brooks, P.C. (2007) Disruption of endothelial cell interactions with the novel HU177 cryptic collagen epitope inhibits angiogenesis. *Clin. Cancer Res.*, **13**, 3068–3078.

39. Rabbany, S.Y., Heissig, B., Hattori, K., and Rafii, S. (2003) Molecular pathways regulating mobilization of marrow-derived stem cells for tissue revascularization. *Trends Mol. Med.*, **9**, 109–117.

40. Hamano, Y., Zeisberg, M., Sugimoto, H., Lively, J.C., Maeshima, Y., Yang, C., Hynes, R.O., Werb, Z., Sudhakar, A., and Kalluri, R. (2003) Physiological levels of tumstatin, a fragment of collagen IV alpha 3 chain, are generated by MMP-9 proteolysis and suppress angiogenesis via alpha V beta 3 integrin. *Cancer Cell.*, **6**, 589–601.

41. Davis, G.E. and Senger, D.R. (2005) Endothelial extracellular matrix: biosynthesis, remodeling, and functions during vascular morphogenesis and neovessel stabilization. *Circ. Res.*, **97**, 1093–1107.

42. Takada, Y., Ye, X., and Simon, S. (2007) The integrins. *Genome Biol.*, **8**, 215.

43. Avraamides, C.J., Garmy-Susini, B., and Varner, J.A. (2008) Integrins in angiogenesis and lymphangiogenesis. *Nat. Rev. Cancer.*, **8**, 604–617.

44. Porter, J.C., Bracke, M., Smith, A., Davies, D., and Hogg, N. (2002) Signaling through integrin LFA-1 leads to filamentous actin polymerization and remodeling resulting in enhanced T cell adhesion. *J. Immunol.*, **68**, 6330–6335.

45. Yang, J.T., Rayburn, H., and Hynes, R.O. (1993) Embryonic mesodermal defects in alpha 5 integrin-deficient mice. *Development*, **119**, 1093–1105.

46. Bader, B.L., Rayburn, H., Crowley, D., and Hynes, R.O. (1998) Extensive vasculogenesis, angiogenesis, and organogenesis precede lethality in mice lacking all alpha v integrins. *Cell*, **95**, 507–519.

47. Bouvard, D., Brakebusch, C., Gustafsson, E., Aszódi, A., Bengtsson, T., Berna, A., and Fässler, R. (2001) Functional consequences of integrin gene mutations in mice. *Circ. Res.*, **89**, 211–223.

48. Hodivala-Dilke, K.M., McHugh, K.P., Tsakiris, D.A., Rayburn, H., Crowley, D., Ullman-Culleré, M., Ross, F.P., Coller, B.S., Teitelbaum, S., and Hynes, R.O. (1999) Beta3-integrin-deficient mice are a model for Glanzmann thrombasthenia showing placental defects and reduced survival. *J. Clin. Invest.*, **103**, 229–238.

49. Mahabeleshwar, G.H., Feng, W., Phillips, D.R., and Byzova, T.V. (2006) Integrin signaling is critical for pathological angiogenesis. *J. Exp. Med.*, **203**, 2495–2507.

50. Weis, S.M., Lindquist, J.N., Barnes, L.A., Lutu-Fuga, K.M., Cui, J., Wood, M.R., and Cheresh, D.A. (2007) Cooperation between VEGF and beta 3 integrin during cardiac vascular development. *Blood*, **109**, 1962–1970.

51. Senger, D.R., Perruzzi, C.A., Streit, M., Koteliansky, V.E., de Fougerolles, A.R., and Detmar, M. (2002) The alpha(1)beta(1) and alpha(2)beta(1) integrins provide critical support for vascular endothelial growth factor signaling, endothelial cell migration, and tumor angiogenesis. *Am. J. Pathol.*, **160**, 195–204.
52. Garmy-Susini, B., Jin, H., Zhu, Y., Sung, R.J., Hwang, R., and Varner, J. (2005) Integrin alpha 4 beta1-VCAM-1-mediated adhesion between endothelial and mural cells is required for blood vessel maturation. *J. Clin. Invest.*, **115**, 1542–1551.
53. Muether, P.S., Dell, S., Kociok, N., Zahn, G., Stragies, R., Vossmeyer, D., and Joussen, A.M. (2007) The role of integrin alpha 5 beta 1 in the regulation of corneal neovascularization. *Exp. Eye. Res.*, **85**, 356–365.
54. Brooks, P.C., Clark, R.A., and Cheresh, D.A. (1994) Requirement of vascular integrin alpha v beta 3 for angiogenesis. *Science*, **264**, 569–571.
55. Friedlander, M., Brooks, P.C., Shaffe, R.W., Kincaid, C.M., Varner, J.A., and Cheresh, D.A. (1995) Definition of two angiogenic pathways by distinct alpha v integrins. *Science*, **270**, 1500–1502.
56. Xiong, J.P., Stehle, T., Zhang, R., Joachimiak, A., Frech, M., Goodman, S.L., and Arnaout, M.A. (2002) Crystal structure of the extracellular segment of integrin alpha V beta 3 in complex with an Arg-Gly-Asp ligand. *Science*, **296**, 151–155.
57. Takagi, J. and Springer, T.A. (2002) Integrin activation and structural rearrangement. *Immunol. Rev.*, **186**, 141–163.
58. Arnaout, M.A., Mahalingam, B., and Xiong, J.P. (2005) Integrin structure, allostery, and bidirectional signaling. *Annu. Rev. Cell Dev. Biol.*, **21**, 381–410.
59. Calderwood, D.A. (2004) Talin controls integrin activation. *Biochem. Soc. Trans.*, **32**, 434–437.
60. Wegener, K.L. and Campbell, I.D. (2008) Transmembrane and cytoplasmic domains in integrin activation and protein-protein interactions. *Mol. Membr. Biol.*, **25**, 76–87.
61. Takagi, J. (2007) Structural basis for ligand recognition by integrins. *Curr. Opin. Cell Biol.*, **19**, 557–564.
62. Somanat, P.R., Ciocea, A., and Byzova, T.V. (2009) Integrin and growth factor receptor alliance in angiogenesis. *Cell Biochem. Biophys.*, **53**, 53–64.
63. Serini, G., Napione, L., and Bussolino, F. (2008) Integrins team up with tyrosine kinase receptors and plexins to control angiogenesis. *Curr. Opin. Hematol.*, **15**, 235–242.
64. Sepulveda, J.L., Gkretsi, V., and Wu, C. (2005) Assembly and signaling of adhesion complexes. *Curr. Top. Dev. Biol.*, **68**, 183–225.
65. Dike, L.E. and Ingber, D.E. (1996) Integrin-dependent induction of early growth response genes in capillary endothelial cells. *J. Cell Sci.*, **109**, 2855–2863.
66. Mettouchi, A., Klein, S., Guo, W., Lopez-Lago, M., Lemichez, E., Westwick, J.K., and Giancotti, F.G. (2001) Integrin-specific activation of Rac controls progression through the G(1) phase of the cell cycle. *Mol. Cell.*, **8**, 15–27.
67. Koyama, H., Raines, E.W., Bornfeldt, K.E., Roberts, J.M., and Ross, R. (1996) Fibrillar collagen inhibits arterial smooth muscle proliferation through regulation of Cdk2 inhibitors. *Cell*, **87**, 1069–1078.
68. Vlahakis, N.E., Young, B.A., Atakilit, A., Hawkridge, A.E., Issaka, R.B., Boudreau, N., and Sheppard, D. (2007) Integrin alpha 9 beta 1 directly binds to vascular endothelial growth factor (VEGF)-A and contributes to VEGF-A-induced angiogenesis. *J. Biol. Chem.*, **282**, 15187–15196.
69. Wang, J.Y., Gualco, E., Peruzzi, F., Sawaya, B.E., Passiatore, G., Marcinkiewicz, C., Staniszewska, I., Ferrante, P., Amini, S., Khalili, K., and Reiss, K. (2007) Interaction between serine phosphorylated IRS-1 and beta1-integrin affects the stability of neuronal processes. *J. Neurosci. Res.*, **85**, 2360–2373.

70. Mori, S., Wu, C.Y., Yamaji, S., Saegusa, J., Shi, B., Ma, Z., Kuwabara, Y., Lam, K.S., Isseroff, R.R., Takada, Y.K., and Takada, Y. (2008) Direct binding of integrin alpha v beta 3 to FGF1 plays a role in FGF1 signaling. *J. Biol. Chem.*, **283**, 18066–18075.
71. Shim, W.S., Ho, I.A., and Wong, P.E. (2007) Angiopoietin: a TIE(d) balance in tumor angiogenesis. *Mol. Cancer Res.*, **5**, 655–665.
72. Chaqour, B. and Goppelt-Struebe, M. (2006) Mechanical regulation of the Cyr61/CCN1 and CTGF/CCN2 proteins. *FASEB J.*, **273**, 3639–3649.
73. Chen, C.C., Chen, N., and Lau, L.F. (2001) The angiogenic factors Cyr61 and connective tissue growth factor induce adhesive signaling in primary human skin fibroblasts. *J. Biol. Chem.*, **276**, 10443–11052.
74. Petitclerc, E., Boutaud, A., Prestayko, A., Xu, J., Sado, Y., Ninomiya, Y., Sarras, M.P., Hudson, B.G., and Brooks, P.C. (2000) New functions for non-collagenous domains of human collagen type IV. Novel integrin ligands inhibiting angiogenesis and tumor growth in vivo. *J. Biol. Chem.*, **275**, 8051–8061.
75. Wickström, S.A., Alitalo, K., and Keski-Oja, J. (2004) An endostatin-derived peptide interacts with integrins and regulates actin cytoskeleton and migration of endothelial cells. *J. Biol. Chem.*, **279**, 20178–20185.
76. Woodal, B.P., Nyström, A., Iozzo, R.A., Eble, J.A., Niland, S., Krieg, T., Eckes, B., Pozzi, A., and Iozzo, R.V. (2008) Integrin alpha 2 beta 1 is the required receptor for endorepellin angiostatic activity. *J. Biol. Chem.*, **283**, 2335–2343.
77. Aidoudi, S., Bujakowska, K., Kieffer, N., and Bikfalvi, A. (2008) The CXC-chemokine CXCL4 interacts with integrins implicated in angiogenesis. *PLoS ONE*, **3**, e2657.
78. Magnon, C., Galaup, A., Mullan, B., Rouffiac, V., Bouquet, C., Bidart, J.M., Griscelli, F., Opolon, P., and Perricaudet, M. (2005) Canstatin acts on endothelial and tumor cells via mitochondrial damage initiated through interaction with alpha v beta 3 and alpha v beta 5 integrins. *Cancer Res.*, **65**, 4353–4361.
79. Sudhakar, A. and Boosani, C.S. (2008) Inhibition of tumor angiogenesis by tumstatin: insights into signaling mechanisms and implications in cancer regression. *Pharm. Res.*, **25**, 2731–2739.
80. Soldi, R., Mitola, S., Strasly, M., Defilippi, P., Tarone, G., and Bussolino, F. (1999) Role of alpha v beta 3 integrin in the activation of vascular endothelial growth factor receptor-2. *EMBO J.*, **8**, 882–892.
81. Toledo, M.S., Suzuki, E., Handa, K., and Hakomori, S. (2005) Effect of ganglioside and tetraspanins in microdomains on interaction of integrins with fibroblast growth factor receptor. *J. Biol. Chem.*, **280**, 16227–16234.
82. Clemmons, D.R. and Maile, L.A. (2005) Interaction between insulin-like growth factor-I receptor and alpha V beta 3 integrin linked signaling pathways: cellular responses to changes in multiple signaling inputs. *Mol. Endocrinol.*, **19**, 1–11.
83. Schneller, M., Vuori, K., and Ruoslahti, E. (1997) Alpha v beta 3 integrin associates with activated insulin and PDGFbeta receptors and potentiates the biological activity of PDGF. *EMBO J.*, **16**, 5600–56007.
84. Rahman, S., Patel, Y., Murray, J., Patel, K.V., Sumathipala, R., Sobel, M., and Wijelath, E.S. (2005) Novel hepatocyte growth factor (HGF) binding domains on fibronectin and vitronectin coordinate a distinct and amplified Met-integrin induced signaling pathway in endothelial cells. *BMC Cell Biol.*, **6**, 8–14.
85. Falcioni, R., Antonini, A., Nisticò, P., Di Stefano, S., Crescenzi, M., Natali, P.G., and Sacchi, A. (1997) Alpha 6 beta 4 and alpha 6 beta 1 integrins associate with ErbB-2 in human carcinoma cell lines. *Exp. Cell Res.*, **236**, 76–85.

86. Cascone, I., Napione, L., Maniero, F., Serini, G., and Bussolino, F. (2005) Stable interaction between alpha 5 beta 1 integrin and Tie2 tyrosine kinase receptor regulates endothelial cell response to Ang-1. *J. Cell Biol.*, **170**, 993–1004.
87. Chung, J., Yoon, S.O., Lipscomb, E.A., and Mercurio, A.M. (2004) The Met receptor and alpha 6 beta 4 integrin can function independently to promote carcinoma invasion. *J. Biol. Chem.*, **279**, 32287–32293.
88. Sahni, A. and Francis, C.W. (2004) Stimulation of endothelial cell proliferation by FGF-2 in the presence of fibrinogen requires alpha v beta 3. *Blood*, **104**, 3635–3641.
89. Oinuma, I., Ishikawa, Y., Katoh, H., and Negishi, M. (2004) The Semaphorin 4D receptor Plexin-B1 is a GTPase activating protein for R-Ras. *Science*, **305**, 862–865.
90. Folgiero, V., Avetrani, P., Bon, G., Di Carlo, S.E., Fabi, A., Nisticò, C., Vici, P., Melucci, E., Buglioni, S., Perracchio, L., Sperduti, I., Rosanò, L., Sacchi, A., Mottolese, M., and Falcioni, R. (2008) Induction of ErbB-3 expression by alpha 6 beta 4 integrin contributes to tamoxifen resistance in ERbeta1-negative breast carcinomas. *PLoS ONE*, **13**, e1592.
91. Mitola, S., Brenchio, B., Piccinini, M., Tertoolen, L., Zammataro, L., Breier, G., Rinaudo, M.T., den Hertog, J., Arese, M., and Bussolino, F. (2006) Type I collagen limits VEGFR-2 signaling by a SHP2 protein-tyrosine phosphatase-dependent mechanism. *Circ. Res.*, **98**, 45–54.
92. Seo, D.W., Li, H., Guedez, L., Wingfield, P.T., Diaz, T., Salloum, R., Wei, B.Y., and Stetler-Stevenson, W.G. (2003) TIMP-2 mediated inhibition of angiogenesis: an MMP-independent mechanism. *Cell*, **114**, 171–180.
93. Wallez, Y. and Huber, P. (2008) Endothelial adherens and tight junctions in vascular homeostasis, inflammation and angiogenesis. *Biochim. Biophys. Acta*, **1778**, 794–809.
94. Gory-Fauré, S., Prandini, M.H., Pointu, H., Roullot, V., Pignot-Paintrand, I., Vernet, M., and Huber, P. (1999) Role of vascular endothelial-cadherin in vascular morphogenesis. *Development*, **126**, 2093–20102.
95. Wang, Y., Jin, G., Miao, H., Li, J.Y., Usam, S., and Chien, S. (2006) Integrins regulate VE-cadherin and catenins: dependence of this regulation on Src, but not on Ras. *Proc. Natl. Acad. Sci. U.S.A.*, **3**, 1774–1179.
96. Imamichi, Y. and Menke, A. (2007) Signaling pathways involved in collagen-induced disruption of the E-cadherin complex during epithelial-mesenchymal transition. *Cells Tissues Organs*, **185**, 80–90.
97. Chartier, N.T., Lainé, M., Gout, S., Pawlak, G., Marie, C.A., Matos, P., Block, M.R., and Jacquier-Sarlin, M.R. (2006) Laminin-5-integrin interaction signals through PI 3-kinase and Rac1b to promote assembly of adherens junctions in HT-29 cells. *J. Cell Sci.*, **119**, 31–46.
98. Wu, J.C., Yan, H.C., Chen, W.T., Chen, W.H., Wang, C.J., Chi, Y.C., and Kao, W.Y. (2008) JNK signaling pathway is required for bFGF-mediated surface cadherin downregulation on HUVEC. *Exp. Cell Res.*, **314**, 421–429.
99. Tremble, P., Damsky, C.H., and Werb, Z. (1995) Components of the nuclear signaling cascade that regulate collagenase gene expression in response to integrin-derived signals. *J. Cell Biol.*, **29**, 1707–1720.
100. Huhtala, P., Humphries, M.J., McCarthy, J.B., Tremble, P.M., Werb, Z., and Damsky, C.H. (1995) Cooperative signaling by alpha 5 beta 1 and alpha 4 beta 1 integrins regulates metalloproteinase gene expression in fibroblasts adhering to fibronectin. *J. Cell Biol.*, **129**, 867–879.
101. Langholz, O., Röckel, D., Mauch, C., Kozlowska, E., Bank, I., Krieg, T., and Eckes, B. (1995) Collagen and collagenase gene expression in three-dimensional collagen lattices are differentially regulated by alpha 1 beta 1 and alpha 2 beta 1 integrins. *J. Cell Biol.*, **131**, 1903–1915.
102. Xie, B., Laouar, A., and Huberman, E. (1998) Fibronectin-mediated cell adhesion is required for induction of

92-kDa type IV collagenase/gelatinase (MMP-9) gene expression during macrophage differentiation. The signaling role of protein kinase C-beta. *J. Biol. Chem.*, **273**, 11576–11582.

103. Faisal Khan, K.M., Laurie, G.W., McCaffrey, T.A., and Falcone, D.J. (2002) Exposure of cryptic domains in the alpha 1-chain of laminin-1 by elastase stimulates macrophages urokinase and matrix metalloproteinase-9 expression. *J. Biol. Chem.*, **277**, 13778–13786.

104. Ylipalosaari, M., Thomas, G.J., Nystrom, M., Salhimi, S., Marshall, J.F., Huotari, V., Tervahartiala, T., Sorsa, T., and Salo, T. (2005) Alpha v beta 6 integrin down-regulates the MMP-13 expression in oral squamous cell carcinoma cells. *Exp. Cell Res.*, **309**, 273–283.

105. Thomas, G.J., Poomsawat, S., Lewis, M.P., Hart, I.R., Speight, P.M., and Marshall, J.F. (2001) Alpha v beta 6 integrin upregulates matrix metalloproteinase 9 and promotes migration of normal oral keratinocytes. *J. Invest. Dermatol.*, **116**, 898–904.

106. Naito, S., Shimizu, S., Matsuu, M., Nakashima, M., Nakayama, T., Yamashita, S., and Sekine, I. (2002) Ets-1 upregulates matrix metalloproteinase-1 expression through extracellular matrix adhesion in vascular endothelial cells. *Biochem. Biophys. Res. Commun.*, **291**, 130–138.

107. Iyer, V., Pumiglia, K., and DiPersio, C.M. (2005) Alpha 3 beta 1 integrin regulates MMP-9 mRNA stability in immortalized keratinocytes: a novel mechanism of integrin-mediated MMP gene expression. *J. Cell Sci.*, **118**, 1185–1195.

108. Hangai, M., Kitaya, N., Xu, J., Chan, C.K., Kim, J.J., Werb, Z., Ryan, S.J., and Brooks, P.C. (2002) Matrix metalloproteinase-9-dependent exposure of a cryptic migratory control site in collagen is required before retinal angiogenesis. *Am. J. Pathol.*, **161**, 1429–1437.

109. Odaka, C., Tanioka, M., and Itoh, T. (2005) Matrix metalloproteinase-9 in macrophages induces thymic neovascularization following thymocyte apoptosis. *J. Immunol.*, **174**, 846–853.

110. Pernasetti, F., Nickel, J., Clark, D., Baeuerle, P.A., Van Epps, D., and Freimark, B. (2006) Novel anti-denatured collagen humanized antibody D93 inhibits angiogenesis and tumor growth: An extracellular matrix-based therapeutic approach. *Int. J. Oncol.*, **29**, 1371–1379.

111. Borrirukwanit, K., Lafleur, M.A., Mercuri, F.A., Blick, T., Price, J.T., Fridman, R., Pereira, J.J., Leardkamonkarn, V., and Thompson, E.W. (2007) The type I collagen induction of MT1-MMP-mediated MMP-2 activation is repressed by alpha V beta 3 integrin in human breast cancer cells. *Matrix Biol.*, **26**, 291–305.

112. Wang, X.Q., Su, P., and Paller, A.S. (2003) Ganglioside GM3 blocks the activation of epidermal growth factor receptor induced by integrin at specific tyrosine sites. *J. Biol. Chem.*, **278**, 48770–48778.

113. Knoblauch, A., Will, C., Goncharenko, G., Ludwig, S., and Wixler, V. (2007) The binding of Mss4 to alpha-integrin subunits regulates matrix metalloproteinase activation and fibronectin remodeling. *FASEB J.*, **21**, 497–510.

114. Wang, D.R., Sato, M., Li, L.N., Miura, M., Kojima, N., and Senoo, H. (2003) Stimulation of pro-MMP-2 production and activation by native forms of extracellular type I collagen in cultured hepatic stellate cells. *Cell Struct. Funct.* **28**, 505–513.

115. Akkawi, S., Nassar, T., Tarshis, M., Cines, D.B., and Higazi, A.A. (2006) LRP and alpha v beta 3 mediate tPA activation of smooth muscle cells. *Am. J. Physiol. Heart Circ. Physiol.*, **291**, 1351–1359.

116. Brooks, P.C., Strömblad, S., Sanders, L.C., von Schalscha, T.L., Aimes, R.T., Stetler-Stevenson, W.G., Quigley, J.P., and Cheresh, D.A. (1996) Localization of matrix metalloproteinase MMP-2 to the surface of invasive cells by interaction with integrin alpha v beta 3. *Cell*, **85**, 683–693.

117. Brooks, P.C., Silletti, S., von Schalscha, T.L., Friedlander, M., and Cheresh, D.A. (1998) Disruption of angiogenesis by PEX, a noncatalytic metalloproteinase fragment with integrin binding activity. *Cell*, **92**, 391–400.
118. Silletti, S., Kessler, T., Goldberg, J., Boger, D.L., and Cheresh, D.A. (2001) Disruption of matrix metalloproteinase 2 binding to integrin alpha v beta 3 by an organic molecule inhibits angiogenesis and tumor growth in vivo. *Proc. Natl. Acad. Sci. U.S.A.*, **98**, 119–224.
119. Bar-Shavit, R., Eskohjido, Y., Fenton, J.W., Esko, J.D., and Vlodavsky, I. (1993) Thrombin adhesive properties: induction by plasmin and heparan sulfate. *J. Cell Biol.*, **123**, 1279–1287.
120. Xue, W., Mizukami, I., Todd, R.F., and Petty, H.R. (1997) Urokinase-type plasminogen activator receptors associate with beta1 and beta3 integrins of fibrosarcoma cells: dependence on extracellular matrix components. *Cancer Res.*, **57**, 1682–1689.
121. Lechner, A.M., Assfalg-Machleidt, I., Zahler, S., Stoeckelhuber, M., Machleidt, W., Jochum, M., and Nägler, D.K. (2006) RGD-dependent binding of procathepsin X to integrin alpha v beta 3 mediates cell-adhesive properties. *J. Biol. Chem.*, **281**, 39588–39597.
122. Gálvez, B.G., Matías-Román, S., Yáñez-Mó, M., Sánchez-Madrid, F., and Arroyo, A.G. (2002) ECM regulates MT1-MMP localization with beta1 or alpha v beta 3 integrins at distinct cell compartments modulating its internalization and activity on human endothelial cells. *J. Cell Biol.*, **159**, 509–521.
123. Stefanidakis, M., Ruohtula, T., Borregaard, N., Gahmberg, C.G., Koivunen, E., Stefanidakis, M., Ruohtula, T., Borregaard, N., Gahmberg, C.G., and Koivunen, E. (2004) Intracellular and cell surface localization of a complex between alpha M beta 2 integrin and promatrix metalloproteinase-9 progelatinase in neutrophils. *J. Immunol.*, **172**, 7060–7068.
124. Morini, M., Mottolese, M., Ferrari, N., Ghiorzo, F., Buglioni, S., Mortarini, R., Noonan, D.M., Natali, P.G., and Albini, A. (2000) The alpha 3 beta 1 integrin is associated with mammary carcinoma cell metastasis, invasion, and gelatinase B (MMP-9) activity. *Int. J. Cancer*, **87**, 336–342.
125. Björklund, M., Heikkilä, P., and Koivunen, E. (2004) Peptide inhibition of catalytic and noncatalytic activities of matrix metalloproteinase-9 blocks tumor cell migration and invasion. *J. Biol. Chem.*, **279**, 29589–29897.
126. Dumin, J.A., Dickeson, S.K., Stricker, T.P., Bhattacharyya-Pakrasi, M., Roby, J.D., Santoro, S.A., and Parks, W.C. (2001) Pro-collagenase-1 (matrix metalloproteinase-1) binds the alpha(2)beta(1) integrin upon release from keratinocytes migrating on type I collagen. *J. Biol. Chem.*, **276**, 29368–29374.
127. Bridges, L.C. and Bowditch, R.D. (2005) ADAM-integrin interactions: potential integrin regulated ectodomain shedding activity. *Curr. Pharm. Des.*, **11**, 837–847.
128. Bergers, G., Brekken, R., McMahon, G., Vu, T.H., Itoh, T., Tamaki, K., Tanzawa, K., Thorpe, P., Itohara, S., Werb, Z., and Hanahan, D. (2000) Matrix metalloproteinase-9 triggers the angiogenic switch during carcinogenesis. *Nat. Cell Biol.*, **2**, 737–744.
129. Du, R., Lu, K.V., Petritch, C., Liu, P., Ganss, R., Passwque, E., Song, H., Vandenberg, S., Johnson, R.S., Werb, Z., and Bergers, G. (2008) HIF1 alpha induces the recruitment of bone-morrow derived vascular modulatory cells to regulate tumor angiogenesis. *Cancer Cell*, **13**, 206–220.
130. Jin, H., Su, J., Garmy-Susini, B., Kleeman, J., and Varner, J. (2006) Integrin alpha 4 beta 1 promotes monocyte trafficking and angiogenesis in tumors. *Cancer Res.*, **66**, 2146–2152.
131. Margadant, C., Raymond, K., Kreft, M., Sachs, N., Janssen, H., and Sonnenberg, A. (2009) Integrin alpha 3 beta 1 inhibits directional migration and wound re-epithelialization in the skin. *J. Cell Sci.*, **122**, 278–288.

132. Kanasaki, K., Kanda, Y., Palmsten, K., Tanjore, H., Lee, S.B., Lebleu, V.S., Gattone, V.H. Jr., and Kalluri, R. (2008) Integrin beta1-mediated matrix assembly and signaling are critical for the normal development and function of the kidney glomerulus. *Dev. Biol.*, **313**, 584–593.
133. Somanath, P.R., Kandel, E.S., Ha, N., and Byzova, T.V. (2007) Akt1 signaling regulates integrin activation, matrix recognition, and fibronectin assembly. *J. Biol. Chem.*, **82**, 22964–22976.
134. Lohikangas, L., Gullberg, D., and Johansson, S. (2001) Assembly of laminin polymers is dependent on beta1-integrins. *Exp. Cell Res.*, **265**, 135–144.
135. Fleischmajer, R.S., Perlish, R.S., MacDonald, E.D., Schechter, A., Murdoch, A.D., Iozzo, R.V., and Yamada, Y. (1998) There is binding of collagen IV to beta 1 integrin during early skin basement membrane assembly. *Ann. N. Y. Acad. Sci.*, **857**, 212–227.
136. Abraham, S., Kogata, N., Fässler, R., and Adams, R.H. (2008) Integrin beta1 subunit controls mural cell adhesion, spreading, and blood vessel wall stability. *Circ. Res.*, **102**, 562–570.
137. Grazioli, A., Alves, C.S., Konstantopoulos, K., and Yang, J.T. (2006) Defective blood vessel development and pericyte/pvSMC distribution in alpha 4 integrin-deficient mouse embryos. *Dev. Biol.*, **293**, 165–177.

8
Signaling Pathways in Tumor Angiogenesis[1]

Cristina Abrahams, Christopher Daly, Alexandra Eichten, Zhe Li, Irene Noguera-Troise, and Gavin Thurston

8.1
Introduction

The growth of solid tumors depends on the growth and development of a vascular system within the tumor, a process known as *tumor angiogenesis* [1, 2]. Because of the dependence of tumors on growing blood vessels, much work in the past two decades has focused on the signaling pathways that mediate tumor angiogenesis. Although tumor vessels are chaotic and abnormal, many of the angiogenic signaling pathways that are required for their growth are also required for the development of the normal vascular system during embryogenesis. Thus, a close parallel has developed in which studies on tumor angiogenesis have shed light on developmental angiogenesis, and vice versa.

In this chapter, we review several of the major signaling pathways that show promise as targets for tumor angiogenesis. We first review the vascular endothelial growth factor (VEGF) pathway, which is a primary upstream activator of angiogenesis and which has been targeted by numerous approved and late-stage therapeutic agents. We next summarize the Delta–Notch and angiopoietin–Tie pathways, which are important signaling pathways within the vasculature, which act, at least in part, to regulate VEGF signals. Several potential therapeutic agents to target these pathways are in preclinical and clinical development. Finally, we briefly summarize the transforming growth factor-β (TGF)-β and ephrin–Eph pathways, two systems that play well-documented roles in developmental angiogenesis and appear to play roles in tumor angiogenesis, yet are earlier in their therapeutic application. This set of signaling pathways includes those that are the most advanced from a therapeutic perspective, but is only a partial list of the available targets, since several other angiogenic pathways, including Slit-Robo and netrin-Unc, are also being intensely studied as potential tumor targets.

1) All authors contributed equally to this manuscript.

Tumor Angiogenesis – From Molecular Mechanisms to Targeted Therapy. Edited by Francis S. Markland, Stephen Swenson, and Radu Minea
Copyright © 2010 WILEY-VCH Verlag GmbH & Co. KGaA, Weinheim
ISBN: 978-3-527-32091-2

8.2
VEGF Pathway

8.2.1
Receptors, Ligands, and Signaling Pathway

The VEGF pathway is the major mediator for *de novo* formation of blood vessels in physiological and pathological conditions. There are five secreted ligands in the mammalian VEGF gene family: VEGF-A, VEGF-B, VEGF-C, VEGF-D, and placental growth factor (PGF, or PlGF) (Figure 8.1a). Within each of these

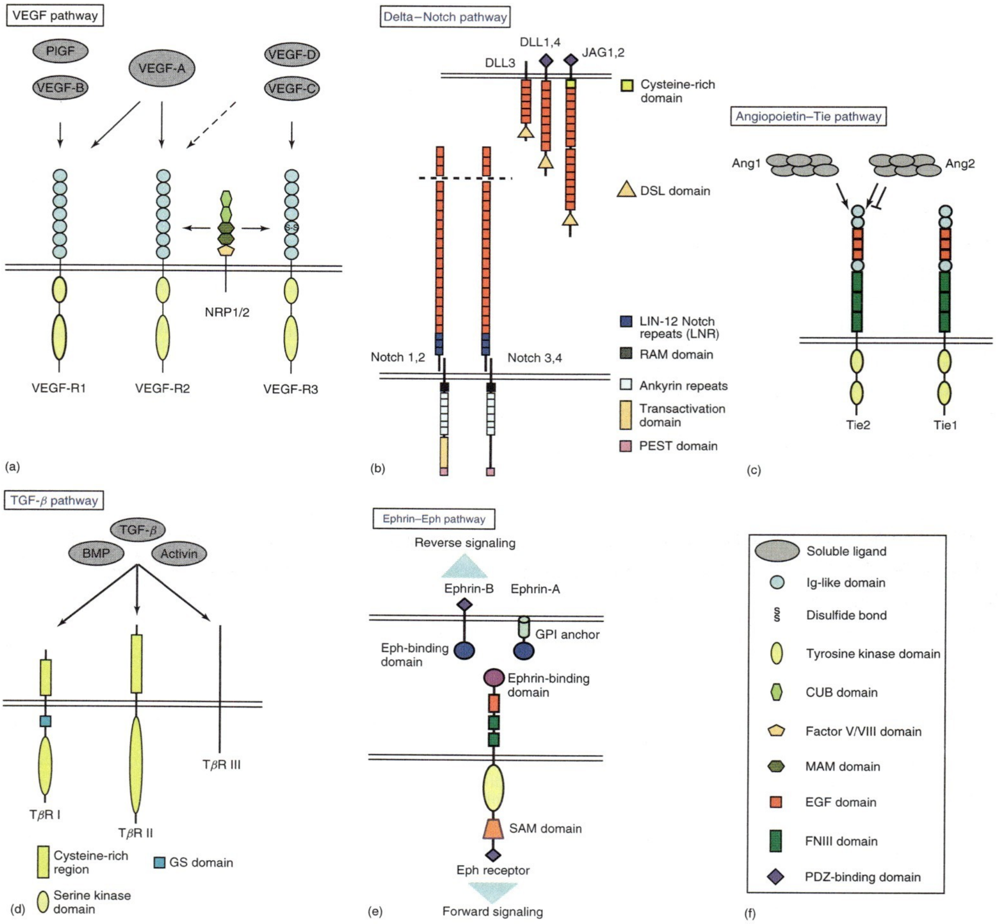

Figure 8.1 Diagram of ligand and receptor components in the angiogenic signaling pathways described in this review: (A) VEGF, (B) Delta-Notch, (C) Angiopoietin-Tie2, (D) TGF-β, and (E) Ephrin. The legend for the symbols used to denote various common protein domains is shown in panel (F).

genes, alternative splicing generates multiple VEGF ligand isoforms with varied biochemical properties, mainly the degree to which the ligands interact with the extracellular matrix. VEGF ligands preferentially form homodimers and exert their angiogenic effects by binding predominantly to three receptors: VEGF-R1 (Flt-1), VEGF-R2 (KDR or Flk-1), and VEGF-R3 (Flt-4). VEGFRs consist of seven immunoglobulin-like domains, a single transmembrane spanning region, and an intracellular portion containing a split tyrosine kinase domain. Unlike the ligands, whose expression is broad and not restricted to any particular cell type, the VEGF receptors are mainly expressed by cells in the vascular system. Expression of VEGF-R1 and VEGF-R2 is most prominent on vascular endothelial cells, while VEGF-R3 is expressed by lymphatic endothelial cells. In general, VEGF-A binds to VEGF-R2 and VEGF-R1, VEGF-C and VEGF-D interact with VEGF-R3, and VEGF-B and PlGF bind to VEGF-R1. However, VEGF ligands can have somewhat overlapping receptor binding properties depending on cellular context and/or proteolytic processing of VEGF isoforms. In addition to the tyrosine kinase receptors, VEGF ligands can also interact with coreceptors Neuropilin 1 (NRP1) and NRP2 [3]. NRPs were originally identified as semaphorin receptors involved in neuronal guidance [4, 5]. It was later shown that NRPs bind certain VEGF isoforms, which in turn form complexes with VEGF receptors to modulate VEGF signaling [6].

VEGF signaling is initiated by ligand binding, which consequently leads to receptor dimerization, phosphorylation on multiple intracellular tyrosine residues, and transmission of the signal into the cell. Intracellular propagation of the signal translates into diverse endothelial cell functions including proliferation, cell survival, transcriptional activation, migration, and vascular leakage. These different biological outcomes are modulated by a variety of mechanisms. At the receptor level, formation of different receptor dimerization partners, bioavailability of certain ligands, and ligand affinity may contribute to altering the signaling response [7]. For example, VEGF-A signals mainly through VEGF-R2 to mediate angiogenesis; however, other VEGF receptors can also play a role in modulating VEGF signaling. VEGF-R1 is thought to be a decoy receptor for VEGF-A; although VEGF-R1 binds VEGF-A with higher affinity than VEGF-R2, it has weak kinase activity. VEGF-R2 and VEGF-R1 can also form heterodimers that interact with VEGF-A. Furthermore, binding of VEGF-A to NRP1 is thought to enhance VEGF-R2 activity [8]. It has been demonstrated that VEGF-R2-induced activation of the Ras/Raf/mitogen-activated protein kinase (MAPK) cascade stimulates proliferation of endothelial cells [9]. Meanwhile, phosphoinositide-3-kinase (PI3K)/Akt signaling mediates survival of endothelial cells [10]. Many other signal transducing molecules have been identified including PLCγ, SHP2, Nck, and STATs [9].

8.2.2
Requirement for VEGF Pathway during Early Embryogenesis

Studies using knockout mice highlight the functional relevance of the VEGF system in vasculogenesis, angiogenesis, and lymphangiogenesis. It is noteworthy that genetic deletion of individual VEGF ligands and receptors often results in embryonic lethality. The significance of VEGF-A is underscored by the fact that disruption of one VEGF-A allele is lethal at embryonic day (E) 11–12 due to vascular abnormalities [11, 12]. Targeted inactivation of both VEGF-A alleles results in more severe defects in vascular development and mice die at E9.5–10.5 [11]. Loss of VEGF-A impinges on most processes during early vascular development including an almost total lack of blood vessels [11, 12]. The requirement for VEGF-A continues from embryonic to perinatal development, but diminishes as mice mature. Similarly, VEGF-R2 null mouse embryos resemble VEGF-A knockout animals, and die at E8.5–9.5 as a result of severely impaired vascular development [13]. Unlike the phenotype of VEGFR2 deficient mice, inactivation of VEGFR1 is embryonically lethal due to excessive growth of endothelial cells, suggesting an opposing role for VEGF-R1 in regulating vessel growth [14]. Mice with targeted genetic deletions of VEGF-C or VEGF-R3 die as a result of lack of lymphatic vessels and cardiovascular failure, respectively [15, 16]. Loss of NRP1 or NRP2 also leads to lethality owing to vascular regression and severe reduction of lymphatic vessels and capillaries [17, 18]. In the case of PlGF, VEGF-B, or VEGF-D, genetic disruptions have little effect on normal vascular or lymphatic development, possibly indicating functional redundancy within members of the VEGF signaling pathway [19–21].

8.2.3
Multiple Roles of VEGF Signaling in Tumor Angiogenesis

In pathological settings including tumors, VEGF-A is a potent angiogenic factor (reviewed in [22]) that stimulates proliferation, migration, and acts a survival factor for VEGF-R2-expressing cells. Further, VEGF alters vascular morphology and increases vascular leakage. In addition, VEGF-A has been reported to play a role in the recruitment of bone marrow derived cells, affecting chemotaxis of monocytes and inflammatory cells to the tumor site. Owing to the pleiotropic biological activities exerted by VEGF-A, it is no surprise that VEGF expression accelerates tumor growth and progression. In fact, elevated expression levels of VEGF-A have been reported in many human cancers and are often associated with poor prognosis [23, 24]. Conversely, tumors that do not produce VEGF-A have fewer blood vessels, resulting in significantly smaller tumors compared to their VEGF-expressing counterparts [25].

The remaining VEGF ligands have been reported to regulate various aspects of tumor vasculature. PlGF has been demonstrated to induce proangiogenic effects including tumor growth, angiogenesis, leukocyte infiltration, and stromal cell migration (review in [26]). These effects may be due to direct actions of PlGF on VEGF-R1, or could be indirect via increasing the bioavailability of VEGF-A for VEGF-R2. The role of VEGF-B in pathological angiogenesis is not well understood, but is suggested to be involved in tumor cell migration and invasion [27]. Numerous studies have established a role of VEGF-C and VEGF-D in stimulating tumor lymphangiogenesis, and consequently facilitating the dissemination of metastatic cells via the lymphatic system (reviewed in [28, 29]). It has also been shown that interaction of NRP2 with VEGF-C signaling may play a role in tumoral lymphangiogenesis [29]. Similar to VEGF-A, expression levels of these other ligands have also been documented to be generally higher in tumors compared to normal tissues [26, 29].

8.2.4 VEGF as a Therapeutic Target

The requisite role of VEGF in tumor angiogenesis and lymphangiogenesis has made it an attractive target for pharmacological inhibition. Multiple strategies have been employed to inhibit VEGF signaling. Monoclonal antibodies, neutralizing peptides, and soluble decoy receptors provide blockade of specific target molecule(s), most commonly VEGF-A or VEGF-R2. On the other hand, broad-spectrum agents like tyrosine kinase inhibitors (TKIs) can block multiple targets such as all three VEGF receptors and other protein kinases. Direct inhibition of VEGF or VEGF-R2 leads to potent inhibition of angiogenesis in many tumor models [30]. Specifically, VEGF-A or VEGF-R2 blockade causes vascular pruning and decreased vascular density, ultimately suppressing tumor growth. In addition to hindering new blood vessel formation, VEGF-A/VEGF-R2 inhibitors also affect vascular function by modulating blood flow and decreasing vascular permeability. Although inhibition of the VEGF-A/VEGF-R2 axis has shown the most promise to date, suggestive results have also been reported for antiangiogenic therapies that target other members of the VEGF signaling pathway. Antibodies against VEGF-R1 [31], PlGF [32], and NRP1 [33] block tumor angiogenesis and abrogate or delay growth progression in certain tumor models. Meanwhile, inhibitors of lymphangiogenesis such as anti-VEGF-C and VEGF-D antibodies [34], VEGF-R3-soluble receptors or TKIs [35], and anti-NRP2 antibodies [36] reduce functional tumor lymphatics, consequently decreasing the incidence of metastasis in tumor models. These lymphangiogenic inhibitors represent a novel class of therapeutics that take aim at decreasing the incidence of tumor metastasis.

So far, three agents that work primarily by inhibiting angiogenesis have obtained FDA approval for the treatment of human cancer: bevacizumab (Avastin, Genentech/Roche), sorafenib (Nexavar, Bayer), and sunitinib (Sutent, Pfizer). The humanized anti-VEGF monoclonal antibody, bevacizumab, was first reported to suppress the growth of human A673 rhabdomyosarcomas and MDA-MD-435 breast carcinomas in nude mice [37]. Today, bevacizumab is approved for patients with metastatic breast cancer, metastatic colorectal cancer, and non–small cell lung carcinoma (NSCLC). Kinase inhibitors (KIs) sorafenib and sunitinib, which target VEGF receptors, are approved as monotherapies for treatment of renal cell carcinomas. Additionally, sunitinib therapy is also approved for gastrointestinal stromal tumors, and sorafenib for hepatocellular carcinomas. Greater antitumor effects are observed in several tumor types when anti-VEGF agents are administered in combination with standard chemotherapy than either treatment alone. Bevacizumab in combination with 5-fluorouracil-based chemotherapy increases median survival and extends progression-free survival of colorectal cancer patients [38]. Improved progression-free survival and higher partial response rates are also observed for patients with NSCLC treated with bevacizumab and paclitaxel [39]. The increased efficacy using combination treatments may be due to the vascular pruning effects of anti-VEGF agents, which are thought to remodel the tortuous, abnormal tumor vessels into "normal" vessels. It is hypothesized that this "vessel normalization" may improve vessel flow and tissue oxygenation, and hence enhance delivery of cytotoxic agents into the tumor mass [40]. As a general class of therapeutic agents, VEGF blockers have been associated with some consistent side effects such as increased blood pressure and proteinuria, as well as some rare but serious side effects including gastrointestinal perforation and thromboembolic events. These may be in part due to anti-VEGF blocking normal physiological functions modulated by VEGF signaling.

It is important to note that anti-VEGF agents have also been shown to fail in some cancer patients. Some tumors do not respond to treatment, and are thus inherently resistant to anti-VEGF therapy. Inherent resistance may be attributed to preexisting microenvironmental conditions or redundant pathways that protect the tumors from anti-VEGF blockers [41]. The ability of tumors to mobilize and recruit myeloid cells is one mechanism that has been suggested to enable resistance to anti-VEGF therapy [42]. In other cases, tumors can lose sensitivity to anti-VEGF agents and reestablish tumor growth and vascularization. Mechanisms of adaptive resistance include upregulation of alternative proangiogenic factors or recruitment of proangiogenic bone marrow–derived vascular progenitor cells (reviewed in [41]). For example, in the RIP-Tag transgenic mouse model of pancreatic islet cell carcinoma, refractoriness to VEGF-R2 blockade may develop as a result of the upregulation of fibroblast growth factor [43]. The efficacy of anti-VEGF treatments and resistance mechanisms probably vary depending on tumor type and/or tumor context. Therefore, development of multimodality approaches to interfere with multiple angiogenesis pathways is of major clinical interest. For example, aflibercept (VEGF Trap, Regeneron) is a soluble decoy receptor with high affinity for VEGF, PlGF, and VEGF-B [44]. Aflibercept has demonstrated potent antitumor effects in various preclinical tumor models including ovarian cancer, renal cell cancer, pancreatic islet carcinomas, gliomas, and melanomas [44–48]. Improved results were also observed when aflibercept was administered with chemotherapy or radiation therapy [49, 50]. Specifically, combination treatment with aflibercept and paclitaxel resulted in a significant decrease in tumor burden and prolonged survival benefit in a mouse model of human ovarian cancer [49]. Combination regiments that block multiple proangiogenic pathways hold promise for improving treatment efficacy and clinical benefit for cancer patients.

8.3 Delta–Notch Pathway

8.3.1 Receptors, Ligands, and Signaling Pathway

The Delta/Jagged–Notch system is a signaling pathway comprising ligands and receptors that are both modular single-pass transmembrane proteins, which together establish short-range communication between adjacent cells. Delta–Notch signaling can occur between cells of the same type (e.g., endothelial–endothelial cell communication) or involve two different cell types (e.g., endothelial-smooth muscle cell communication). The signals transduced through this pathway have an essential role in cell fate decisions during embryonic development and in normal tissue homeostasis. In general, Notch signaling regulates cell fate decisions, such as promoting cell differentiation or inducing proliferation and maintaining an undifferentiated cellular fate.

Genes for Notch ligands and receptors are found in all metazoa, from sea urchins to humans. In this review we will focus on the mammalian genes/proteins and, for simplicity, only on the well-established canonical signaling pathway. There are five ligands (Delta-like ligand (Dll) 1, 3, and 4 and Jagged 1 and 2) and four receptors (Notch 1–4) (Figure 8.1b). All ligands are modular proteins containing a cysteine-rich Delta, Serrate, Lag2 (DSL) domain, flanked by a short N-terminal domain (also referred to as module at N-termini of Notch ligands (MNNLs)) followed by a variable number of epidermal growth factor (EGF-like repeats, 6–8 for Delta's and 14 for Jagged's) motifs that precede the transmembrane region. The Jagged proteins also have an additional cysteine-rich region with sequence similar to a domain found in von Willebrand factor type C domain [51]. The DSL region together with the N-terminal and the first two EGF-like domains appears to be responsible for binding to the Notch receptor [51, 52]. The C-terminal portion of the ligands comprises a short intracellular region that is the most divergent among the Notch ligands, with many, but not all, having a stretch of lysine residues and a postsynaptic density, discs large, zonula occludens-1, (PDZ)-binding motif that is proposed to interact with the cytoskeleton [51].

The Notch receptors are large proteins that contain an extracellular domain composed of a large number of EGF-like repeats (29–36). In mammals, the minimal region of Notch that is necessary and sufficient for interaction with the ligands is not well defined, but apparently includes EGF-like repeats 1–15, as has been reported for a soluble Jagged 1 and an N-terminal fragment of Notch 2 [52]. EGF-domains 21–30 compose a region where many mutations have been found that appear to result in constitutive activation of Notch signaling. Between the ligand-binding domain and the transmembrane region, Notch receptors have a negative regulatory region (NRR) consisting of three LIN-12 Notch repeats (LNRs) and a heterodimerization domain (HD) preceding the transmembrane region. The LNR region "protects" Notch from accidental activation in the absence of ligand. The HD domain contains furin and metalloprotease cleavage sites that are exposed once the receptor binds ligands. In T-cell acute lymphoblastic leukemia, Notch mutations found have been identified within the NRR that allow constitutive activation [53].

Notch receptors are positioned at the cell surface as cleaved heterodimers with the extracellular domain still associated to a membrane-tethered intracellular domain. Direct binding of Delta/Jagged ligands to Notch receptors triggers two regulated proteolytic events: an A Disintegrin and Metalloproteinase (ADAM)/TACE mediated cleavage followed by a γ-secretase mediated cleavage, with subsequent release of the intracellular domain of the Notch receptor (NICD) [54]. After proteolytic cleavage, the NICD translocates into the nucleus where it assembles into a transcription activation complex together with recombination signal-binding protein for immunoglobulin kappa J region (RBPJ) and further recruiting of MAML1. The resultant complex promotes transcription of target genes such as the basic helix loop helix family of transcription factors Hes1 (hairy and enhancer of split 1) and Hey1/2 (hairy or enhancer of split related with YRPW motif proteins). The Notch target genes can apparently be divided into a set of "consensus" target genes, which are common across many cell types, and a set of "specific" target genes, which depend on the cell type or context. Further work is needed to fully characterize these gene sets *in vivo*.

8.3.2
Requirement for Delta–Notch Pathway during Embryogenesis

Various Delta/Jagged Notch genes are essential for normal vascular development, as demonstrated by gene ablation studies that have produced severe vascular defects and embryonic lethality in the mouse and zebrafish. From these studies it has become apparent that the Notch pathway displays its actions after the formation of the primitive vascular plexus, during the process of vessel remodeling needed to form a hierarchic vascular network. Notch1 might also have an earlier role by driving cell fate decisions in the hemangioblast stage affecting endothelial versus hematopoietic fates. Genetic deletion of various Notch pathway genes, including Notch1, Dll4, ADAM10, RBPJ, and Hey1 plus Hey2 double knockout, all lead to vascular defects and lethality at E9.5–10.5 [55]. Further, Notch4 null mice show no developmental defects in spite of prominent vascular expression. However, double-mutant Notch1 + Notch4 embryos display more pronounced vascular defects than the Notch1 null, suggesting that indeed Notch4 might have a role in vascular development. Moreover, abnormal overexpression of Notch 4 results in embryos with severely dilated vessels and disorganized vascular networks, indicating that the precise dosage of Notch4 is essential for the establishment of a normal vasculature. In addition, Notch2, Dll1, and Jagged1 seem to play a role later in development since embryonic lethality is observed at E11.5–12 at late stages of vessel remodeling [55]. Deletion of Notch3 in mice does not result in embryonic lethality, but instead these mice display abnormal differentiation of vascular arterial smooth muscle cells and thus still reveal an important role in establishing a functional vasculature. Among the Notch receptors and ligands, Dll4 displays restricted vascular expression [56]. Similar to the Notch1 mutants, Dll4 knockout mice die at E9.5, mainly because of severe vascular defects that include failure to remodel the primitive vascular plexus in embryonic and extraembryonic tissues (placenta and yolk sac). These mice also show defective arterial branching and stenosis of large arteries.

8.3.3
Role of Delta–Notch Signaling in Tumor Angiogenesis

In human tumors and in preclinical mouse tumor models, Dll4 is specifically and strongly expressed in the tumor vasculature compared to that of mature vessels in the surrounding normal tissue. Examples of enhanced expression of Dll4 in human tumors have been reported in clear cell renal carcinomas and bladder tumors [57, 58]. Moreover, manipulation of Dll4 in tumors resulted in striking vascular changes. Activation of Notch signaling, by overexpression of the full-length active Dll4 in tumor cells, produced relatively straight and unbranched vessels that lacked sprouts [59]. Reciprocally, inhibition of Dll4–Notch signaling produced aberrant, nonproductive tumor vessels, which was associated with reduced tumor perfusion and growth. These experiments provide rationale for the pharmacologic inhibition of Dll4–Notch in tumor vessels, as a means to treat solid tumors.

8.3.4
Delta–Notch as a Therapeutic Target

There are several approaches to inhibit Notch signaling. One approach is to inhibit Notch–ligand interactions, using either decoy ligands/receptors [59, 60] or specific monoclonal antibodies that bind to the ligands and block their interaction with the receptors [59, 61, 62]. Another approach is to develop antibodies that bind to Notch receptors and block the conformational change necessary for efficient cleavage/activation of the receptor [63]. Another approach is through the use of small-molecule γ-secretase inhibitors, which are currently being tested in clinical trials with tamoxifen in early and advanced breast cancers (e.g., developed by Merck, MK-0752) [64]. The potency and specificity of these approaches will need to be evaluated in preclinical and early clinical studies.

8.4
Angiopoietin–Tie Pathway

8.4.1
Receptors, Ligands, and Signaling Pathway

Angiopoietins are a family of secreted ligands that signal through the endothelial cell–specific receptor tyrosine kinase Tie2. A structurally related endothelial cell–specific kinase, Tie1, does not directly interact with angiopoietins but it appears to modulate angiopoietin signaling through Tie2. Tie1 and Tie2 have a similar extracellular domain structure (overall amino acid sequence identity is ~30%) consisting of two immunoglobulin-like domains at the N-terminus followed by three EGF repeats, a third immunoglobulin-like domain, and then three fibronectin type III domains [65]. The cytoplasmic region of both receptors contains a split tyrosine kinase domain (amino acid identity in the cytoplasmic domains is ~80%).

There are three genuine angiopoietin proteins that bind to Tie2 receptors (Figure 8.1c): in humans, these are known as *Ang-1*, *Ang-2*, and *Ang-4* (the mouse ortholog of the latter is known as *Ang-3*). This discussion will be limited to Ang-1 and Ang-2, which are much better characterized. Ang-1 was isolated by expression cloning as an activating ligand for Tie2 that is widely expressed by cells surrounding blood vessels [66]. The ~500 amino acid Ang-1 protein contains an N-terminal coiled coil domain that mediates Ang-1 multimerization followed by a C-terminal fibrinogen-like domain that mediates binding to Tie2.

Subsequent to the identification of Ang-1, Ang-2 was cloned by homology screening [67]. The Ang-2 protein has the same overall domain structure as Ang-1 and is ~60% identical. However, although Ang-2 binds Tie2 with an affinity similar to that of Ang-1, it does not, in general, elicit robust phosphorylation of Tie2 on endothelial cells, and when present in excess, Ang-2 can inhibit Ang-1-dependent Tie2 activation [67]. In addition to different Tie2 activation properties, Ang-2 also exhibits a very different expression pattern from that of Ang-1. Ang-2 is expressed primarily in endothelial cells themselves rather than in vascular support cells, and its expression is rather restricted during development [67]. In the adult, Ang-2 does not appear to be expressed by most endothelial cells, but its expression is highly upregulated in angiogenic vessels, for example, in tumors and the remodeling ovary [68]. The combination of the Ang-2 expression data and its apparent function as a Tie2 antagonist led to the hypothesis that Ang-2 is induced in remodeling endothelial cells to weaken the vessel stabilizing effects of Ang-1/Tie2 signaling, thus allowing for subsequent vessel sprouting or regression [68]. Several lines of evidence have challenged this original model and will be discussed further below.

Like other receptor tyrosine kinases, Tie2 undergoes phosphorylation on tyrosine following ligand binding. Studies with angiopoietin ligands engineered to be of precise multimeric structure indicate that Tie2 activation requires at least a ligand tetramer; this is consistent with the finding that native Ang-1 and Ang-2 exist as a mixture of tetramers and higher-order multimers [69]. The different Tie2 activation properties of Ang-1 and Ang-2 do not appear to reflect different multimeric structures.

Once phosphorylated, Tie2 recruits and activates a variety of cytoplasmic signaling proteins, initiating the process of signal transduction. The primary cell biological effect of Tie2 activation appears to be inhibition of apoptosis via strong activation of the PI3K/Akt survival pathway [70, 71]. Interestingly, when compared directly to VEGF and bFGF, Ang-1 is a more potent activator of Akt but is a weaker activator of ERK [72], consistent with Ang-1 having strong prosurvival effects while being a weak stimulator of proliferation. Two recent papers make the provocative suggestion that downstream signaling of Tie2 activation is influenced by the subcellular localization of the receptor, such that Ang-1 elicits stronger Akt signaling when cells are confluent and Tie2 is localized at cell–cell junctions, and stronger ERK signaling when cells are sparse and Tie2 is localized at sites of cell–substratum contacts [73, 74]. Thus, the effects of Tie2 activation on endothelial cell function may depend on the nature of the cell–cell and cell–matrix contacts.

While Tie1 does not directly bind Ang-1 or Ang-2 or possess a highly active kinase domain [75], it does become phosphorylated following Ang-1 stimulation of endothelial cells [76]. This may reflect the formation of Tie1–Tie2 complexes and Tie2-dependent phosphorylation of Tie1. Knockdown of Tie1 using siRNA indicates that it is not required for Ang-1-dependent activation of major signaling pathways like Akt and extracellular signal-regulated kinases (ERK) [77]. Thus, the functional significance of Tie1 in angiopoietin signaling remains unclear.

8.4.2
Requirement for Angiopoietin–Tie Pathway during Embryogenesis

Mice lacking Tie2 die by ~E10.5 and exhibit prominent heart defects and defective remodeling/maturation of the primary vascular plexus, indicating a role for Tie2 in embryonic angiogenesis [78]. Further, while supporting a critical role for Tie2 in regulation of blood vessel morphology, humans with an activating point mutation in the Tie2 kinase domain exhibit venous malformations characterized by enlarged vessels with an increased number of endothelial cells [79]. In Tie1-null mice, vascular patterning is not noticeably affected, but these mice do exhibit edema and hemorrhage, suggesting a role for Tie1 in promoting vascular integrity [78, 80].

Consistent with a role for Ang-1 as an essential Tie2 activator, Ang-1 null mice die at ~E12.5 and exhibit defects in vascular remodeling/maturation that are reminiscent of those seen in Tie2 knockout mice [81]. In contrast, and consistent with the lack of widespread expression of Ang-2 during development, genetic studies have shown that Ang-2 is not required for embryonic vascular development, but is required for subsequent vascular remodeling in the postnatal retina [82] and for proper formation of a functional lymphatic vasculature. Apparently confirming the role of Ang-2 as a Tie2 antagonist, transgenic mice in which Ang-2 is overexpressed in endothelial cells die at E9.5–10.5 and exhibit vascular defects similar to those seen in Tie2 and Ang-1 null mice [67]. Overall, the genetic data has clearly established that Ang-1/Tie2 signaling is critical for later stages of embryonic vascular development.

8.5
Ang-1 as a Promoter of Vascular Stability/Quiescence

While VEGF is known to promote sprouting angiogenesis and a leaky vasculature, Ang-1 promotes vessel enlargement in the absence of sprouting. The vessel enlarging effect of Ang-1 is limited in most tissues to the first postnatal month and appears to reflect increased endothelial cell proliferation (whether the proliferation is a direct effect of Ang-1 is unclear, however it does not require VEGF) [83]. Importantly, vessels exposed to Ang-1 are resistant to leak induced by VEGF or inflammatory agents [84]. The antileak effect is observed upon acute administration of Ang-1, well before the vessel enlargement effect is apparent; thus, these effects are separable [85]. In addition to its antipermeability effect, Ang-1 appears to limit vascular activation/dysfunction in multiple settings. For example, Ang-1 inhibits leukocyte infiltration in multiple animal models (diabetic retinopathy [86], sepsis [87], cardiac allograft arteriosclerosis [88], acute lung injury [89]) and promotes endothelial cell survival (and animal survival) in response to whole body irradiation [90]. In addition, there is evidence that Ang-1 has beneficial effects in a monocrotaline-induced model of pulmonary hypertension [91] and that it lowers blood pressure in spontaneously hypertensive rats [92]. Together, these findings suggest a role for Ang-1 in maintaining a quiescent, well-functioning vasculature.

Although the mechanisms underlying these effects of Ang-1 remain to be fully elucidated, there are some hints in the literature. Potent activation of the Akt pathway is likely to underly the *in vivo* survival effects of Ang-1 and it may also be relevant to the hypotensive effects, since Akt can lead to activation of eNOS and production of NO in endothelial cells [93]. The effects of Ang-1 on leukocyte infiltration in inflammatory models may reflect, at least partly, inhibition of endothelial cell adhesion molecule expression. Decreased expression of these proteins (ICAM-1, VCAM-1, E-selectin) has also been attributed to activation of the PI3K/Akt pathway [94]. With respect to permeability, several reports suggest that Ang-1 works by inhibiting calcium influx and/or activation of the small GTPase Rho [95, 96]. Both calcium flux and Rho are believed to promote permeability through effects on the actin cytoskeleton.

8.5.1
Is Ang-2 a Proinflammatory, Vascular Destabilizing Factor or a Protective Factor?

The initial hypothesis that Ang-2 functions as a Tie2 antagonist and a vessel destabilizing factor has been challenged by several findings. First, Ang-2 can clearly promote Tie2 phosphorylation (albeit relatively weakly) in cultured endothelial cells. This has been shown by addition of exogenous Ang-2 [97, 98] as well as in a model where expression of endogenous Ang-2 is induced in endothelial cells by the transcription factor FOXO1 [99]. Since FOXO1 is phosphorylated and inhibited by Akt, it induces gene expression when Akt activity is low. FOXO1-induced Ang-2 functions as an autocrine activator of Tie2/Akt signaling and promotes endothelial cell survival [99]. Thus, Ang-2 induction, in general, might be a compensatory response to low Akt activity, for example, when Ang-1/Tie2 signaling is weak. Conversely, we have shown that strong Akt activation in response to Ang-1 leads to inhibition of Ang-2 expression [72]. An interesting possibility suggested by these findings is that transcriptional pathways that switch on Ang-2 expression might also regulate the expression of a protein(s) that control the response to Ang-2. Thus, endothelial cells that actively produce Ang-2 might also respond to it with a robust activation of Tie2, whereas cells not producing Ang-2 would remain unresponsive.

In addition to the *in vitro* data, there are several examples where Ang-1 and Ang-2 have similar effects *in vivo*. For example, both Ang-1 and Ang-2 can inhibit inflammation in a cardiac allograft model [100] and can inhibit leak in ear skin [99]. In addition, both Ang-1 and Ang-2 rapidly activate Tie2/Akt signaling in mouse heart and inhibit expression of FOXO1 target genes [99].

However, there are instances in which Ang-2 has effects that are seemingly inconsistent with the proposal that it is a Tie2 agonist. For example, inflammation and vessel leak in models of peritonitis and acute lung injury are reduced in Ang-2 knockout mice [101, 102]. Conversely, administration of Ang-2 can promote vascular leak and/or leukocyte extravasation in these settings [101–103]. These findings suggest that in these models Ang-2 is a proinflammatory factor, presumably via inhibition of Tie2 signaling (although effects on Tie2 phosphorylation have not been documented). In addition, in a mouse model of revascularization after ischemia, transgenic delivery of Ang-2 impairs collateral vessel growth; this was attributed to inhibition of Tie2 phosphorylation, which was observed in the lungs of Ang-2 expressing mice [104]. Thus, the controversy regarding the role of Ang-2 as an agonist/antagonist remains. The most likely explanation for the conflicting findings is that endothelial cell responsiveness to Ang-2 is regulated in a tissue-specific manner by mechanisms that are yet to be elucidated. Furthermore, it should be stressed that an inability of the vessels in a particular tissue to respond to exogenous Ang-2 does not mean that endogenous Ang-2, when expressed at that site, will behave in a similar manner.

8.5.2
Angiopoietins in Tumor Angiogenesis

As noted above, Ang-2 expression is highly upregulated in tumor vessels, suggesting a potential role in tumor angiogenesis [68]. Experiments involving implantation of tumor cells engineered to overexpress Ang-2 have yielded conflicting results, with some studies suggesting a proangiogenic, protumor effect, and other studies suggesting the opposite (conflicting results have also been obtained regarding the role of Ang-1 in tumors). The situation has been clarified to some extent with the recent development of reagents that can specifically interfere with the function of endogenous Ang-2 produced by tumor endothelial cells. An antibody that specifically blocks Ang-2 function is able to significantly inhibit the growth of multiple human tumor lines in mouse xenograft studies [105]. More recently, it was demonstrated that the growth of subcutaneous tumors is inhibited in Ang-2 null mice, indicating a positive influence of Ang-2 on tumor progression [106]. Although the precise mechanisms through which Ang-2 promotes tumor growth remain to be elucidated, these results suggest the possibility that Ang-2 may represent another target for antiangiogenic strategies in solid tumors. Outstanding questions include "is Ang-2 behaving as a Tie2 agonist in tumors" and "what are the effects of Ang-2 blockade on tumor vessel morphology?" The application of potent inhibitors of angiopoietins to other disease models (e.g., inflammation) should help clarify their role in other settings of angiogenesis and vascular remodeling.

8.6 TGF-β Pathway

8.6.1 Receptors, Ligands, and Signaling Pathway

The TGF-β family is composed of multifunctional ligands including three TGF-β isoforms (TGF-β1–3), activins, and bone morphogenetic proteins (BMPs) (Figure 8.1d), all of which are involved in many different physiological and pathological processes [107]. TGF-β family members signal via two related single transmembrane threonine/serine kinase receptors, type I and type II [108]. In mammals, seven type I receptors, also called activin receptor-like kinase (ALK) 1–7 and five type II receptors have been identified [109]. Upon ligand interaction with the receptors, a heterodimeric complex composed of two type I and two type II receptors forms and the type II kinase phosphorylates the type I receptor [110]. In addition, a third family of TGF-β receptors exists, which is composed of the TGF-β type III receptors (TβRIII) endoglin and betaglycan. Type III receptors are transmembrane proteins with short intracellular domains lacking an enzymatic motif, which can form heterodimeric complexes with TGF-β type II receptor (TβRII) [111]. Each TGF-β family member interacts with a specific set of receptors. For example, TGF-β interacts mostly with TβRII and TGF-β type I receptor (TβRI – ALK5) [112], except in endothelial cells where it signals mainly via ALK1 [113]. Upon receptor activation, signals are transduced from the membrane to the nucleus via intracellular effector molecules named Smads [114]. Two major Smad pathways are induced upon ligand binding: the Smad 2/3 pathway, which is induced upon activation of ALK4, 5, and 7, and the Smad 1/5/8 pathway, which is activated by ALK 1, 2, 3, and 6 signaling [114].

8.6.2 Requirement for TGF-β Pathway during Embryogenesis

Gene targeting studies in mice have shown that loss of TGF-β signaling components leads to abnormal differentiation and maturation of the primitive vascular plexus of the embryo, oftentimes resulting in embryonic or perinatal lethality [115]. Since ALK1 and endoglin are the only two members that are mainly expressed on endothelial cells and involved in tumor angiogenesis, these will be the focus of this section. Homozygous knockout of either endoglin or ALK1 is embryonically lethal due to vascular defects [116, 117]. More specifically, while endothelial cells differentiated normally in endoglin-deficient mice, vascular organization was affected [116]. In addition, a defect in development of vascular smooth muscle cells (VSMCs), which are closely associated with endothelial cells and impact vascular stability, was observed before defects in the endothelial cells became apparent. These observations suggest that endoglin-deficient VSMCs contribute to the embryonic lethal vascular defect [116]. The phenotype of endoglin-deficient mice is highly reminiscent of ALK1-deficient mice, thereby suggesting that endoglin also plays a role in ALK1 signaling in angiogenesis [117]. Heterozygous endoglin as well as ALK1 mice are viable, but a percentage of those develop a phenotype similar to that observed in hereditary hemorrhagic telangiectasia (HHT) patients [118, 119]. HHT is characterized by telangiectases and arteriovenous malformations (AVMs) typically found in the skin and mucocutaneous tissues [120, 121], which clearly suggests a role for ALK1 and endoglin in angiogenesis.

8.6.3 Preclinical and Clinical Data on TGF-β Pathway in Tumor Angiogenesis

Endoglin is expressed in tumor associated endothelium in many solid cancers, including breast, prostate, and cervical cancer [111]. Targeting of endoglin in murine tumor models using antiendoglin antibodies or endoglin plasmid DNA and recombinant protein resulted in decreased tumor growth and decreased tumor angiogenesis [122]. Endoglin expression on tumor endothelium appears to have diagnostic value based on detecting its expression *in vivo* in both patients and animal models using conjugated antiendoglin antibodies [122]. Intratumor microvessel density, as determined by antiendoglin staining, and circulating levels of soluble endoglin have prognostic significance in cancer [111]. In addition to using endoglin as a diagnostic or prognostic tool, investigators developed a human/mouse chimeric antibody to directly target endoglin with encouraging pharmacokinetic and toxicity data in nonhuman primates [123], which may be pursued in clinical trials. Altogether, these preclinical and clinical data suggest that antiendoglin therapy could be beneficial in cancer patients. However, caution should be taken as endoglin is also weakly expressed on normal endothelium [111].

ALK1 expression is low in normal vasculature and induced in preexisting feeding arteries and newly forming arterial vessels during tumor angiogenesis [124]. These findings suggest that ALK1 is as good a target as endoglin. To date, no data showing the effect of blocking ALK1 in preclinical models have been published. However, recent announcements suggest that a biologic inhibitor of ALK1 is in phase 1 trials in solid tumors (PF-3446962, Pfizer).

8.7 Ephrin–Eph Pathway

8.7.1 Receptors and Ligands

Eph stands for erythropoietin-producing human hepatocellular carcinoma, the cell line from which the first gene encoding a member of this receptor family was cloned in late 1980s [125]. Eph receptors comprise the largest family of transmembrane tyrosine kinase receptors. The unique feature of this family of receptor tyrosine kinases is that they are involved in transducing bidirectional intercellular signals through interaction with their corresponding ligands, ephrins (Eph family receptor interacting proteins) [126–131].

The Eph receptor family contains 16 members that are structurally conserved. Eph receptors are transmembrane proteins in which the extracellular domain consists of an N-terminal ephrin-binding domain, followed by an EGF-like motif and two fibronectin type III repeats (Figure 8.1e). The intracellular tyrosine kinase domain is connected to the extracellular domain via the intrajuxtamembrane domain that is conserved among Eph receptors. The tyrosine kinase domain is followed by a sterile alpha-motif (SAM) domain and the C-terminal PDZ-binding domain. The Eph receptors are classified into two categories [132], EphAs (EphA1–EphA10) and EphBs (EphB1–EphB6), based on their sequence similarity as well as binding specificities to two structurally distinct classes of a total of nine ephrin ligands, namely type A ephrins and type B ephrins, respectively. Like their receptors, ephrins are also membrane-anchored proteins. While all ephrins contain an N-terminal extracellular Eph-binding domain, type A ephrins (ephrinA1–A6) are glycosylphosphatidylinositol (GPI)-anchored cell surface membrane proteins lacking intracellular domains, whereas type B ephrins (ephrinB1–B3) contain a short cytoplasmic tail ending with a C-terminal PDZ-binding domain. In most cases, EphA receptors interact with ephrinA ligands and EphB receptors interact with ephrinB ligands, although cross talk between A and B family members has also been reported [129].

8.7.2
Eph–Ephrin Receptor–Ligand Bidirectional Signaling

In general, Eph–ephrin interaction transduces bidirectional signaling between adjacent cells. The signaling event mediated by ephrin-induced Eph receptor activation is referred to as *forward signaling*. In such a process, Eph receptors behave as typical RTKs, which, upon ligand binding, undergo receptor oligomerization, receptor kinase activation, and autophosphorylation, followed by recruiting and/or phosphorylation of downstream signaling molecules. Conversely, the presence of the intracellular domain in the ephrinB proteins enables them to transduce a so-called "reverse signal", via recruiting intracellular signaling effector proteins upon binding to their EphB receptors [127]. Although some common signaling effectors such as Src kinases, PI3K, MAPK, and Rho GTPases are involved Eph–ephrin pathways, cell content–dependent, and ligand–receptor-specific downstream signaling determines the cellular responses triggered.

Relatively little is known about the signaling events downstream of ephrin–Eph complexes during angiogenesis and vascular development. Both EphA2 and EphB4 receptors have been shown to regulate endothelial cell migration via PI3K/Rac signaling [133, 134]. In addition, EphA2 reportedly associates with focal adhesion kinase (FAK) and p130 (Cas) to regulate ephrinA1-induced cytoskeleton reorganization [135]. On the other hand, ephrinA1 binding also induces binding to EphA2 by SHP-2, which in turn dephosphorylates FAK and paxillin [136].

8.7.2.1 EphrinA-EphA

The first implication of involvement of Eph proteins in blood vessel development came from the fact that ephrinA1 was cloned from human umbilical vein endothelial cells (HUVECs) in early studies [137]. Later studies showed that administration of recombinant ephrinA1 triggered chemotaxis of EphA2-expressing bovine adrenal capillary endothelial cells *in vitro* as well as angiogenesis in rat cornea *in vivo* [138, 139]. Further, EphA2 was shown to regulate endothelial cell migration and vascular assembly *in vitro* [133]. Although these results support a role for the ephrinA1/EphA2 signaling in angiogenic remodeling, this pathway appears to be dispensable during embryonic vascularization since EphA2$^{-/-}$ mice are viable [140, 141]. However, complementary expression of ephrinA1 in tumor cells and EphA2 in blood vessel endothelium was reported in a xenograft tumor model [142]. Further, ephrinA1 was shown to be regulated by VEGF [143]. However, another report suggests that EphA2 may inhibit many VEGF-dependent angiogenic activities [136], indicating a potential role for this pathway at sites of neovascularization. Therefore, targeting the ephrinA1/EphA2 pathway might be a useful therapeutic approach for tumor and other angiogenesis-associated diseases.

8.7.2.2 EphrinB-EphB

It is now clear that ephrinB2/EphB4 signaling is essential in the development of the vascular system. This ligand-receptor pair shows a striking complementary vascular-specific expression pattern in mice [144–146]. In particular, ephrin-B2 is predominantly expressed in arterial endothelial cells (and some collecting lymphatics), whereas its receptor EphB4 is expressed in veins (and throughout the lymphatic vasculature) [147]. Genetic inactivation of either ephrinB2 or EphB4 in mice results in embryonic lethality by E10.5–11.5 [146, 148, 149]. The defects in these knockout embryos are very similar, with disrupted angiogenic remodeling, perturbed arteriovenous specification, as well as disorganized capillary network [146, 148, 149]. The vascular-specific role of ephrinB2/EphB4 was further confirmed by the fact that vascular-specific inactivation of ephrinB2 phenocopies the ephrinB2 null mice [150]. Interestingly, no significant vascular phenotype was observed in mice lacking either EphB2 or EphB3 receptors, which can also bind ephrinB2 [148]. However, a proportion of the mice carrying null mutations in both genes die around E11.5, with a primary vascular plexus remodeling defect [148], suggesting that EphB4 alone is not sufficient for vascular development. In addition, several lines of cell biology data support the notion that ephrinB2-mediated reverse signaling triggers propulsive activities during angiogenesis, whereas forward signaling through EphB4 mediates repulsive responses [134, 151].

Because of the clear involvement of ephrinB2/EphB4 in vascular development, efforts are underway to inhibit this pathway in tumor angiogenesis. One study found that a soluble form of EphB4 inhibited cell migration and invasion of cells from Kaposi's sarcoma [152]. Beyond these insightful observations, very limited knowledge has been obtained on the definitive role of ephrinB2/EphB4 (and/or other ephrin/Eph members) signaling in tumor angiogenesis.

8.8
Summary and Conclusions

The chapter has summarized five major signaling systems that are involved in tumor angiogenesis. A key initiator of tumor angiogenesis is the hypoxia-mediated expression of VEGF ligands by tumor cells. These VEGF ligands provide upstream *trans*-activation from the tumor cells to the vasculature, by interacting with VEGF receptors on endothelial cells. Subsequently, much of the signaling described in this chapter is mediated in "*cis*", that is, via localized or even cell–cell signals acting within the vascular compartment. These signals include the Delta–Notch, angiopoietin–Tie2, TGF-β, and ephrin pathways, which can act among endothelial cells or between endothelial cells and perivascular cells. For each pathway, multiple ligands and/or receptors are involved in tumor angiogenesis. Thus, while it is possible that additional pathways will prove to be suitable targets for tumor angiogenesis, much work in the next few years is likely to characterize the effects of these particular pathways on tumor vessel growth and function, to examine the specific role of different growth factors and receptors within these pathways, and to begin to understand how these pathways interact in the dynamic and imperfect process of tumor angiogenesis.

References

1. Folkman, J. (1971) Tumor angiogenesis: therapeutic implications. *N. Engl. J. Med.*, **285** (21), 1182–1186.
2. Folkman, J. and Shing, Y. (1992) Angiogenesis. *J. Biol. Chem.*, **267** (16), 10931–10934.
3. Soker, S., Takashima, S., Miao, H.Q., Neufeld, G., and Klagsbrun, M. (1998) Neuropilin-1 is expressed by endothelial and tumor cells as an isoform-specific receptor for vascular endothelial growth factor. *Cell*, **92** (6), 735–745.
4. He, Z. and Tessier-Lavigne, M. (1997) Neuropilin is a receptor for the axonal chemorepellent Semaphorin III. *Cell*, **90** (4), 739–751.
5. Kolodkin, A.L. *et al.* (1997) Neuropilin is a semaphorin III receptor. *Cell*, **90** (4), 753–762.
6. Pellet-Many, C., Frankel, P., Jia, H., and Zachary, I. (2008) Neuropilins: structure, function and role in disease. *Biochem. J.*, **411** (2), 211–226.
7. Olsson, A.K., Dimberg, A., Kreuger, J., and Claesson-Welsh, L. (2006) VEGF receptor signalling – in control of vascular function. *Nat. Rev. Mol. Cell Biol.*, **7** (5), 359–371.
8. Guttmann-Raviv, N. *et al.* (2006) The neuropilins and their role in tumorigenesis and tumor progression. *Cancer Lett.*, **231** (1), 1–11.
9. Meadows, K.N., Bryant, P., and Pumiglia, K. (2001) Vascular endothelial growth factor induction of the angiogenic phenotype requires Ras activation. *J. Biol. Chem.*, **276** (52), 49289–49298.
10. Fujio, Y. and Walsh, K. (1999) Akt mediates cytoprotection of endothelial cells by vascular endothelial growth factor in an anchorage-dependent manner. *J. Biol. Chem.*, **274** (23), 16349–16354.
11. Carmeliet, P. *et al.* (1996) Abnormal blood vessel development and lethality in embryos lacking a single VEGF allele. *Nature*, **380** (6573), 435–439.
12. Ferrara, N. *et al.* (1996) Heterozygous embryonic lethality induced by targeted inactivation of the VEGF gene. *Nature*, **380** (6573), 439–442.

13. Shalaby, F. *et al.* (1995) Failure of blood-island formation and vasculogenesis in Flk-1-deficient mice. *Nature*, **376** (6535), 62–66.
14. Fong, G.H., Zhang, L., Bryce, D.M., and Peng, J. (1999) Increased hemangioblast commitment, not vascular disorganization, is the primary defect in flt-1 knock-out mice. *Development*, **126** (13), 3015–3025.
15. Dumont, D.J. *et al.* (1998) Cardiovascular failure in mouse embryos deficient in VEGF receptor-3. *Science*, **282** (5390), 946–949.
16. Karkkainen, M.J. *et al.* (2004) Vascular endothelial growth factor C is required for sprouting of the first lymphatic vessels from embryonic veins. *Nat. Immunol.*, **5** (1), 74–80.
17. Kawasaki, T. *et al.* (1999) A requirement for neuropilin-1 in embryonic vessel formation. *Development*, **126** (21), 4895–4902.
18. Yuan, L. *et al.* (2002) Abnormal lymphatic vessel development in neuropilin 2 mutant mice. *Development*, **129** (20), 4797–4806.
19. Baldwin, M.E. *et al.* (2005) Vascular endothelial growth factor D is dispensable for development of the lymphatic system. *Mol. Cell Biol.*, **25** (6), 2441–2449.
20. Bellomo, D. *et al.* (2000) Mice lacking the vascular endothelial growth factor-B gene (Vegfb) have smaller hearts, dysfunctional coronary vasculature, and impaired recovery from cardiac ischemia. *Circ. Res.*, **86** (2), E29–E35.
21. Carmeliet, P. *et al.* (2001) Synergism between vascular endothelial growth factor and placental growth factor contributes to angiogenesis and plasma extravasation in pathological conditions. *Nat. Med.*, **7** (5), 575–583.
22. Ferrara, N. (2004) Vascular endothelial growth factor: basic science and clinical progress. *Endocr. Rev.*, **25** (4), 581–611.
23. Dvorak, H.F. (2002) Vascular permeability factor/vascular endothelial growth factor: a critical cytokine in tumor angiogenesis and a potential target for diagnosis and therapy. *J. Clin. Oncol.*, **20** (21), 4368–4380.
24. Poon, R.T., Fan, S.T., and Wong, J. (2001) Clinical implications of circulating angiogenic factors in cancer patients. *J. Clin. Oncol.*, **19** (4), 1207–1225.
25. Viloria-Petit, A. *et al.* (2003) Contrasting effects of VEGF gene disruption in embryonic stem cell-derived versus oncogene-induced tumors. *EMBO J.*, **22** (16), 4091–4102.
26. Fischer, C., Mazzone, M., Jonckx, B., and Carmeliet, P. (2008) FLT1 and its ligands VEGFB and PlGF: drug targets for anti-angiogenic therapy? *Nat. Rev. Cancer*, **8** (12), 942–956.
27. Fan, F. *et al.* (2005) Expression and function of vascular endothelial growth factor receptor-1 on human colorectal cancer cells. *Oncogene*, **24** (16), 2647–2653.
28. Alitalo, K., Tammela, T., and Petrova, T.V. (2005) Lymphangiogenesis in development and human disease. *Nature*, **438** (7070), 946–953.
29. Thiele, W. and Sleeman, J.P. (2006) Tumor-induced lymphangiogenesis: a target for cancer therapy? *J. Biotechnol.*, **124** (1), 224–241.
30. Ellis, L.M. and Hicklin, D.J. (2008) Pathways mediating resistance to vascular endothelial growth factor-targeted therapy. *Clin. Cancer Res.*, **14** (20), 6371–6375.
31. Wu, Y. *et al.* (2006) Anti-vascular endothelial growth factor receptor-1 antagonist antibody as a therapeutic agent for cancer. *Clin. Cancer Res.*, **12** (21), 6573–6584.
32. Fischer, C. *et al.* (2007) Anti-PlGF inhibits growth of VEGF(R)-inhibitor-resistant tumors without affecting healthy vessels. *Cell*, **131** (3), 463–475.
33. Pan, Q. *et al.* (2007) Blocking neuropilin-1 function has an additive effect with anti-VEGF to inhibit tumor growth. *Cancer Cell*, **11** (1), 53–67.
34. Achen, M.G. *et al.* (2000) Monoclonal antibodies to vascular endothelial growth factor-D block its interactions with both VEGF receptor-2 and VEGF receptor-3. *Eur. J. Biochem.*, **267** (9), 2505–2515.
35. Kubo, H. *et al.* (2002) Blockade of vascular endothelial growth factor

receptor-3 signaling inhibits fibroblast growth factor-2-induced lymphangiogenesis in mouse cornea. *Proc. Natl. Acad. Sci. U.S.A.*, **99** (13), 8868–8873.

36. Caunt, M. *et al.* (2008) Blocking neuropilin-2 function inhibits tumor cell metastasis. *Cancer Cell*, **13** (4), 331–342.

37. Presta, L.G. *et al.* (1997) Humanization of an anti-vascular endothelial growth factor monoclonal antibody for the therapy of solid tumors and other disorders. *Cancer Res.*, **57** (20), 4593–4599.

38. Hurwitz, H. *et al.* (2004) Bevacizumab plus irinotecan, fluorouracil, and leucovorin for metastatic colorectal cancer. *N. Engl. J. Med.*, **350** (23), 2335–2342.

39. Sandler, A. *et al.* (2006) Paclitaxel-carboplatin alone or with bevacizumab for non-small-cell lung cancer. *N. Engl. J. Med.*, **355** (24), 2542–2550.

40. Jain, R.K. (2005) Normalization of tumor vasculature: an emerging concept in antiangiogenic therapy. *Science*, **307** (5706), 58–62.

41. Bergers, G. and Hanahan, D. (2008) Modes of resistance to anti-angiogenic therapy. *Nat. Rev. Cancer*, **8** (8), 592–603.

42. Shojaei, F. *et al.* (2007) Tumor refractoriness to anti-VEGF treatment is mediated by CD11b+Gr1+ myeloid cells. *Nat. Biotechnol.*, **25** (8), 911–920.

43. Casanovas, O., Hicklin, D.J., Bergers, G., and Hanahan, D. (2005) Drug resistance by evasion of antiangiogenic targeting of VEGF signaling in late-stage pancreatic islet tumors. *Cancer Cell*, **8** (4), 299–309.

44. Holash, J. *et al.* (2002) VEGF-Trap: a VEGF blocker with potent antitumor effects. *Proc. Natl. Acad. Sci. U.S.A.*, **99** (17), 11393–11398.

45. Byrne, A.T. *et al.* (2003) Vascular endothelial growth factor-trap decreases tumor burden, inhibits ascites, and causes dramatic vascular remodeling in an ovarian cancer model. *Clin. Cancer Res.*, **9** (15), 5721–5728.

46. Huang, J. *et al.* (2003) Regression of established tumors and metastases by potent vascular endothelial growth factor blockade. *Proc. Natl. Acad. Sci. U.S.A.*, **100** (13), 7785–7790.

47. Inai, T. *et al.* (2004) Inhibition of vascular endothelial growth factor (VEGF) signaling in cancer causes loss of endothelial fenestrations, regression of tumor vessels, and appearance of basement membrane ghosts. *Am. J. Pathol.*, **165** (1), 35–52.

48. Frischer, J.S. *et al.* (2004) Effects of potent VEGF blockade on experimental Wilms tumor and its persisting vasculature. *Int. J. Oncol.*, **25** (3), 549–553.

49. Hu, L. *et al.* (2005) Vascular endothelial growth factor trap combined with paclitaxel strikingly inhibits tumor and ascites, prolonging survival in a human ovarian cancer model. *Clin. Cancer Res.*, **11** (19 Pt 1), 6966–6971.

50. Wachsberger, P.R. *et al.* (2007) VEGF trap in combination with radiotherapy improves tumor control in u87 glioblastoma. *Int. J. Radiat. Oncol. Biol. Phys.*, **67** (5), 1526–1537.

51. D'Souza, B., Miyamoto, A., and Weinmaster, G. (2008) The many facets of Notch ligands. *Oncogene*, **27** (38), 5148–5167.

52. Gordon, W.R., Arnett, K.L., and Blacklow, S.C. (2008) The molecular logic of Notch signaling–a structural and biochemical perspective. *J. Cell Sci.*, **121** (Pt 19), 3109–3119.

53. Weng, A.P. *et al.* (2004) Activating mutations of NOTCH1 in human T cell acute lymphoblastic leukemia. *Science*, **306** (5694), 269–271.

54. Weinmaster, G. and Kopan, R. (2006) A garden of Notch-ly delights. *Development*, **133** (17), 3277–3282.

55. Hofmann, J.J. and Iruela-Arispe, M.L. (2007) Notch signaling in blood vessels: who is talking to whom about what? *Circ. Res.*, **100** (11), 1556–1568.

56. Gale, N.W. *et al.* (2004) Haploinsufficiency of delta-like 4 ligand results in embryonic lethality due to major defects in arterial and vascular development. *Proc. Natl. Acad. Sci. U.S.A.*, **101** (45), 15949–15954.

57. Mailhos, C. *et al.* (2001) Delta4, an endothelial specific notch ligand expressed at sites of physiological and tumor angiogenesis. *Differentiation*, **69** (2-3), 135–144.

58. Patel, N.S. *et al.* (2005) Up-regulation of delta-like 4 ligand in human tumor vasculature and the role of basal expression in endothelial cell function. *Cancer Res.*, **65** (19), 8690–8697.
59. Noguera-Troise, I. *et al.* (2006) Blockade of Dll4 inhibits tumour growth by promoting non-productive angiogenesis. *Nature*, **444** (7122), 1032–1037.
60. Dufraine, J., Funahashi, Y., and Kitajewski, J. (2008) Notch signaling regulates tumor angiogenesis by diverse mechanisms. *Oncogene*, **27** (38), 5132–5137.
61. Ridgway, J. *et al.* (2006) Inhibition of Dll4 signalling inhibits tumour growth by deregulating angiogenesis. *Nature*, **444** (7122), 1083–1087.
62. Yamanda, S. *et al.* (2009) Role of ephrinB2 in non-productive angiogenesis induced by Delta-like 4 blockade. *Blood*, **113** (15), 3631–3639.
63. Li, K. *et al.* (2008) Modulation of Notch signaling by antibodies specific for the extracellular negative regulatory region of NOTCH3. *J. Biol. Chem.*, **283** (12), 8046–8054.
64. Miele, L. (2008) Rational targeting of Notch signaling in breast cancer. *Expert Rev. Anticancer Ther.*, **8** (8), 1197–1202.
65. Sato, T.N., Qin, Y., Kozak, C.A., and Audus, K.L. (1993) Tie-1 and tie-2 define another class of putative receptor tyrosine kinase genes expressed in early embryonic vascular system. *Proc. Natl. Acad. Sci. U.S.A.*, **90** (20), 9355–9358.
66. Davis, S. *et al.* (1996) Isolation of angiopoietin-1, a ligand for the TIE2 receptor, by secretion-trap expression cloning. *Cell*, **87** (7), 1161–1169.
67. Maisonpierre, P.C. *et al.* (1997) Angiopoietin-2, a natural antagonist for Tie2 that disrupts in vivo angiogenesis. *Science*, **277** (5322), 55–60.
68. Holash, J., Wiegand, S.J., and Yancopoulos, G.D. (1999) New model of tumor angiogenesis: dynamic balance between vessel regression and growth mediated by angiopoietins and VEGF. *Oncogene*, **18** (38), 5356–5362.
69. Davis, S. *et al.* (2003) Angiopoietins have distinct modular domains essential for receptor binding, dimerization and superclustering. *Nat. Struct. Biol.*, **10** (1), 38–44.
70. Kim, I. *et al.* (2000) Angiopoietin-1 regulates endothelial cell survival through the phosphatidylinositol 3'-Kinase/Akt signal transduction pathway. *Circ. Res.*, **86** (1), 24–29.
71. Papapetropoulos, A. *et al.* (2000) Angiopoietin-1 inhibits endothelial cell apoptosis via the Akt/survivin pathway. *J. Biol. Chem.*, **275** (13), 9102–9105.
72. Daly, C. *et al.* (2004) Angiopoietin-1 modulates endothelial cell function and gene expression via the transcription factor FKHR (FOXO1). *Genes Dev.*, **18** (9), 1060–1071.
73. Fukuhara, S. *et al.* (2008) Differential function of Tie2 at cell-cell contacts and cell-substratum contacts regulated by angiopoietin-1. *Nat. Cell Biol.*, **10** (5), 513–526.
74. Saharinen, P. *et al.* (2008) Angiopoietins assemble distinct Tie2 signalling complexes in endothelial cell-cell and cell-matrix contacts. *Nat. Cell Biol.*, **10** (5), 527–537.
75. Marron, M.B., Hughes, D.P., Edge, M.D., Forder, C.L., and Brindle, N.P. (2000) Evidence for heterotypic interaction between the receptor tyrosine kinases TIE-1 and TIE-2. *J. Biol. Chem.*, **275** (50), 39741–39746.
76. Saharinen, P. *et al.* (2005) Multiple angiopoietin recombinant proteins activate the Tie1 receptor tyrosine kinase and promote its interaction with Tie2. *J. Cell Biol.*, **169** (2), 239–243.
77. Yuan, H.T. *et al.* (2007) Activation of the orphan endothelial receptor Tie1 modifies Tie2-mediated intracellular signaling and cell survival. *FASEB J.*, **21** (12), 3171–3183.
78. Sato, T.N. *et al.* (1995) Distinct roles of the receptor tyrosine kinases Tie-1 and Tie-2 in blood vessel formation. *Nature*, **376** (6535), 70–74.
79. Vikkula, M. *et al.* (1996) Vascular dysmorphogenesis caused by an activating mutation in the receptor tyrosine kinase TIE2. *Cell*, **87** (7), 1181–1190.

80. Puri, M.C., Rossant, J., Alitalo, K., Bernstein, A., and Partanen, J. (1995) The receptor tyrosine kinase TIE is required for integrity and survival of vascular endothelial cells. *EMBO J.*, **14** (23), 5884–5891.
81. Suri, C. *et al.* (1996) Requisite role of angiopoietin-1, a ligand for the TIE2 receptor, during embryonic angiogenesis. *Cell*, **87** (7), 1171–1180.
82. Gale, N.W. *et al.* (2002) Angiopoietin-2 is required for postnatal angiogenesis and lymphatic patterning, and only the latter role is rescued by Angiopoietin-1. *Dev. Cell*, **3** (3), 411–423.
83. Thurston, G. *et al.* (2005) Angiopoietin 1 causes vessel enlargement, without angiogenic sprouting, during a critical developmental period. *Development*, **132** (14), 3317–3326.
84. Thurston, G. *et al.* (1999) Leakage-resistant blood vessels in mice transgenically overexpressing angiopoietin-1. *Science*, **286** (5449), 2511–2514.
85. Thurston, G. *et al.* (2000) Angiopoietin-1 protects the adult vasculature against plasma leakage. *Nat. Med.*, **6** (4), 460–463.
86. Joussen, A.M. *et al.* (2002) Suppression of diabetic retinopathy with angiopoietin-1. *Am. J. Pathol.*, **160** (5), 1683–1693.
87. Witzenbichler, B., Westermann, D., Knueppel, S., Schultheiss, H.P., and Tschope, C. (2005) Protective role of angiopoietin-1 in endotoxic shock. *Circulation*, **111** (1), 97–105.
88. Nykanen, A.I. *et al.* (2003) Angiopoietin-1 protects against the development of cardiac allograft arteriosclerosis. *Circulation*, **107** (9), 1308–1314.
89. McCarter, S.D. *et al.* (2007) Cell-based angiopoietin-1 gene therapy for acute lung injury. *Am. J. Respir. Crit. Care Med.*, **175** (10), 1014–1026.
90. Cho, C.H. *et al.* (2004) Designed angiopoietin-1 variant, COMP-Ang1, protects against radiation-induced endothelial cell apoptosis. *Proc. Natl. Acad. Sci. U.S.A.*, **101** (15), 5553–5558.
91. Zhao, Y.D., Campbell, A.I., Robb, M., Ng, D., and Stewart, D.J. (2003) Protective role of angiopoietin-1 in experimental pulmonary hypertension. *Circ. Res.*, **92** (9), 984–991.
92. Lee, J.S. *et al.* (2008) Angiopoietin-1 prevents hypertension and target organ damage through its interaction with endothelial Tie2 receptor. *Cardiovasc. Res.*, **78** (3), 572–580.
93. Babaei, S. *et al.* (2003) Angiogenic actions of angiopoietin-1 require endothelium-derived nitric oxide. *Am. J. Pathol.*, **162** (6), 1927–1936.
94. Kim, I., Moon, S.O., Park, S.K., Chae, S.W., and Koh, G.Y. (2001) Angiopoietin-1 reduces VEGF-stimulated leukocyte adhesion to endothelial cells by reducing ICAM-1, VCAM-1, and E-selectin expression. *Circ. Res.*, **89** (6), 477–479.
95. Jho, D. *et al.* (2005) Angiopoietin-1 opposes VEGF-induced increase in endothelial permeability by inhibiting TRPC1-dependent Ca2 influx. *Circ. Res.*, **96** (12), 1282–1290.
96. Li, X. *et al.* (2004) Role of protein kinase Czeta in thrombin-induced endothelial permeability changes: inhibition by angiopoietin-1. *Blood*, **104** (6), 1716–1724.
97. Kim, I. *et al.* (2000) Angiopoietin-2 at high concentration can enhance endothelial cell survival through the phosphatidylinositol 3'-kinase/Akt signal transduction pathway. *Oncogene*, **19** (39), 4549–4552.
98. Teichert-Kuliszewska, K. *et al.* (2001) Biological action of angiopoietin-2 in a fibrin matrix model of angiogenesis is associated with activation of Tie2. *Cardiovasc. Res.*, **49** (3), 659–670.
99. Daly, C. *et al.* (2006) Angiopoietin-2 functions as an autocrine protective factor in stressed endothelial cells. *Proc. Natl. Acad. Sci. U.S.A.*, **103** (42), 15491–15496.
100. Nykanen, A.I. *et al.* (2006) Common protective and diverse smooth muscle cell effects of AAV-mediated angiopoietin-1 and -2 expression in rat cardiac allograft vasculopathy. *Circ. Res.*, **98** (11), 1373–1380.
101. Bhandari, V. *et al.* (2006) Hyperoxia causes angiopoietin 2-mediated acute

lung injury and necrotic cell death. *Nat. Med.*, **12** (11), 1286–1293.

102. Fiedler, U. *et al.* (2006) Angiopoietin-2 sensitizes endothelial cells to TNF-alpha and has a crucial role in the induction of inflammation. *Nat. Med.*, **12** (2), 235–239.

103. Parikh, S.M. *et al.* (2006) Excess circulating angiopoietin-2 may contribute to pulmonary vascular leak in sepsis in humans. *PLoS Med.*, **3** (3), e46.

104. Reiss, Y. *et al.* (2007) Angiopoietin-2 impairs revascularization after limb ischemia. *Circ. Res.*, **101** (1), 88–96.

105. Oliner, J. *et al.* (2004) Suppression of angiogenesis and tumor growth by selective inhibition of angiopoietin-2. *Cancer Cell*, **6** (5), 507–516.

106. Nasarre, P. *et al.* (2009) Host-derived angiopoietin-2 affects early stages of tumor development and vessel maturation but is dispensable for later stages of tumor growth. *Cancer Res.*, **69** (4), 1324–1333.

107. ten Dijke, P. and Arthur, H.M. (2007) Extracellular control of TGFbeta signalling in vascular development and disease. *Nat. Rev. Mol. Cell Biol.*, **8** (11), 857–869.

108. Schmierer, B. and Hill, C.S. (2007) TGFbeta-SMAD signal transduction: molecular specificity and functional flexibility. *Nat. Rev. Mol. Cell Biol.*, **8** (12), 970–982.

109. Piek, E., Heldin, C.H., and Ten Dijke, P. (1999) Specificity, diversity, and regulation in TGF-beta superfamily signaling. *FASEB J.*, **13** (15), 2105–2124.

110. Wrana, J.L., Attisano, L., Wieser, R., Ventura, F., and Massague, J. (1994) Mechanism of activation of the TGF-beta receptor. *Nature*, **370** (6488), 341–347.

111. ten Dijke, P., Goumans, M.J., and Pardali, E. (2008) Endoglin in angiogenesis and vascular diseases. *Angiogenesis*, **11** (1), 79–89.

112. Franzen, P. *et al.* (1993) Cloning of a TGF beta type I receptor that forms a heteromeric complex with the TGF beta type II receptor. *Cell*, **75** (4), 681–692.

113. Goumans, M.J. *et al.* (2002) Balancing the activation state of the endothelium via two distinct TGF-beta type I receptors. *EMBO J.*, **21** (7), 1743–1753.

114. ten Dijke, P. and Hill, C.S. (2004) New insights into TGF-beta-Smad signalling. *Trends Biochem. Sci.*, **29** (5), 265–273.

115. Goumans, M.J. and Mummery, C. (2000) Functional analysis of the TGF-beta receptor/Smad pathway through gene ablation in mice. *Int. J. Dev. Biol.*, **44** (3), 253–265.

116. Li, D.Y. *et al.* (1999) Defective angiogenesis in mice lacking endoglin. *Science*, **284** (5419), 1534–1537.

117. Urness, L.D., Sorensen, L.K., and Li, D.Y. (2000) Arteriovenous malformations in mice lacking activin receptor-like kinase-1. *Nat. Genet.*, **26** (3), 328–331.

118. Srinivasan, S. *et al.* (2003) A mouse model for hereditary hemorrhagic telangiectasia (HHT) type 2. *Hum. Mol. Genet.*, **12** (5), 473–482.

119. Torsney, E. *et al.* (2003) Mouse model for hereditary hemorrhagic telangiectasia has a generalized vascular abnormality. *Circulation*, **107** (12), 1653–1657.

120. van den Driesche, S., Mummery, C.L., and Westermann, C.J. (2003) Hereditary hemorrhagic telangiectasia: an update on transforming growth factor beta signaling in vasculogenesis and angiogenesis. *Cardiovasc. Res.*, **58** (1), 20–31.

121. Marchuk, D.A., Srinivasan, S., Squire, T.L., and Zawistowski, J.S. (2003) Vascular morphogenesis: tales of two syndromes. *Hum. Mol. Genet.*, **12** (Spec No.1), R97–112.

122. Dallas, N.A. *et al.* (2008) Endoglin (CD105): a marker of tumor vasculature and potential target for therapy. *Clin. Cancer Res.*, **14** (7), 1931–1937.

123. Shiozaki, K. *et al.* (2006) Antiangiogenic chimeric anti-endoglin (CD105) antibody: pharmacokinetics and immunogenicity in nonhuman primates and effects of doxorubicin. *Cancer Immunol. Immunother.*, **55** (2), 140–150.

124. Seki, T., Yun, J., and Oh, S.P. (2003) Arterial endothelium-specific activin receptor-like kinase 1 expression suggests its role in arterialization and

vascular remodeling. *Circ. Res.*, **93** (7), 682–689.

125. Hirai, H., Maru, Y., Hagiwara, K., Nishida, J., and Takaku, F. (1987) A novel putative tyrosine kinase receptor encoded by the eph gene. *Science*, **238** (4834), 1717–1720.
126. Heroult, M., Schaffner, F., and Augustin, H.G. (2006) Eph receptor and ephrin ligand-mediated interactions during angiogenesis and tumor progression. *Exp. Cell Res.*, **312** (5), 642–650.
127. Himanen, J.P., Saha, N., and Nikolov, D.B. (2007) Cell-cell signaling via Eph receptors and ephrins. *Curr. Opin. Cell Biol.*, **19** (5), 534–542.
128. Kuijper, S., Turner, C.J., and Adams, R.H. (2007) Regulation of angiogenesis by Eph-ephrin interactions. *Trends Cardiovasc. Med.*, **17** (5), 145–151.
129. Lackmann, M. and Boyd, A.W. (2008) Eph, a protein family coming of age: more confusion, insight, or complexity? *Sci. Signal.*, **1** (15), re2.
130. Pasquale, E.B. (2008) Eph-ephrin bidirectional signaling in physiology and disease. *Cell*, **133** (1), 38–52.
131. Surawska, H., Ma, P.C., and Salgia, R. (2004) The role of ephrins and Eph receptors in cancer. *Cytokine Growth Factor Rev.*, **15** (6), 419–433.
132. Gale, N.W. *et al.* (1996) Eph receptors and ligands comprise two major specificity subclasses and are reciprocally compartmentalized during embryogenesis. *Neuron*, **17** (1), 9–19.
133. Brantley-Sieders, D.M. *et al.* (2004) EphA2 receptor tyrosine kinase regulates endothelial cell migration and vascular assembly through phosphoinositide 3-kinase-mediated Rac1 GTPase activation. *J. Cell Sci.*, **117** (Pt 10), 2037–2049.
134. Marston, D.J., Dickinson, S., and Nobes, C.D. (2003) Rac-dependent trans-endocytosis of ephrinBs regulates Eph-ephrin contact repulsion. *Nat. Cell Biol.*, **5** (10), 879–888.
135. Carter, N., Nakamoto, T., Hirai, H., and Hunter, T. (2002) EphrinA1-induced cytoskeletal re-organization requires FAK and p130(cas). *Nat. Cell Biol.*, **4** (8), 565–573.
136. Miao, H., Burnett, E., Kinch, M., Simon, E., and Wang, B. (2000) Activation of EphA2 kinase suppresses integrin function and causes focal-adhesion-kinase dephosphorylation. *Nat. Cell Biol.*, **2** (2), 62–69.
137. Holzman, L.B., Marks, R.M., and Dixit, V.M. (1990) A novel immediate-early response gene of endothelium is induced by cytokines and encodes a secreted protein. *Mol. Cell Biol.*, **10** (11), 5830–5838.
138. Daniel, T.O. *et al.* (1996) ELK and LERK-2 in developing kidney and microvascular endothelial assembly. *Kidney Int. Suppl.*, **57**, S73–S81.
139. Pandey, A., Shao, H., Marks, R.M., Polverini, P.J., and Dixit, V.M. (1995) Role of B61, the ligand for the Eck receptor tyrosine kinase, in TNF-alpha-induced angiogenesis. *Science*, **268** (5210), 567–569.
140. Brantley-Sieders, D.M. *et al.* (2005) Impaired tumor microenvironment in EphA2-deficient mice inhibits tumor angiogenesis and metastatic progression. *FASEB J.*, **19** (13), 1884–1886.
141. Guo, H. *et al.* (2006) Disruption of EphA2 receptor tyrosine kinase leads to increased susceptibility to carcinogenesis in mouse skin. *Cancer Res.*, **66** (14), 7050–7058.
142. Brantley, D.M. *et al.* (2002) Soluble Eph A receptors inhibit tumor angiogenesis and progression in vivo. *Oncogene*, **21** (46), 7011–7026.
143. Cheng, N. *et al.* (2002) Blockade of EphA receptor tyrosine kinase activation inhibits vascular endothelial cell growth factor-induced angiogenesis. *Mol. Cancer Res.*, **1** (1), 2–11.
144. Gale, N.W. *et al.* (2001) Ephrin-B2 selectively marks arterial vessels and neovascularization sites in the adult, with expression in both endothelial and smooth-muscle cells. *Dev. Biol.*, **230** (2), 151–160.
145. Shin, D. *et al.* (2001) Expression of ephrinB2 identifies a stable genetic difference between arterial and venous vascular smooth muscle as well

as endothelial cells, and marks subsets of microvessels at sites of adult neovascularization. *Dev. Biol.*, **230** (2), 139–150.

146. Wang, H.U., Chen, Z.F., and Anderson, D.J. (1998) Molecular distinction and angiogenic interaction between embryonic arteries and veins revealed by ephrin-B2 and its receptor Eph-B4. *Cell*, **93** (5), 741–753.

147. Makinen, T. *et al.* (2005) PDZ interaction site in ephrinB2 is required for the remodeling of lymphatic vasculature. *Genes Dev.*, **19** (3), 397–410.

148. Adams, R.H. *et al.* (1999) Roles of ephrinB ligands and EphB receptors in cardiovascular development: demarcation of arterial/venous domains, vascular morphogenesis, and sprouting angiogenesis. *Genes Dev.*, **13** (3), 295–306.

149. Gerety, S.S., Wang, H.U., Chen, Z.F., and Anderson, D.J. (1999) Symmetrical mutant phenotypes of the receptor EphB4 and its specific transmembrane ligand ephrin-B2 in cardiovascular development. *Mol. Cell*, **4** (3), 403–414.

150. Gerety, S.S. and Anderson, D.J. (2002) Cardiovascular ephrinB2 function is essential for embryonic angiogenesis. *Development*, **129** (6), 1397–1410.

151. Fuller, T., Korff, T., Kilian, A., Dandekar, G., and Augustin, H.G. (2003) Forward EphB4 signaling in endothelial cells controls cellular repulsion and segregation from ephrinB2 positive cells. *J. Cell Sci.*, **116** (Pt 12), 2461–2470.

152. Scehnet, J.S. *et al.* (2009) The role of Ephs, Ephrins, and growth factors in Kaposi sarcoma and implications of EphrinB2 blockade. *Blood*, **113** (1), 254–263.

Part IV
Therapeutic Approaches and Angiogenesis

Tumor Angiogenesis – From Molecular Mechanisms to Targeted Therapy. Edited by Francis S. Markland,
Stephen Swenson, and Radu Minea
Copyright © 2010 WILEY-VCH Verlag GmbH & Co. KGaA, Weinheim
ISBN: 978-3-527-32091-2

9
Development of an Integrin-Targeted Antiangiogenic Agent

Stephen Swenson, Radu Minea, and Francis S. Markland

Angiogenesis is a critical process in tumor and disease progression. A number of features are central to both tumor growth and development, and the recruitment and invasion of a growing vascular network supplying the tumor with nutrients and a mechanism of escape to form metastatic foci. One class of molecules important to both processes is a family of cell surface receptors identified as integrins. Integrins are α/β heterodimeric glycoproteins in which the different α subunits combine with distinct β subunits resulting in a range of specificities toward various extracellular matrix (ECM) proteins. Several classes of integrins recognize the RGD sequence present in ECM proteins, allowing integrins to link cytoskeletal proteins with the ECM and to be involved in bidirectional signaling that alters cellular functions. Among these interactions are the adhesion of both endothelial cells and cancer cells to ECM proteins, interactions that are integral to metastasis, tumor growth, and angiogenesis. Function blocking antibodies targeted to integrins have been shown to retard tumor growth and subsequent tumor-induced angiogenesis. The problem with this approach is that a single integrin is targeted, which may allow the tumor to circumvent this type of blockage. A more broad-spectrum agent that binds to and blocks the function of several different integrins at a time is available, these agents are known as *disintegrins*. Originally purified from the venom of Viperidae family of snakes, disintegrins role in nature is presumably to block platelet aggregation following envenomation based on interaction of an integrin on the activated platelet surface with an RGD sequence in the disintegrin. It has been observed that integrins overexpressed on some tumor types and angiogenic vasculature have similar affinity for RGD motifs found in ECM proteins. On the basis of disintegrin structure, we have developed a recombinant form of a snake venom disintegrin, which we call *vicrostatin* (VN or VCN). VN is a potent antiangiogenic/antitumor agent in *in vitro* and *in vivo* laboratory studies. Further development of the recombinant venom–derived disintegrin along with new technology looking at stand-alone disintegrin domains from human A Disintegrin and Metalloproteinase (ADAM) family members protein may offer a novel therapeutic approach in targeting tumor-induced angiogenesis.

Tumor Angiogenesis – From Molecular Mechanisms to Targeted Therapy. Edited by Francis S. Markland, Stephen Swenson, and Radu Minea
Copyright © 2010 WILEY-VCH Verlag GmbH & Co. KGaA, Weinheim
ISBN: 978-3-527-32091-2

9.1 Toward Molecular Treatment of Cancer

At a cellular level, the attachment of metastatic cells is highly dependent on the successful induction by the transformed cell of normal cell–cell and cell–matrix interactions, so that the newly established colonies can integrate with circulatory and other physiological systems. These interactions are extremely complex and involve surface interactions between tumor cells and surrounding tissues [1], and the integrin class of molecules is particularly important to these interactions. Disintegrins are small, disulfide-rich peptides, many of which contain an Arg-Gly-Asp (RGD) sequence, that bind to integrins on the surface of normal and malignant cells. Potent natural disintegrins have been identified and evaluated for their ability to limit cancer progression including the limitation of tumor-induced angiogenesis. The focus of this chapter is to give the description of the development of an integrin-targeted antiangiogenic cancer therapeutic. The exemplar molecules used in this description are the powerful natural snake venom–purified disintegrin contortrostatin (CN) and its recombinant derivative VN, both of which are potent inhibitors of integrin binding [2–4].

Angiogenesis is defined as the formation of new blood vessels, a process that requires new capillaries to sprout from existing blood vessels. It is essential in a number of physiological processes, including development, reproduction, and wound repair. Under these conditions, angiogenesis is highly regulated and the normally quiescent vasculature is activated only for a brief period. However, persistent unregulated angiogenesis has been implicated in several pathological conditions, notably in tumor development [5]. The importance of angiogenesis for tumor growth and the therapeutic implications of this phenomenon were first elucidated by Dr. Judah Folkman as early as 1971 [6]. Subsequently, considerable research in this field has shown that the growth of new blood vessels is required for tumor growth and metastasis. The newly developed vasculature is used by tumors to obtain oxygen and nutrients and discard waste products. In addition, the blood vessels embedded within the tumor allow tumor cells to escape the primary tumor by entering the circulatory system and spreading to distant sites (metastasis).

Integrins are α/β heterodimeric glycoproteins [7], the different α subunits combine with distinct β subunits resulting in a range of specificities toward various ECM proteins [8]. Several classes of integrins recognize the RGD sequence present in ECM proteins [9], allowing integrins to link cytoskeletal proteins with the ECM and to be involved in bidirectional signaling that alters cellular functions. Among these interactions are the adhesion of both endothelial cells and cancer cells to ECM proteins [9], interactions that are integral to metastasis, and tumor growth. The importance of vitronectin receptors (integrins $\alpha v\beta 3$ and $\alpha v\beta 5$) in angiogenesis is well known [10, 11]. A monoclonal antibody to integrin $\alpha v\beta 3$, as well as a cyclic RGD-containing peptide [12], perturbed angiogenesis and produced regression of a human cancer growing on CAM (chorioallantoic membrane). Administration of antagonists of $\alpha v\beta 3$ apparently causes apoptosis of vascular endothelial cells responsible for angiogenesis, following selective activation of p53 and increased expression of the p53-inducible cell cycle inhibitor $p21^{WAF1/CIP1}$ *in vivo* [13]. Thus, *in vivo* survival of vascular endothelial cells depends on attachment and spreading on the ECM. Integrin $\alpha v\beta 5$ has also been found to play a role in angiogenesis.

An anti-$\alpha v\beta 3$ antibody blocked angiogenesis induced by basic fibroblast growth factor (bFGF), whereas an anti-$\alpha v\beta 5$ antibody blocked vascular endothelial growth factor (VEGF)-induced angiogenesis [14]. This suggests that downstream signal transduction pathways of the two growth factors are distinct, and indicates that there are two angiogenic pathways mediated by distinct αv integrins. Integrin $\alpha 5\beta 1$ antagonists inhibit tumor angiogenesis, and block metastases in an animal tumor model [15]. Both $\alpha 5\beta 1$ and $\alpha v\beta 3$ have been shown to participate in important angiogenic pathways [16]; inhibition of endothelial cell lumen formation was achieved only by blocking both $\alpha v\beta 3$ and $\alpha 5\beta 1$.

Yet, integrins are not only central to angiogenesis. It was shown in early studies that RGD-containing peptides inhibited experimental peritoneal seeding of human cancer cells [17]. Metastasis to remote sites, including bone, lungs, liver, and brain, involves dissemination of breast cancer cells via the bloodstream and is the primary cause of death in human breast cancer. A significant component of metastasis involves adhesion of cancer cells within the vasculature. Cancer cell adhesion depends on integrins and is dependent on integrin activation. Integrin $\alpha v\beta 3$ supports breast cancer cell attachment under blood flow conditions in an activation-dependent manner [18]. Other findings suggest that integrin $\alpha v\beta 3$ expression in tumor cells accelerates the development of bone lesions, presumably through increased invasion of and adhesion to bone [19]. Bone metastasis was common *in vivo* following introduction of $\alpha v\beta 3$-positive breast cancer cells, but not with $\alpha v\beta 3$ negative cells. Thus, $\alpha v\beta 3$ expression appears to play a key role in the development of bone metastasis from breast cancer [20]. Other integrins, however, appear to be involved in cancer invasion as well, particularly $\alpha 4\beta 6$ [21]. Pathological examination indicates that the level of αv expression in breast cancer is directly related to migration or metastasis [18, 19, 22, 23], with higher levels of integrin associated with a more metastatic phenotype. Because more than one integrin inhibitor was required to block migration of breast cancer (BC) cell lines, BC therapy based on integrin antagonists would most likely require concomitant use of multiple agents [23]. Antibodies to integrins interfere with many of the cell–tissue interactions of the tumorigenic and metastatic processes, including migration, attachment, and angiogenesis [24]. Since a single antagonist, targeting a broad range of integrins, can be envisioned to disrupt all of these activities, integrins present a very attractive target for anticancer drug development. In evolution, this class of molecule has also proven to be an attractive target for interspecies warfare. Disintegrins [25], a class of proteins originally described in snake venoms, avidly attach to a range of integrins, disrupting normal cellular functions. Thus, agents such as CN and VN are ideally suited to block breast cancer progression since they block multiple integrin-mediated pathways.

9.2 Disintegrins as Molecular Weapons against Cancer

Disintegrins are small, disulfide-rich, Arg-Gly-Asp (RGD)-containing peptides that bind to integrins on the surface of normal and malignant cells [26–28]. Disintegrins have been characterized from many snake venoms and were originally characterized as platelet aggregation inhibitors [29–32]. Nuclear magnetic resonance has been used to determine the structures of several disintegrins [33, 34]; they display little secondary structure, and are characterized by an RGD sequence at the tip of a 13–amino acid flexible loop held by disulfide bonds and protruding from the main body of the peptide chain [30].

Integrins are the natural binding partner of disintegrins and a number of disintegrins have been evaluated as antiangiogenic agents. Many disintegrins including echistatin [35], salmosin [36], alternagin [37], and accutin [38] have been evaluated *in vitro* for their ability to inhibit processes critical to angiogenesis. These disintegrins are all purified from different species of snakes from locations around the world. Not all of the venom-derived disintegrins possess RGD at the tip of the flexible loop as evidenced by obtustatin [39]. CN and now VN are the most well studied of the disintegrins in *in vivo* models of different forms of cancer including breast, ovarian, prostate, and glioma [2, 4, 40–44].

CN was purified from southern copperhead venom [32], and is a homodimer. It has a molecular mass (Mr) of 13 500; each chain has an RGD sequence and an Mr of 6700 [32, 45]. The X-ray crystal structure of a closely related disintegrin acostatin reveals unequivocally that the two chains are aligned in an antiparallel array with the RGD motifs at either end of the dimer and about 69.5 Å apart [46]. CN binds competitively to integrins of the $\beta 1$, $\beta 3$, and $\beta 5$ subclasses, including receptors for fibronectin ($\alpha 5\beta 1$), vitronectin ($\alpha v\beta 3$, $\alpha v\beta 5$), and fibrinogen ($\alpha IIb\beta 3$) [32, 45, 47]. The lack of integrin specificity by CN is an advantage in controlling breast cancer growth and dissemination. CN disrupts angiogenesis induced by both bFGF and VEGF, consistent with the observation that a cyclic RGD-peptide had no selectivity among αv family members [14]; in animal models of cancer, CN demonstrates antiinvasive, antiangiogenic, and antimetastatic activities [44, 45, 48, 49]. For example, in a recent study [40], human umbilical vein endothelial cells (HUVEC) were exposed to CN for zero to three days and then tested for migration, invasion, and tube formation. All three activities were inhibited by up to 90%. Further, immunochemical staining showed both actin and VE (vascular endothelial)-cadherin organization to be disrupted by CN in HUVEC [40].

While *in vitro* the natural purified form of CN is fully functional in biological assays, when administered *in vivo* the naked protein displays no activity at the milligrams per kilogram levels we have evaluated. In an effort to extend the circulatory half-life as well as develop a passive targeting system to the growing tumor, we have developed a nanosomal encapsulation formulation of CN. Nanosomes are submicroscopic spheres composed of thin but durable membranes made primarily of phospholipids and cholesterol. The composition, number of lipid layers, size, charge, and permeability of the membrane can be altered to enhance delivery of a variety of therapeutic agents encapsulated inside the nanosome [50, 51]. Advantages associated with lipid encapsulation include (i) enhanced drug delivery to the desired site; (ii) prolonged drug half-life and thus reduced dosing frequency; and (iii) reduced drug-related toxicities. In the next section we describe our success in utilizing a nanosomal formulation of VN liposomal vicrostatin (LVN) as an effective therapeutic in both prostate and breast cancer. In addition, our previous studies with nanosomal formulations of snake venom–derived and recombinant disintegrins showed undetectable immunogenicity, extended circulatory half-life, and lack of nontarget effects [4].

In vivo studies have shown that CN is well tolerated and can be infused without detrimental effect on blood pressure, body temperature, or other physiological parameters when evaluated in a canine model of arterial occlusion. CN-treated animals did not exhibit changes in heart rate, EKG, or blood coagulation parameters, and CN showed no evidence of toxicity at the doses tested [52]. In chronic studies in nude mice, CN was administered in a nanosomal formulation by both daily intratumor (IT) and twice weekly intravenous (IV) injection over seven weeks. There were no visible side effects or signs of internal bleeding, indicating that mice tolerate chronic administration of CN quite well [4, 44].

9.3 Recombinant Expression of a Venom-Derived Disintegrin

For a number of years our laboratory has worked with CN, a disintegrin isolated from *Agkistrodon contortrix contortrix* (southern copperhead) snake venom. A major obstacle in the pathway to clinical development of CN was the supply of the protein itself; for purification it exists only as a very small fraction of the total venom protein (~0.01%), and for recombinant production, its peculiar structure stabilized by numerous disulfide bonds makes its expression in commonly employed recombinant systems a very difficult task. In addition, it has recently been reported that there is variation in the protein composition of the venom based on geographical location of the individual snake as well as the age of the snake. This would make quality control of clinical material an issue in moving CN to the clinic. Nonetheless, we have successfully employed a recombinant expression system for which we developed a proprietary production method capable of generating

```
Contortrostatin   DAPANPCCDAATCKLTTGSQCADGLCCDQCKFMKEGTVCRRARGDDLDDYCNGISAGCPRNPFHA---
                   ************************************************************** .:.
Vicrostatin       GDAPANPCCDAATCKLTTGSQCADGLCCDQCKFMKEGTVCRRARGDDLDDYCNGISAGCPRNPHKGPAT
```

Figure 9.1 Vicrostatin is a C-terminal modified form of a monomer of contortrostatin. The recombinant disintegrin VN is based on the sequence of CN with the addition of a portion of the C-terminal sequence of echistatin in an effort to enhance the affinity of the recombinant molecule for integrin $\alpha5\beta1$.

approximately 200 mg of purified active recombinant disintegrin from 1 l of bacterial culture in small-scale laboratory conditions. To generate recombinant disintegrins, we have successfully adapted a commercially available *Escherichia coli* expression system consisting of the Origami B (DE3) expression host in combination with the pET32a vector (Novagen) for our production needs. A sequence-engineered form of CN, called *VN* (Figure 9.1), has been directionally cloned into pET32a expression vector under the control of an isopropyl β-D-1-thiogalactopyranoside (IPTG) inducible promoter. In addition, a unique tobacco etch virus (TEV) protease cleavage site has been incorporated into the construct, which facilitates the removal of the thioredoxin fusion partner from the expressed VN. Briefly, for recombinant VN production, multiple colonies of transformed Origami B cells were used to establish primary cultures by inoculating 5 ml LB broth batches containing carbenicillin (100 $\mu g\ ml^{-1}$), tetracycline (12.5 $\mu g\ ml^{-1}$), and kanamycin (15 $\mu g\ ml^{-1}$). The primary cultures were grown overnight at 37 °C and 250 rpm in a shaker incubator and used to seed secondary cultures. Batches of fresh 500 ml LB broth in the presence of the three antibiotics were then inoculated with the previously established primary cultures and grown at 37 °C and 250 rpm to an OD600 of 0.6–1.0. At this point, the cells were induced with IPTG added to a final concentration of 1 mM and cultured for another 4–5 h at either 25 °C or 37 °C and 250 rpm. At the end of the induction period, the cultures were centrifuged at 4000 g and bacterial pellets lysed by a scalable homogenization method; cells were homogenized in a microfluidizer (Microfluidics M-110L, Microfluidics, Newton, MA) at room temperature by resuspending the cell pellets in five volumes of water before commencing the homogenization. The operating conditions of the homogenizer included applied pressures of 14 000–18 000 psi, bacterial slurry flow rates of 300–400 ml min^{-1} and multiple passes of the slurry through the processor. The insoluble cellular debris was then removed by centrifuging the bacterial lysate at 40 000 g and the soluble cell lysate was collected and further analyzed by SDS-PAGE for recombinant protein expression. The expressed fusion protein, thioredoxinA-VN (Trx-VN), was proteolyzed by adding recombinant TEV protease to the soluble cell lysates according to the manufacturer's protocol (Invitrogen). TEV protease treatment efficiently cleaved VN from its TrxA fusion partner; the status of proteolytic cleavage was monitored by SDS-PAGE. When proteolysis was complete the proteolyzed lysate was passed through a 0.22 μm filter, diluted 100-fold in water, and ultrafiltrated through a 50 kDa molecular weight cutoff (MWCO) cartridge (Biomax50, Millipore, MA) in a tangential flow ultrafiltration device (Labscale TFF system, Millipore, MA) that removed most of the higher molecular weight bacterial proteins. The resulting ultrafiltrates were then reconcentrated against a 5 kDa MWCO cartridge (Biomax5, Millipore MA) using the same tangential flow ultrafiltration device. VN was further purified by reverse phase HPLC. The recombinant disintegrin we have produced through this system is recognized by polyclonal antisera raised against native CN and inhibits ADP-induced platelet aggregation in a dose-dependent manner with an IC_{50} almost identical to native CN (~60 nM) [2]. Moreover, VN inhibits tumor cell adhesion to vitronectin and fibronectin, inhibits endothelial cell and tumor cell invasion through a laminin-rich reconstituted basement membrane, and inhibits endothelial cell tube formation in a manner indistinguishable from native venom-derived CN. In conclusion, due to its robustness and reproducibility, we believe that our recombinant production method will be easily translatable for scale-up and cGMP production.

9.4 Functional *In vitro* Evaluation of VN

9.4.1 Integrin Binding

In order to examine VN binding to different breast cancer cell lines, we characterized the integrin profile of several of these cell lines by fluorescence-activated cell sorting (FACS) analysis using fluorescently labeled anti-integrin monoclonal antibodies (Table 9.1). We have shown that VN binds with different affinities to a panel of human breast cancer cell lines dependent on the integrin display status of the individual cell line. Using an FACS-based assay we evaluated the ability of fluorescently labeled VN to bind to HUVEC and the breast cancer cell lines MDA-MB-435 and MDA-MB-231 cells (Figure 9.2). This study revealed that while each of the cells has a different display of integrins, VN binds to both tumor cell lines as well as endothelial cells. In further experiments we observed differential binding of VN to different subclones of MDA-MB-231 (data not shown), which we attribute to their slightly altered integrin profiles. These two subclones, when injected intraventricularly in nude mice, are shown to preferentially metastasize to different locations in the body, bone (MDA-MB-231BO), or brain (MDA-MB-231BR) [53].

Table 9.1 Integrin profiles of selected breast cancer cell lines.

	Relative integrin surface expression				
	αv	$\alpha v\beta 3$	$\alpha 5\beta 1$	$\beta 1$	$\alpha 2\beta 1$
Breast cancer cell lines					
MDA-MB-231 (parent)	+++	+	+++	+++	+++
MDA-MB-231BO	+++	+	++	+++	+++++
MDA-MB-231BR	+++	+	++	++	+++++
MDA-MB-468	−	−	+	++	+
MCF-7	+++	−	+	+	+++
Reference cell line					
MDA-MB-435	+++	+++	++	++	+

− Means no detectable expression; + to +++++ indicates increasing levels of expression. MDA-MB-231BO and MDA-MB-231BR are cell lines that have been shown to metastasize preferentially to either bone (BO) or brain (BR).

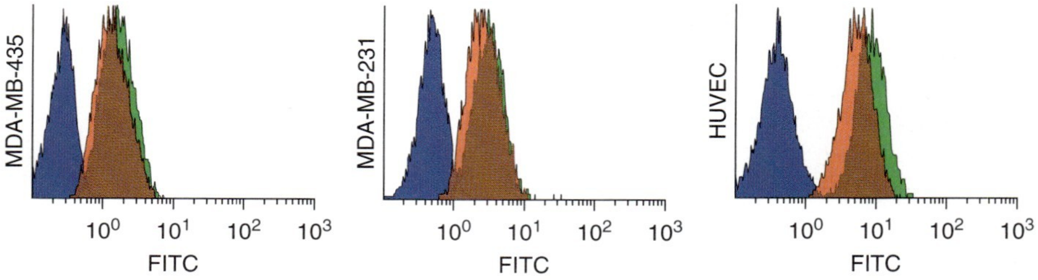

Figure 9.2 Binding analysis of FITC-VN using FACS analysis. MDA-MB-435, MDA-MB-231, and HUVEC were incubated with FITC alone (blue), FITC-CN (red), and FITC-VN (green). The direct binding of labeled protein was assessed by flow cytometry. As shown in the above representative flow-cytometric plots, FITC-labeled VN binds avidly to MDA-MB-435, MDA-MB-231, and HUVEC.

The results of these studies support our hypothesis that the promiscuous nature of integrin binding by VN allows for broad targeting toward many tumor types.

9.4.2 Binding Affinities of VN to Integrins

To assess the binding affinities of VN with soluble functional integrins, two approaches were evaluated. First, the disintegrins were immobilized on the surface of a BiaCore surface Plasmon resonance chip. The soluble integrin was then allowed to interact with the immobilized disintegrin. This approach was unsuccessful with VN, presumably because the integrins had limited access to the disintegrin. This was confirmed by flowing an anti-VN antibody across the surface and this also displayed limited binding. It is believed that the disintegrin interaction with the carboxymethyl cellulose surface causes VN to be buried and unavailable for binding. In an attempt to solve this issue fluorescence polarization (FP) was used to determine binding kinetics. In this method, differing concentrations of functional integrin were incubated with a constant amount of FITC (fluorescein isothiocyanate)-labeled VN. As VN is a small molecule it rapidly depolarizes the excitation light. Upon binding to the large integrin, the fluorescent tag on VN tumbles in solution at a slower rate resulting in increased levels of polarization. The measured FP value is a weighted average of FP values of the bound and free fluorescent VN and is therefore a direct measure of the fraction bound. Data generated in these experiments can be analyzed like standard radioligand binding, and kinetics of binding can be determined as with Scatchard analysis using a nonlinear curve fit. From this set of experiments, we determined the dissociation constants for VN and CN with integrins $\alpha v\beta 3$, $\alpha 5\beta 1$, and $\alpha v\beta 5$ (Table 9.2). Recombinant VN was purposely designed with a carboxy terminal extension, which was expected to enhance affinity for $\alpha 5\beta 1$. This was confirmed as CN and VN exhibit nearly identical affinities for $\alpha v\beta 3$ and similar affinity for $\alpha v\beta 5$, while there is over an order of magnitude difference in the K_d values for binding to $\alpha 5\beta 1$ when comparing VN (higher affinity binding) to CN.

Table 9.2 Binding affinities of VN and CN for selected integrins.

Disintegrin	Integrin K_d		
	$\alpha v\beta 3$	$\alpha 5\beta 1$	$\alpha v\beta 5$
CN (nM)	6.6	191.3	19.5
VN (nM)	7.4	15.2	41.2

9.4.3
Inhibition of Cellular Processes Critical to Tumor Progression

To assess the ability of VN to block processes critical to tumor survival, progression, and angiogenesis – adhesion, migration, and invasion, we measured its inhibitory effect on different tumor cell lines, as well as endothelial cells. Inhibition of adhesion is evaluated through the ability of VN to block cell attachment to a number of different ECM proteins. The problem with this type of assay is that despite increasing the concentration of VN, complete inhibition of adhesion will never be achieved since cells use a panel of integrins to attach to ECM proteins including some that are not targeted by VN. Nonetheless, we observed a dose dependence in the inhibition of adhesion of MDA-MB-231 and MDA-MB-435 cells, to both vitronectin and fibronectin, ECM proteins that are ligands for integrins targeted by VN (data not shown). Another process, cellular migration, is inhibited by VN in a dose-dependent manner. To evaluate tumor and endothelial cell migration, a phagokinetic tracking assay is employed. In this assay cells are plated on a collagen coated coverslip with a substratum of colloidal gold. As the cells move they displace or injest the colloidal gold leaving tracks on the surface of the coverslip. Then, using dark-field microscopy the tracks can be visualized and photographed. Using image analysis software the area of the tracks in a photographed field can be determined and a "migration index" can be calculated as a percentage of the field lacking gold. Comparison of the migration indices between the different treatment groups and the untreated controls reveals the inhibition of migration by VN. Untreated cells will migrate and denude a significant portion of the vision field; however, following treatment by increasing concentrations of VN the migration of both tumor and endothelial cells is significantly limited (Figure 9.3). Finally, the ability of cells to invade through the ECM was evaluated using modified Boyden chambers. These chambers contain a Matrigel coated porous membrane (pore size 8 μm). A chemoattractant is placed in the lower chamber and untreated cells invade through the membrane toward the attractant. As seen in Figure 9.4, VN blocks the invasion of endothelial (HUVEC) and tumor cells (MDA-MB231) in a dose-dependent manner. A positive control cytochalasin D (CytoD, a cell permeable and potent inhibitor of actin polymerization) was included. These results show that VN has essentially identical activity in inhibiting invasion of endothelial and breast cancer (MDA-MB-231) cells. The results also convincingly demonstrate one

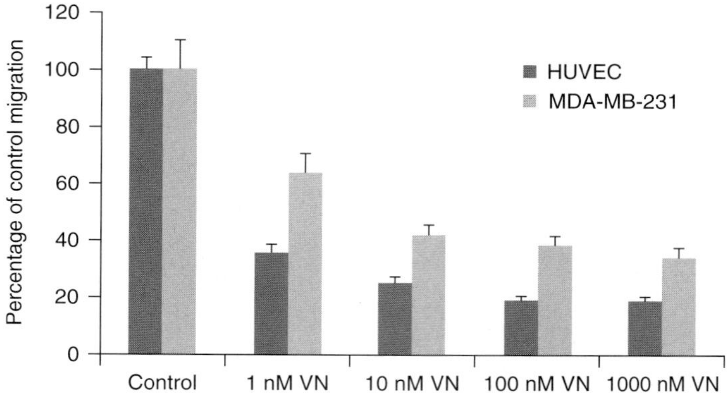

Figure 9.3 VN inhibits endothelial and tumor cell migration. Utilizing a colloidal gold cell tracking assay we evaluated the efficacy of VN, at different concentrations, to inhibit cellular migration. With the breast cancer cell line MDA-MB-231, a gradual decrease in migration was observed. With HUVEC the inhibition of migration was more immediate with 1 nM VN inhibiting migration by >65% and increasing concentrations of VN further inhibited migration.

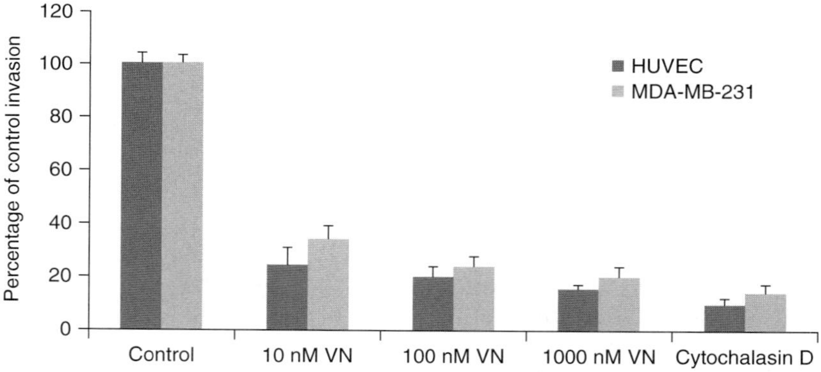

Figure 9.4 VN inhibits endothelial and tumor cell invasion. Using a Boyden chamber assay and fluorescent quantitation of invaded cells, VN inhibit invasion of vascular endothelial cells (HUVEC) as well as a breast cancer cell line (MDA-MB-231) at low nanomolar concentrations. A positive control, cytochalasinD (CytoD), was used to identify the level of fluorescence for complete actin polymerization disruption.

of the important attributes of VN that it inhibits endothelial cell as well as tumor cell invasion in the low nanomolar range.

9.4.4 Inhibition of HUVEC Tube Formation

To assess the ability of VN to interfere with tube formation, HUVEC were maintained in EGM-2 complete media and grown to confluency. The HUVEC were

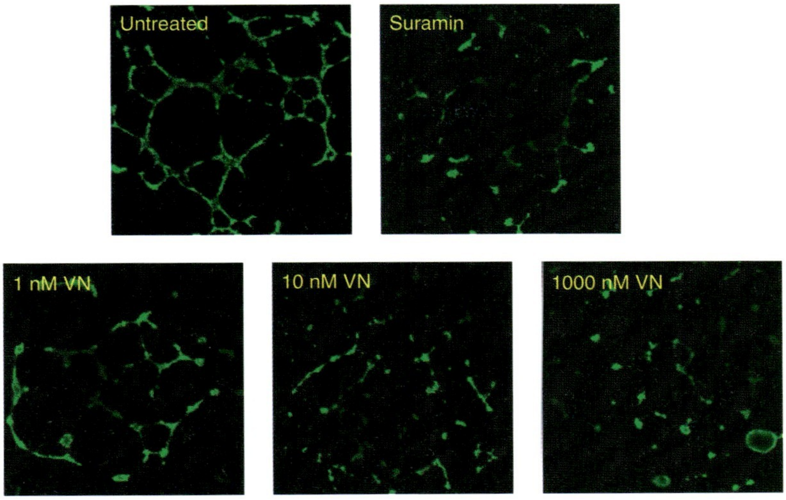

Figure 9.5 Inhibition of HUVEC tube formation by VN. HUVEC were plated on "Endothelial Cell Tube Formation "plates (BD Biosciences) in the presence of various concentrations of VN (0–1000 nM), or a known tube formation inhibitor Suramin (used as a positive control). Representative figures from independent experiments are shown above. Cells were stained with Calcein AM and imaged using confocal microscopy.

then harvested by brief trypsinization, washed in the presence of soybean trypsin inhibitor (1 mg ml^{-1}), and resuspended in basal media. After being maintained in suspension for 15–30 min, cells were seeded onto Endothelial Cell Tube Formation plates (BD BioCoat™ Angiogenesis System), an *in vitro* endothelial tubulogenesis system, at a concentration of 25 000 cells per well and immediately treated with various concentrations of VN, CN (positive control), or Suramin (supplier provided positive control) and incubated for 18 h at 37 °C. At the end of the incubation period, cells were washed twice with phosphate buffered saline (PBS) and then stained with 8 μg ml^{-1} Calcein AM in PBS at 37 °C. After 30 min the cells were washed again two times with PBS and then imaged (Figure 9.5) using confocal microscopy at 2.5× and 10× magnification. On the captured images, the total length of tubes was quantitated with Zeiss LSM image software and data were plotted against the total length of tubes (in micrometers) generated by untreated cells. Representative tubes from six different wells were measured by three individuals and averaged to form each data point (Figure 9.6). VN exhibits potent dose-dependent inhibition of tube formation at low nanomolar concentrations.

9.4.5
Nanosomal Encapsulation of VN

As previously described naked disintegrins display no therapeutic efficacy when administered intravenously in animal models of human cancer. Previously we had encapsulated CN within a unilamellar nanosphere composed of phospholipids and

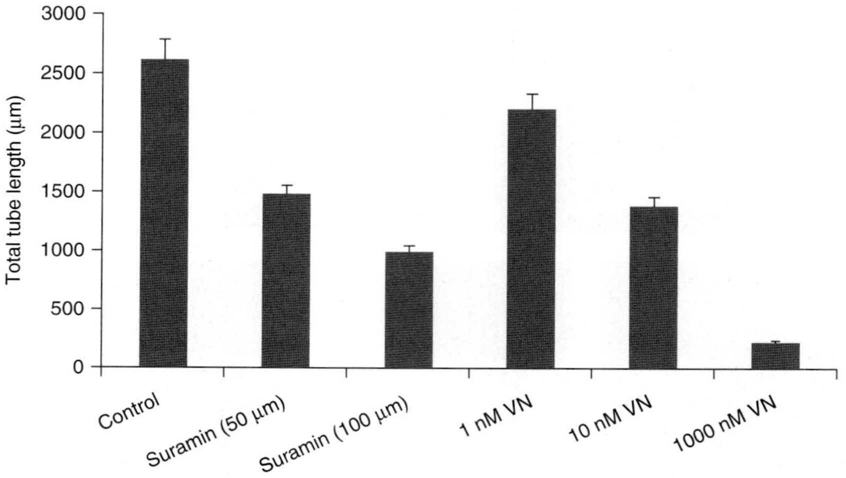

Figure 9.6 Quantitation of tube formation inhibition by varying concentrations of VN. The tubes formed by HUVECs were quantitated in multiple fields collected from three repeated experiments by computing the total tube length with Zeiss LSM image software and averaged to form each data point. The data shown above was assembled from multiple independent experiments.

cholesterol. VN was encapsulated in a similar manner yielding the therapeutic agent to be delivered to the study animals. Briefly, to prepare LVN, stock solutions of phospholipids and cholesterol were prepared by dissolving each lipid in a chloroform/methanol solvent mixture (liposomal formulation of CN and VN was prepared by Molecular GPS Technologies, Gary Fujii, Ph.D., CEO, Rancho Dominguez, CA). Thin lipid films were created by pipetting aliquots of the lipid solutions into round bottom glass tubes followed by solvent evaporation at 65 °C under a stream of nitrogen gas. The dried lipids and cholesterol were further dried under vacuum for 48 h. This process yielded lipid powder mixtures that were used to prepare LVN. VN was dissolved in a hydration buffer (10 mM sodium phosphate and 262 mM sucrose, pH 7.2) and added to the dried lipids. The lipid dispersion was incubated for 5 min at 50 °C. The LVN was formed by homogenization by passing the material through a microfluidizer (M110L; Microfluidics, Newton, MA). The material was processed between 10 000 and 18 000 psi while maintaining an elevated temperature (45–65 °C). Samples of the nanosome batch were taken during the process and the size distribution of LVN was determined – 60–80-nm diameter with an average of 72 nm, with an ultrafine particle analyzer (UPA150) (Microtrac, North Largo, FL). After processing, unencapsulated VN was removed by ultrafiltration using an Amicon UF membrane of 100 000 MWCO and the LVN sterilized by filtration through a 0.2 μM PVDF (polyvinylidene fluoride) filter. Since homogenization has been used by others in the industry for nanosome preparations, we are encouraged that LVN can be scaled to volumes necessary for commercialization.

9.5
Functional *In vivo* Evaluation of VN

9.5.1
Circulatory Half-Life of LVN

Previously we had determined the circulatory half-life of CN and nanosomal CN (LCN). We repeated these studies for VN and nanosomal VN (LVN). To carry out this study, blood samples were taken at different times out to 72 h following IV administration of ^{125}I-VN or L-^{125}I-VN. Gamma counting of collected blood samples revealed that there was a rapid decrease to <0.5% of the administered counts in the blood at 6 h after IV injection of ^{125}I-VN (Figure 9.7). However, in animals given L-^{125}I-VN, the percentage of total injected counts in the blood drops to a level of 63% of the injected counts 6 h postinjection and gradually decreases over the following 66 h. By plotting the decrease in radioactivity in blood over time following IV administration in tumor-free mice, we observed a circulatory half-life of 0.4 h for ^{125}I-VN and 20.4 h for L-^{125}I-VN. Thus, encapsulation of VN in nanosomes not only protects the protein but also maintains it in the circulation for a much longer period than native unencapsulated VN, enabling more effective access to the tumor.

9.5.2
In vivo Biological Assay

We examined the effect of treatment of breast cancer with nanosomal VN (LVN). To determine the functional efficacy of two preparations of homogenized LVN, we

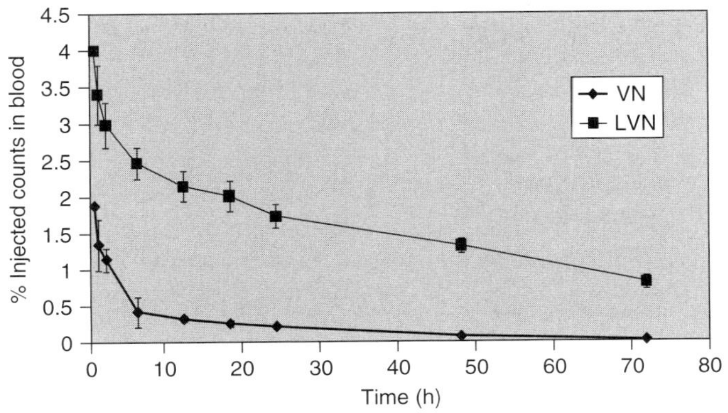

Figure 9.7 Nanosomal VN has an extended circulatory half-life as compared to unencapsulated VN. Five non-tumor-bearing mice per group were injected IV with 20 μCi of native or nanosomal ^{125}I-VN, and 50 μl of blood was drawn from the retroorbital sinus at the times indicated. Nanosomal VN has circulatory $t^1/_2$ = 20.4 h. Native VN is rapidly eliminated from the circulation with $t^1/_2$ = 0.4 h.

used a mouse model of human cancer. For these studies, human metastatic cancer cells (MDA-MB-435, 5 × 10^5 cells) were implanted in the mammary fat pads of five-week-old nude mice. At the time of these studies controversy arose as to the tissue of origin of the MDA-MB-435 cell line. On the basis of the analysis of gene expression, it has been proposed that the cell line is of melanoma origin, not breast cancer cell line [54–56], but more recent evidence suggests that MDA-MB-435 is indeed a poorly differentiated aggressive breast tumor line that expresses both epithelial and melanocytic markers [57]. Nonetheless, the tumor studies were completed with this cell line with future studies utilizing a different cell line without the cloudy background. Tumors were allowed to grow until palpable (14 days) at which time drug administration was initiated. The preparations of homogenized LVN were administered twice weekly via intravenous injection (105 μg per dose for seven weeks), a PBS control was included. Tumors were measured weekly via caliper in a blind fashion. At the end of the treatment period there is a significant inhibitory effect on tumor growth (~80% inhibition) as compared to the PBS-treated control with both the homogenized and research-scale LVN preparations (Figure 9.8). Since the lineage of MDA-MB-435 cells as being of breast cancer origin has come under question [57, 58], these studies were repeated

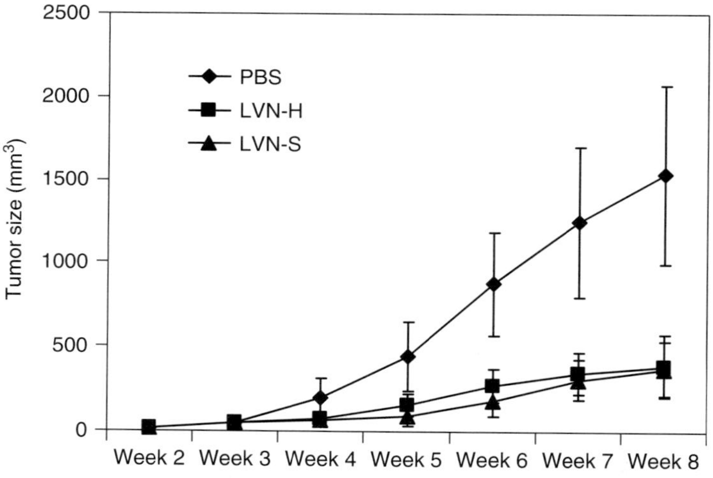

Figure 9.8 Nanosomal disintegrins inhibit the growth of MDA-MB-435 xenografts in a mouse orthotopic model. Nude mice were inoculated orthotopically (mammary fat pads; 5 × 105 MDA-MB-435 cells per mouse). At two weeks following implantation, tumors became palpable and the treatment was commenced. The animals were treated for seven weeks with liposomal formulated disintegrins, prepared by homogenization LVN-H or prepared using the research-scale sonication method (LVN-S) (100 μg twice weekly, intravenous tail vein administration), and compared to a control group (PBS). This animal model was repeated three times and the tumor growth inhibition was quantitated. A significant reduction of tumor growth was observed with both the homogenized formulation and the material prepared by sonication.

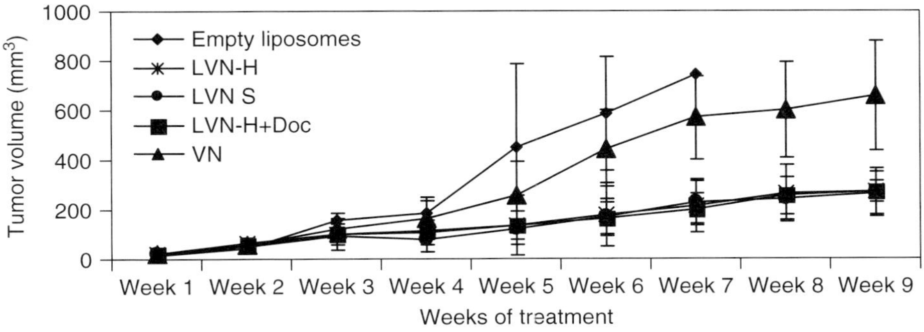

Figure 9.9 Liposomal formulations of VN inhibit the growth of MDA-MB-231 xenografts in a mouse orthotopic model. Nude mice were inoculated orthotopically (mammary fat pads; 2.5 × 106 MDA-MB-231 cells per mouse). At 20 days following implantation, tumors became palpable and the treatment was commenced. The groups of five animals were treated for nine weeks with nanosomal formulated disintegrin prepared by homogenization (LVN-H). In addition, naked VN was administered. All samples were delivered by 100 μg twice weekly, intravenous tail vein injections, and compared to a control group (empty liposomes). The LVN formulation slows tumor growth while the empty liposomes do not affect tumor progression and growth. Docetaxel (DOC, 4 μg g^{-1}) intraperitoneally once per week was used as the chemotherapeutic agent, and in combination with LVN (LVN-H plus DOC) using the same LVN dosing regimen as described. The empty liposome control mice began to die in week 5 and all were dead before measurement in week 8.

using the human breast cancer MDA-MB-231 cell line that does not have the questionable background of the 435 line. In the experiment with MDA-MB-231 cells two different nanosomal formulations of VN were investigated with the addition of groups receiving naked VN as well as a combination of LVN with docetaxel (DOC) (Figure 9.9). There is a similar response to that observed with the 435 cells treated with LVN. In addition, survival is enhanced by treatment with LVN since ≥80% of mice treated with any of the LVN formulations survived for >70 days of treatment while those receiving empty nanosomes died on or before day 58 (Figure 9.10). Figure 9.9 clearly shows that unencapsulated VN has virtually no tumor suppressive activity as compared to LVN. The combination of DOC with LVN did not appear to show any additive activity. However, when animal survival was examined, there was a clear benefit of the combination (Figure 9.10) since at day 73 all animals treated with combination therapy were still alive. In additional combination studies we evaluated the efficacy of LVN in combination with Avastin (data not shown). We used a therapeutic dose of both Avastin and LVN and observed no additive effect on inhibition of tumor growth. These studies are ongoing using different dosing regimens in which lower doses of each agent is being employed; we are monitoring tumor growth suppression as well as animal survival for indications of a beneficial effect of the combination therapy.

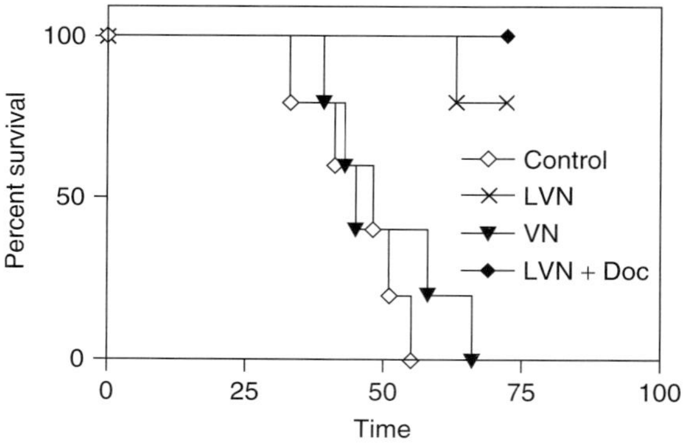

Figure 9.10 Survival of animals treated with homogenized LVN. In the MDA-MB-231 model shown previously survival of the animals was evaluated. In the treatment groups, survival was significantly extended, while the control animals were all dead prior to the eighth week of treatment. In addition, the naked VN–treated animals were all dead by the ninth week of treatment. Treatment groups are the same as shown previously (Figure 9.9).

9.6 Antiangiogenic Effect of LVN Therapy

9.6.1 Tumor Tissue Harvest, Sectioning, Fixation, and Staining

When the tumors became too large (~1.5 cm × 1.5 cm) or the mice became too sick in the above therapeutic efficacy study, we sacrificed the mice. Tumors were excised and placed into an embedding base mold (Tissue Tek Hatfield, PA). The tumors were then immersed in optimal cutting temperature (OCT) to ensure a uniform freezing without ice crystal formation (Tissue Tek Hatfield, PA). Once the tissues were in the mold, and embedded in OCT, they were placed on dry ice for 10–15 min. The tumor blocks were then stored at −80 °C. The frozen blocks were sectioned on a cryostat. The sections were taken from three regions of the tumor, the first, second, and third 20 μm segment of the tumor. The thickness per section was 5 μm. This was done to make sure that the staining was representative of the entire tumor, and not due to the specific portion of the tumor tested. Deeper sections were not taken because they contained necrotic regions, which were devoid of any stainable tissue; therefore, sections were taken only from the first 60 μm of the tumor.

9.6.2 Immunohistochemical (IHC) Staining

The sections were placed on super plus slides (Full Moon Biosystems, Sunnyvale, CA), and fixed in acetone for 10 min. Acetone fixation precipitates proteins, stopping cellular degradation that can destroy antigens. After acetone fixation, the slides were placed on the staining tray and covered with PBS to prevent the sample from drying out.

The first step of immunohistochemical (IHC) staining is blocking in order to stop any nonspecific binding by the secondary antibody. Goat serum (5%) was placed on the tumor sections in order to prevent nonspecific binding of the goat anti-rat secondary antibody. The goat serum blocking solution was made in PBS and placed on the slides for 15 min, then removed and the primary antibody was applied.

The primary antibody was a polyclonal rat anti-mouse CD31/PECAM (BD Biosciences, San Diego, CA), which detects the PECAM antigen on endothelial cells – a specific marker for tumor angiogenesis [59]. Following primary antibody incubation, the slides were washed with PBS three times to remove any unbound antibody, and the secondary antibody – biotinylated goat anti-rat antibody (Vector Laboratories, Burlingame, CA) – at a dilution of 1 : 100 with PBS was placed on the slides (200 μl) and incubated for 45 min. The slides were then washed in PBS as described above. Subsequently, the avidin binding complex (ABC), prepared by diluting 50 μl into 2 ml of PBS (Vector Laboratories, Burlingame, CA), was added to the slides and allowed to incubate for 30 min at room temperature. The slides were washed again in PBS, and then treated with 3-amino-9-ethylcarbazole (AEC) chromogen solution. The red staining identifies the antigen–antibody reaction. The AEC precipitation provides the pigment necessary for visualization. Since the AEC is light sensitive, AEC-treated slides were covered to prevent saturation of the chromogen. After approximately 15 min, the slides were washed in water and counterstained (1.5 min) with Mayer's Hematoxylin to visualize the nuclei. The slides were then mounted with mounting gel and coverslipped (VWR, Westchester, PA).

9.6.3
Evaluation of Tumor-Induced Blood Vessel Growth

To quantitate the blood vessels the stained sections were subject to "random field" analysis [60]. Four randomly assigned fields were chosen and captured on an Olympus E20N digital camera (Olympus America, Melville, NY). To ensure consistency and accuracy, the same magnification of 200× was used, along with the same microscope and digital camera throughout the study. To insure objectivity, the pictures were taken "blindly" without prior knowledge of treatment group. Four images per section and three sections per tumor (one from 0 to 20 μm, one from 20 to 40 μm, and one from 40 to 60 μm) were typically analyzed, 12 images per each tumor. Between 2 and 4 tumors per animal group were analyzed to obtain the microvessel density average in each of the experimental groups.

Once the images were captured, and loaded onto the computer, they were quantitated using simple PCI (Hamimatsu Seuickley, PA). *Blood vessels* were defined as any endothelial cell or endothelial cell cluster, which was distinctly separated from tumor cells or connective tissue [59]. The simple PCI imaging software allows the observer to pick a range of pixel colors denoting blood vessels. Those pixels are then quantified as both a raw number and percent pixels per field of view. The images were analyzed in a blind fashion to reduce bias. After the raw data were compiled, it was analyzed for significance.

As illustrated in Figure 9.11, one can see the microvessel density present in the control (PBS) group stains with ~12% of the total pixels in a field being positive. This means that on average 12% of the viewable field was occupied with positively stained blood vessels. Treatment of the animals with the LVN resulted in a dramatic decrease in tumor-associated microvessel density to 1–2% of the visualized fields (Figure 9.11). The statistical significance between the PBS group and the LVN treatment group reached *P* values on the order of 10^{-18} denoting highly significant findings. In previous studies with MDA-MB-435 tumors the differences between the treatment groups – LVN, Avastin, DOC, and combinations of LVN with both Avastin and DOC – were not statistically significant, demonstrating that LVN can compete as an antiangiogenic with today's standard advanced breast cancer treatments at least in human breast cancer models in the mouse. Another interesting finding is that combining different therapies (e.g., DOC and avastin) had no noticeable additional effect on microvessel density, thus combination treatment did not provide additional advantages in this animal model study. The images shown in Figure 9.11 contain a representative image from the control (PBS) and the LVN treatment group. The control group, PBS, contains larger and thicker blood vessels when compared to the stained tumor sections from the treated group. The tumor section for the control group also has more blood vessels per field of view than the tumor sections from the treated groups. Blood vessels are clearly demarcated by the CD31 staining and can be seen scattered across the sections of the various treated animal tumor sections. Some necrotic regions are evident in the center of the tumors, a characteristic of these tumors after effective therapy. This data further shows the antiangiogenic potential of VN. Treatment groups containing LVN monotherapy has a pronounced inhibitory effect on blood vessel

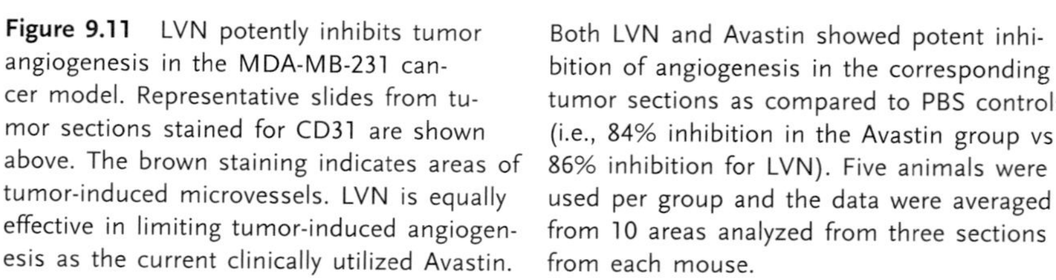

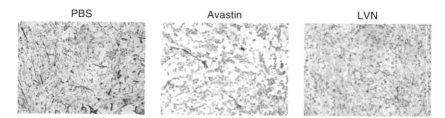

Figure 9.11 LVN potently inhibits tumor angiogenesis in the MDA-MB-231 cancer model. Representative slides from tumor sections stained for CD31 are shown above. The brown staining indicates areas of tumor-induced microvessels. LVN is equally effective in limiting tumor-induced angiogenesis as the current clinically utilized Avastin. Both LVN and Avastin showed potent inhibition of angiogenesis in the corresponding tumor sections as compared to PBS controls (i.e., 84% inhibition in the Avastin group vs 86% inhibition for LVN). Five animals were used per group and the data were averaged from 10 areas analyzed from three sections from each mouse.

growth with similar blood vessel profiles as those containing Avastin, a clinically utilized antiangiogenic agent.

9.7 Toxicology Studies

The toxicity of a single intravenous dose of VN or LVN was evaluated using a rat model. Female Wistar rats (120–130 g) were assigned to 12 groups of 3 animals each, including 2 control groups (PBS and empty liposome control) and 10 experimental groups that received VN or LVN preparations (3, 10, 25, and 75 mg kg^{-1} for VN and 1, 3, 10, 25, 50, and 75 mg kg^{-1} for LVN). Animals were given a single administration of the test agents and evaluated for signs of physical toxicity or stress over 14 days, and then sacrificed at day 14. Signs of toxicity were monitored via physical status, activity, and total body weight; following sacrifice, gross and microscopic pathology was performed and hematological properties were analyzed. There were no adverse effects observed in any of the treated animals. Animals in all treatment groups thrived and gained weight indistinguishable from the control groups. There were no observed changes in behavior immediately following agent administration or throughout the 14 day study. Gross examination following sacrifice revealed no changes in tissue or organs between control and treated animals. There were no significant differences in hematological parameters between even the highest dose (75 mg kg^{-1}) and the controls (Table 9.3). Microscopic examination of major body organs (lung, liver, kidney, pancreas, eyes, heart, and bladder) revealed that there was no observable inflammation, no significant cellular alterations, and no vascular changes in the microscopic sections.

9.8 Summary

Antiangiogenic agents currently in clinical use or clinical trials generally target a single process or molecule involved in the angiogenic pathway such as monoclonal antibodies, that is, bevacizumab, which specifically inhibits the VEGF-A signaling cascade. While this type of specific targeting strategy may be initially effective, it has been shown that tumors can acquire resistance to antiangiogenic therapies, as reviewed by Kerbel *et al.* [61]. Redundant signaling mechanisms may replace the molecule being targeted, for example, that VEGF-A can be replaced by other VEGF family members or pathways unrelated to VEGF during tumor progression. These alternate pathways allow the tumor to bypass the blockage of the function of VEGF-A. In addition, targeting a single pathway applies a positive selective pressure upon tumor cells such that resistant variants may be generated through mutation. An additional selection pressure can be made by the targeted antiangiogenic agent on the tumor through the emergence of hypoxia-resistant mutant tumor cells that

Table 9.3 Average hematologic values for VN LVN toxicity test.

Agent	Dose ($mg\ kg^{-1}$)	Weight (g)	WBC ($\times 10^3\ \mu l^{-1}$)	RBC ($\times 10^6\ \mu l^{-1}$)	HGB ($g\ dl^{-1}$)	MCV (10^{-15} l)	MCHC ($g\ dl^{-1}$)	PLT ($\times 10^3\ \mu l^{-1}$)	MPV (10–15 l)
Saline	–	186 ± 7	5.7 ± 0.6	5.9 ± 0.5	12.2 ± 0.4	55.9 ± 0.6	34.7 ± 1.0	988 ± 39	5.4 ± 0.6
Liposomes	–	179 ± 9	7.3 ± 1.3	7.1 ± 0.6	13.9 ± 0.8	58.9 ± 0.9	35.0 ± 0.9	1028 ± 44	5.4 ± .05
VN	3	183 ± 7	4.7 ± 0.8	6.8 ± 0.4	13.1 ± 0.9	55.8 ± 0.5	34.6 ± 1.1	1160 ± 83	5.1 ± 0.7
	75	176 ± 9	4.4 ± 0.7	6.5 ± 0.6	12.7 ± .06	56.2 ± 0.8	35.2 ± 1.2	1193 ± 64	5.2 ± 0.5
LVN	3	182 ± 6	5.7 ± 0.6	6.3 ± 0.7	13.0 ± 0.7	55.5 ± 0.6	35.0 ± 1.1	976 ± 59	5.4 ± 0.4
	75	188 ± 7	5.4 ± 0.9	6.6 ± 0.4	13.6 ± 0.5	57.3 ± 0.9	35.1 ± 0.9	1022 ± 46	5.3 ± 0.3

Weight = animal weight in grams (g); WBC = white blood cell count; RBC = red blood cell count; HGB = hemoglobin; MCV = mean corpuscular volume; MCHC = mean cell hemoglobin concentration; PLT =platelet count; MPV = mean platelet volume.

are less dependent on angiogenesis. Alterations caused by the targeted agent in the structure of tumor vasculature toward a more mature phenotype might also promote resistance. Therefore, the development of drug resistance is a significant concern in the design of novel antiangiogenic therapies.

As previously discussed, key pathways in tumor progression, including cell migration, invasion, adhesion, metastasis, and angiogenesis, all rely on integrin function. Through drugs that target integrins, it is expected that an impact on all of these pathways will be observed causing a more broad-spectrum inhibitory effect on both tumor and endothelial cells. Such an approach circumvents the problem of drug resistance since multiple pathways are being simultaneously targeted. Owing to the small size of the disintegrin class of integrin antagonists, potential side effects are minimized in that they do not activate the immune system. Clinically relevant methods for the delivery of these agents have been developed and disintegrins have proven to be effective antiangiogenic and antitumor agents in a number of mouse models without significant toxicity. As described previously, a fully functional recombinant form of CN, VN, has been expressed in a highly modified *E. coli* expression system [2]. The expression system is based on introducing and creating both an oxidizing and reducing environment within the bacterial host yielding the ability to form the disulfide bonds needed for proper presentation of the RGD motif. While the interchain disulfide bonds found in CN are not formed, the intrachain bonds are formed allowing for effective presentation of the important RGD motif in VN. This advance in the production and validation of a recombinant monomeric variant of CN solves an important supply problem and will allow further development as a clinical agent. Although current therapies have been promising, disintegrins represent an attractive alternative to conventional antiangiogenic agents.

While a number of venom-purified disintegrins have been described, no disintegrins have been identified as stand-alone polypeptides in humans. Disintegrin-like domains have been identified in a class of protein identified as ADAM proteins. These ADAM proteins are large transmembrane multidomain proteins. The ADAM family members are unique among cell surface proteins in possessing both a potential adhesive domain and a potential protease domain [62, 63]. The diversity of biological properties that are associated with disintegrin metalloproteases implicate them in a variety of important physiological and pathological processes, and suggest potential therapeutic utility in areas including cardiovascular disease, tissue remodeling, tumor suppression, and growth control [64, 65]. Their disintegrin domain contains an integrin-binding sequence in a loop that mediates cell–cell interactions [63]. All family members potentially possess a functional cell-adhesion domain based on amino acid sequence analysis; however, only a few of the human ADAMs carry the amino acid consensus sequence (HEXXHXXGXXH) that indicates the presence of an active metalloproteinase, which is required to mediate the proteolytic cleavage of ECM components and/or cell surface receptors [66]. The levels of expression of an additional class of ADAM proteins, A Disintegrin and Metalloproteinase with Thrombospondin motifs (ADAMTS), have been evaluated in non–small cell lung carcinoma [67, 68]. In these reports, it was observed that the level of ADAMTS-1 is significantly reduced and the level of ADAM12 increased in the tumor samples, as compared to control, indicating that the proteases play different roles in tumor progression. A number of attempts have been made to express and evaluate the activity of the disintegrin domains of the ADAM proteins, but at levels unacceptable for therapeutic evaluation in animal models [69, 70]. In a recent report [71] the disintegrin domain of ADAM15 (also known as *metargidin*), the only human disintegrin that contains the RGD motif, was expressed in a generic *E. coli* strain and shown to inhibit the *in vitro* adhesion and invasion of endothelial cells through a reconstituted basement membrane. Because the amounts of recombinant ADAM5 disintegrin generated in this study were minimal, Trochon-Joseph *et al.* evaluated the anticancer efficacy of this agent *in vivo* by inserting the ADAM15 disintegrin gene into a vector containing a tetracycline-inducible promoter that was further electrotransferred into tumor-bearing nude mice muscle. The *in vivo* results showed that the *in situ* expressed ADAM15 disintegrin domain strongly inhibited cancer progression and vascular growth into the tumor [71]. These *in vivo* results were impressive but, nonetheless, difficult to translate into a clinical therapeutic due to the disappointing recombinant expression levels. In order for recombinant human ADAM disintegrin-like domains (modified ADAM polypeptides, MAPs) to be effectively translated into clinical agents, a recombinant expression system must yield a protein that is not only properly folded but also produced in sufficient amounts for *in vitro* and/or *in vivo* experimentation. As previously described, we have developed a uniquely engineered *E. coli* system suitable for disintegrin expression that yields a fully functional (*in vivo* and *in vitro*) recombinant snake venom disintegrin [2]. Moreover, our recent data show that the system can be successfully employed for ADAM-derived disintegrin domain production as well. Development and implementation of this expression system have overcome a major hurdle in the production of human ADAM-derived disintegrin domains for use as potential antiangiogenic agents.

References

1. Liotta, L.A. and Kohn, E.C. (2001) The microenvironment of the tumour-host interface. *Nature*, **411**, 375–379.
2. Minea, R., Swenson, S., Costa, F., Chen, T.C., and Markland, F.S. (2005) Development of a novel recombinant disintegrin, contortrostatin, as an effective anti-tumor and anti-angiogenic agent. *Pathophysiol. Haemost. Thromb.*, **34**, 177–183.
3. Swenson, S., Costa, F., Ernst, W., Fujii, G., and Markland, F.S. (2005) Contortrostatin, a snake venom disintegrin with anti-angiogenic and anti-tumor activity. *Pathophysiol. Haemost. Thromb.*, **34**, 169–176.
4. Swenson, S., Costa, F., Minea, R., Sherwin, R.P., Ernst, W., Fujii, G., Yang, D., and Markland, F.S. Jr. (2004) Intravenous liposomal delivery of the snake venom disintegrin contortrostatin limits breast cancer progression. *Mol. Cancer Ther.*, **3**, 499–511.
5. Carmeliet, P. (2005) Angiogenesis in life, disease and medicine. *Nature*, **438**, 932–936.
6. Folkman, J. (1971) Tumor angiogenesis: therapeutic implications. *N. Engl. J. Med.*, **285**, 1182–1186.
7. Cheresh, D.A. (1992) Structural and biologic properties of integrin-mediated cell adhesion. *Clin. Lab. Med.*, **12**, 217–236.

8. Hynes, R.O. (1992) Integrins: versatility, modulation, and signaling in cell adhesion. *Cell*, **69**, 11–25.
9. Ruoslahti, E. (1991) Integrins. *J. Clin. Invest.*, **87**, 1–5.
10. Brooks, P.C., Stromblad, S., Klemke, R., Visscher, D., Sarkar, F.H., and Cheresh, D.A. (1995) Antiintegrin alpha v beta 3 blocks human breast cancer growth and angiogenesis in human skin. *J. Clin. Invest.*, **96**, 1815–1822.
11. Brooks, P.C., Clark, R.A., and Cheresh, D.A. (1994) Requirement of vascular integrin alpha v beta 3 for angiogenesis. *Science*, **264**, 569–571.
12. Brooks, P.C., Montgomery, A.M., Rosenfeld, M., Reisfeld, R.A., Hu, T., Klier, G., and Cheresh, D.A. (1994) Integrin alpha v beta 3 antagonists promote tumor regression by inducing apoptosis of angiogenic blood vessels. *Cell*, **79**, 1157–1164.
13. Stromblad, S., Becker, J.C., Yebra, B.C.B.P., and Cheresh, D.A. (1996) Suppression of p53 activity and p21WAF1/CIP1 expression by vascular cell integrin avb3 during angiogenesis. *J. Clin. Invest.*, **98**, 426–433.
14. Friedlander, M., Brooks, P.C., Shaffer, R.W., Kincaid, C.M., Varner, J.A., and Cheresh, D.A. (1995) Definition of two angiogenic pathways by distinct alpha v integrins. *Science*, **270**, 1500–1502.
15. Stoeltzing, O., Liu, W., Reinmuth, N., Fan, F., Parry, G.C., Parikh, A.A., McCarty, M.F., Bucana, C.D., Mazar, A.P., and Ellis, L.M. (2003) Inhibition of integrin alpha5beta1 function with a small peptide (ATN- 161) plus continuous 5-FU infusion reduces colorectal liver metastases and improves survival in mice. *Int. J. Cancer*, **104**, 496–503.
16. Bayless, K.J., Salazar, R., and Davis, G.E. (2000) RGD-dependent vacuolation and lumen formation observed during endothelial cell morphogenesis in three-dimensional fibrin matrices involves the avb3 and a5b1 integrins. *Am. J. Pathol.*, **156**, 1673–1683.
17. Yamamoto, K., Murae, M., and Yasuda, M. (1991) RGD-containing peptides inhibit experimental peritoneal seeding of human ovarian cancer cells. *Acta Obstet. Gynaecol. Jpn.*, **43**, 1687–1692.
18. Felding-Habermann, B., O'Toole, T.E., Smith, J.W., Fransvea, E., Ruggeri, Z.M., Ginsberg, M.H., Hughes, P.E., Pampori, N., Shattil, S.J., Saven, A., and Mueller, B.M. (2001) Integrin activation controls metastasis in human breast cancer. *Proc. Natl. Acad. Sci. U.S.A.*, **98**, 1853–1858.
19. Pecheur, I., Peyruchaud, O., Serre, C.M., Guglielmi, J., Voland, C., Bourre, F., Margue, C., Cohen-Solal, M., Buffet, A., Kieffer, N., and Clezardin, P. (2002) Integrin alpha(v)beta3 expression confers on tumor cells a greater propensity to metastasize to bone. *FASEB J.*, **16**, 1266–1268.
20. Takayama, S., Ishii, S., Ikeda, T., Masamura, S., Doi, M., and Kitajima, M. (2005) The relationship between bone metastasis from human breast cancer and integrin alpha(v)beta3 expression. *Anticancer Res.*, **25**, 79–83.
21. Yoon, S.O., Shin, S., and Lipscomb, E.A. (2006) A novel mechanism for integrin-mediated ras activation in breast carcinoma cells: the alpha6beta4 integrin regulates ErbB2 translation and transactivates epidermal growth factor receptor/ErbB2 signaling. *Cancer Res.*, **66**, 2732–2739.
22. Fujii, Y., Okuda, D., Fujimoto, Z., Horii, K., Morita, T., and Mizuno, H. (2003) Crystal structure of trimestatin, a disintegrin containing a cell adhesion recognition motif RGD. *J. Mol. Biol.*, **332**, 1115–1122.
23. Bartsch, J.E., Staren, E.D., and Appert, H.E. (2003) Adhesion and migration of extracellular matrix-stimulated breast cancer. *J. Surg. Res.*, **110**, 287–294.
24. Clyman, R.I., Mauray, F., and Kramer, R.H. (1992) b1 and b3 Integrins have different roles in the adhesion and migration of vascular smooth muscle cells on extracellular matrix. *Exp. Cell Res.*, **200**, 272–284.
25. Gould, R.J., Polokoff, M.A., Friedman, P.A., Huang, T.F., Holt, J.C., Cook, J.J., and Niewiarowski, S. (1990) Disintegrins: a family of integrin inhibitory proteins from viper venoms. *Proc. Soc. Exp. Biol. Med.*, **195**, 168–171.
26. Huang, T.F., Holt, J.C., Kirby, E.R., and Niewiarowski, S. (1989) Trigramin: primary structure and its inhibition of

von Willebrand factor binding to glycoprotein IIb-IIIa complex on human platelets. *Biochemistry*, **28**, 661–666.

27. Dennis, M.S., Henzel, W.J., Pitti, R.M., Lipari, M.T., Napier, M.A., Deisher, T.A., Bunting, S., and Lazarus, R.A. (1990) Platelet glycoprotein IIb/IIIa protein antagonists from snake venoms: evidence for a family of platelet-aggregation inhibitors. *Proc. Natl. Acad. Sci. U.S.A.*, **87**, 2471–2475.
28. Scarborough, R.M., Rose, J.W., Naughton, M.A., Phillips, D.R., Nannizzi, L., Arfsten, A., Campbell, A.M., and Charo, I.F. (1993) Characterization of the integrin specificities of disintegrins isolated from American pit viper venoms. *J. Biol. Chem.*, **268**, 1058–1065.
29. McLane, M.A., Marcinkiewicz, C., Vijay-Kumar, S., Wierzbicka-Patynowski, I., and Niewiarowski, S. (1998) Viper venom disintegrins and related molecules. *Proc. Soc. Exp. Biol. Med.*, **219**, 109–119.
30. Niewiarowski, S., McLane, M.A., Kloczewiak, M., and Stewart, G.J. (1994) Disintegrins and other naturally occurring antagonists of platelet fibrinogen receptors. *Semin. Hematol.*, **31**, 289–300.
31. Phillips, D.R., Charo, I.F., and Scarborough, R.M. (1991) GP IIb-IIIa: the responsive integrin. *Cell*, **65**, 359–362.
32. Trikha, M., Rote, W.E., Manley, P.J., Lucchesi, B.R., and Markland, F.S. (1994) Purification and characterization of platelet aggregation inhibitors from snake venoms. *Thromb. Res.*, **73**, 39–52.
33. Adler, M., Lazarus, R.A., Dennis, M.S., and Wagner, G. (1991) Solution structure of kistrin, a potent platelet aggregation inhibitor and GP IIb-IIIa antagonist. *Science*, **253**, 445–448.
34. Saudek, V., Atkinson, R.A., Lepage, P., and Pelton, J.T. (1991) The secondary structure of Echistatin from 1H-NMR. *Eur. J. Biochem.*, **202**, 329–338.
35. Kumar, C.C., Nie, H., Rogers, C.P., Malkowski, M., Maxwell, E., Catino, J.J., and Armstrong, L. (1997) Biochemical characterization of the binding of echistatin to integrin alphavbeta3 receptor. *J. Pharmacol. Exp. Ther.*, **283**, 843–853.
36. Kang, I.C., Lee, Y.D., and Kim, D.S. (1999) A novel disintegrin salmosin inhibits tumor angiogenesis. *Cancer Res.*, **59**, 3754–3760.
37. Cominetti, M.R., Terruggi, C.H., Ramos, O.H., Fox, J.W., Mariano-Oliveira, A., De Freitas, M.S., Figueiredo, C.C., Morandi, V., and Selistre-de-Araujo, H.S. (2004) Alternagin-C, a disintegrin-like protein, induces vascular endothelial cell growth factor (VEGF) expression and endothelial cell proliferation in vitro. *J. Biol. Chem.*, **279**, 18247–18255.
38. Yeh, C.H., Peng, H.C., and Huang, T.F. (1998) Accutin, a new disintegrin, inhibits angiogenesis in vitro and in vivo by acting as integrin alphavbeta3 antagonist and inducing apoptosis. *Blood*, **92**, 3268–3276.
39. Monleon, D., Moreno-Murciano, M.P., Kovacs, H., Marcinkiewicz, C., Calvete, J.J., and Celda, B. (2003) Concerted motions of the integrin-binding loop and the C-terminal tail of the non-RGD disintegrin obtustatin. *J. Biol. Chem.*, **278**, 45570–45576.
40. Golubkov, V., Hawes, D., and Markland, F.S. (2003) Antiangiogenic activity of contortrostatin, a disintegrin from Agkistrodon contortrix contortrix snake venom. *Angiogenesis*, **6**, 213–224.
41. Pyrko, P., Wang, W., Markland, F.S., Swenson, S.D., Schmitmeier, S., Schonthal, A.H., and Chen, T.C. (2005) The role of contortrostatin, a snake venom disintegrin, in the inhibition of tumor progression and prolongation of survival in a rodent glioma model. *J. Neurosurg.*, **103**, 526–537.
42. Zhou, Q., Ritter, M., and Markland, F.S. (1996) Contortrostatin, a snake venom protein, which is an inhibitor of breast cancer progression. *Mol. Biol. Cell*, **7** (Suppl), 425a.
43. Zhou, Q., Arnold, C., Leder, C., and Markland, F.S. (1998) Contortrostatin inhibits Kaposi's sarcoma invasion and angiogenesis. *J. Acquir. Immune Defic. Syndr. Hum. Retrovirol.*, **17**, A18.

44. Markland, F.S., Shieh, K., Zhou, Q., Golubkov, V., Sherwin, R.P., Richters, V., and Sposto, R. (2001) A novel snake venom disintegrin that inhibits human ovarian cancer dissemination and angiogenesis in an orthotopic nude mouse model. *Haemostasis*, **31**, 183–191.
45. Trikha, M., De Clerck, Y.A., and Markland, F.S. (1994) Contortrostatin, a snake venom disintegrin, inhibits beta 1 integrin-mediated human metastatic melanoma cell adhesion and blocks experimental metastasis. *Cancer Res.*, **54**, 4993–4998.
46. Moiseeva, N., Bau, R., Swenson, S.D., Markland, F.S., Choe, J.Y., Liu, Z.J., and Allaire, M. Jr. (2008) Structure of acostatin, a dimeric disintegrin from Southern copperhead (Agkistrodon contortrix contortrix), at 1.7 A resolution. *Acta Crystallogr. D: Biol. Crystallogr.*, **64**, 466–470.
47. Zhou, Q., Nakada, M.T., Brooks, P.C., Swenson, S.D., Ritter, M.R., Argounova, S., Arnold, C.A., and Markland, F.S. (2000) Contortrostatin, a homodimeric disintegrin, binds to integrin avb5. *Biochem. Biophys. Res. Commun.*, **267**, 350–355.
48. Schmitmeier, S., Markland, F.S., and Chen, T.C. (2000) Anti-invasive effect of contortrostatin, a snake venom disintegrin, and TNF-alpha on malignant glioma cells. *Anticancer Res.*, **20**, 4227–4233.
49. Zhou, Q., Sherwin, R.P., Parrish, C., Richters, V., Groshen, S.G., Tsao-Wei, D., and Markland, F.S. (2000) Contortrostatin, a dimeric disintegrin from Agkistrodon contortrix contortrix, inhibits breast cancer progression. *Breast Cancer Res. Treat.*, **61**, 249–260.
50. Woodle, M.C. (1993) Surface-modified liposomes: assessment and characterization for increased stability and prolonged blood circulation. *Chem. Phys. Lipids*, **64**, 249–262.
51. Fujii, G. (1996) *Liposomal Amphotericin B(AmBisome): Realization of the Drug Delivery Concept*, Marcel Dekker, New York, p. 491.
52. Cousins, G.R., Sudo, Y., Friedrichs, G.R., Markland, F.S., and Lucchesi, B.R. (1995) Contortrostatin prevents reocclusion after thrombolytic therapy in a canine model of carotid artery thrombosis. *FASEB J.*, **9**, A938.
53. Yoneda, T., Williams, P.J., Hiraga, T., Niewolna, M., and Nishimura, R. (2001) A bone-seeking clone exhibits different biological properties from the MDA-MB-231 parental human breast cancer cells and a brain-seeking clone in vivo and in vitro. *J. Bone Miner. Res.*, **16**, 1486–1495.
54. Christgen, M. and Lehmann, U. (2007) MDA-MB-435: the questionable use of a melanoma cell line as a model for human breast cancer is ongoing. *Cancer Biol. Ther.*, **6**, 1355–1357.
55. Ellison, G., Klinowska, T., Westwood, R.F., Docter, E., French, T., and Fox, J.C. (2002) Further evidence to support the melanocytic origin of MDA-MB-435. *Mol. Pathol.*, **55**, 294–299.
56. Lacroix, M. (2008) Persistent use of "false" cell lines. *Int. J. Cancer*, **122**, 1–4.
57. Chambers, A.F. (2009) MDA-MB-435 and M14 cell lines: identical but not M14 melanoma? *Cancer Res.*, **69**, 5292–5293.
58. Lacroix, M. (2009) MDA-MB-435 cells are from melanoma, not from breast cancer. *Cancer Chemother. Pharmacol.*, **63**, 567.
59. Fox, S.B. and Harris, A.L. (2004) Histological quantitation of tumour angiogenesis. *Apmis*, **112**, 413–430.
60. Protopapa, E., Delides, G.S., and Revesz, L. (1993) Vascular density and the response of breast carcinomas to mastectomy and adjuvant chemotherapy. *Eur. J. Cancer*, **29A**, 1391–1393.
61. Gately, S. and Kerbel, R. (2001) Antiangiogenic scheduling of lower dose cancer chemotherapy. *Cancer J.*, **7**, 427–436.
62. Wolfsberg, T.G., Straight, P.D., Gerena, R.L., Huovila, A.P., Primakoff, P., Myles, D.G., and White, J.M. (1995) ADAM, a widely distributed and developmentally regulated gene family encoding membrane proteins with a disintegrin and metalloprotease domain. *Dev. Biol.*, **169**, 378–383.
63. Wolfsberg, T.G., Primakoff, P., Myles, D.G., and White, J.M. (1995) ADAM,

a novel family of membrane proteins containing a disintegrin and metalloprotease domain: multipotential functions in cell-cell and cell-matrix interactions. *J. Cell Biol.*, **131**, 275–278.

64. Stone, A.L., Kroeger, M., and Sang, Q.X. (1999) Structure-function analysis of the ADAM family of disintegrin-like and metalloproteinase-containing proteins (review). *J. Protein Chem.*, **18**, 447–465.

65. Blobel, C.P. (2005) ADAMs: key components in EGFR signalling and development. *Nat. Rev. Mol. Cell Biol.*, **6**, 32–43.

66. Black, R.A. and White, J.M. (1998) ADAMs: focus on the protease domain. *Curr. Opin. Cell Biol.*, **10**, 654–659.

67. Rocks, N., Paulissen, G., El Hour, M., Quesada, F., Crahay, C., Gueders, M., Foidart, J.M., Noel, A., and Cataldo, D. (2008) Emerging roles of ADAM and ADAMTS metalloproteinases in cancer. *Biochimie*, **90**, 369–379.

68. Rocks, N., Paulissen, G., Quesada Calvo, F., Polette, M., Gueders, M., Munaut, C., Foidart, J.M., Noel, A., Birembaut, P., and Cataldo, D. (2006) Expression of a disintegrin and metalloprotease (ADAM and ADAMTS) enzymes in human non-small-cell lung carcinomas (NSCLC). *Br. J. Cancer*, **94**, 724–730.

69. Lu, D., Xie, S., Sukkar, M.B., Lu, X., Scully, M.F., and Chung, K.F. (2007) Inhibition of airway smooth muscle adhesion and migration by the disintegrin domain of ADAM-15. *Am. J. Respir. Cell Mol. Biol.*, **37**, 494–500.

70. Zigrino, P., Steiger, J., Fox, J.W., Loffek, S., Schild, A., Nischt, R., and Mauch, C. (2007) Role of ADAM-9 disintegrin-cysteine-rich domains in human keratinocyte migration. *J. Biol. Chem.*, **282**, 30785–30793.

71. Trochon-Joseph, V., Martel-Renoir, D., Mir, L.M., Thomaidis, A., Opolon, P., Connault, E., Li, H., Grenet, C., Fauvel-Lafeve, F., Soria, J., Legrand, C., Soria, C., Perricaudet, M., and Lu, H. (2004) Evidence of antiangiogenic and antimetastatic activities of the recombinant disintegrin domain of metargidin. *Cancer Res.*, **64**, 2062–2069.

10
Anti-VEGF Approaches and Newer Antiangiogenic Approaches Which Are Already in Clinical Use

Chandu Vemuri and Steven K. Libutti

10.1
Introduction

The treatment of patients with malignancy has changed dramatically in the past century. Advances in medical and surgical oncology have dramatically increased survival of patients with cancer while decreasing morbidity and mortality. Dr. Judah Folkman, through rigorous application of the scientific method to answering novel questions, discovered a new field of cancer therapeutics which he termed *antiangiogenesis*. The current field of antiangiogenic therapy for malignancy is rapidly expanding with thousands of ongoing clinical trials (Table 10.1). In this chapter, we will discuss the discovery of vascular endothelial growth factor (VEGF), in essence the discovery of the field of angiogenesis-based chemotherapy, and the therapeutics which are approved and in clinical use: monoclonal antibodies (bevacizumab, cetuximab), small molecule tyrosine kinase inhibitors (smTKIs: sorafenib, sunitinib, erlotinib), and inhibitors of the mammalian target of rapamycin (mTOR; temsirolimus, everolimus).

10.2
Discovery of VEGF

The concept that tumors can synthesize and secrete factors to perpetuate their survival through the promotion of neovascularization can be traced back to as early as 1939 [1]. Work done by Harry Green demonstrated that the successful implantation and growth of transplanted tumors was directly correlated with the degree of neovascularization [2]. Furthermore, the intensity and persistence of neovascularization was later shown to be much greater with implanted tumors than with implantation of normal tissues or in association with wound healing [3]. Later, two independent groups found that tumor cells were able to promote neovascularization even when a filter was placed between the tumor and the host, thereby supporting the theory that secreted factors from tumors were angiogenic [4, 5].

Tumor Angiogenesis – From Molecular Mechanisms to Targeted Therapy. Edited by Francis S. Markland, Stephen Swenson, and Radu Minea
Copyright © 2010 WILEY-VCH Verlag GmbH & Co. KGaA, Weinheim
ISBN: 978-3-527-32091-2

Table 10.1 Antiangiogenic therapies currently in clinical use.

Agent	US FDA-approved clinical uses
Anti-VEGF	
Bevacizumab	Metastatic colorectal cancer First-line therapy in combination with irinotecan, 5-FU, and leucovorin Second-line therapy in combination with 5-FU-based therapies Metastatic renal cell cancer in conjunction with interferon α Non–small cell lung cancer with carboplatin and paclitaxel in select patients Recurrent glioblastoma multiforme in patients who had progressed on prior therapies Breast cancer in combination with paclitaxel for untreated patients with metastatic, HER2 negative breast cancer
Small molecule tyrosine kinase inhibitors	
Sorafenib	Advanced renal cell carcinoma Unresectable hepatocellular carcinoma
Sunitinib	Advanced renal cell carcinoma Gastrointestinal stromal tumor
Erlotinib	Locally advanced or metastatic non–small cell lung cancer Pancreatic cancer
Monoclonal antibody against EGFR	
Cetuximab	Metastatic colorectal cancer without K-ras mutations Squamous cell cancer of the head and neck
Inhibitors of mTOR	
Temsirolimus	Metastatic renal cell carcinoma
Everolimus	Advanced renal cell carcinoma in patients who have progressed sorafenib or sunitinib

Dr. Folkman's landmark studies published in the 1970s elegantly expounded on the relationship between tumor growth and angiogenesis. His group found that tumors grew slowly in avascular corneas until vascularization was achieved, after which the growth was rapid [6]. Tumors implanted on the CAM of chick embryos also grew slowly until neovascularization at 72 h, after which growth was accelerated. Later histologic studies revealed that capillary sprouts occurred 24 h after implantation, but vascular penetration of the tumor graft occurred around 72 h [7]. Tumors implanted into the anterior chamber of a rabbit's eye caused neovascularization of the iris, and the rate of tumor growth was directly related to proximity of these vessels in the iris [8, 9]. Tumors in the vitreous of rabbit eyes grew slowly until they reached the retinal surface where neovascularization occurred and growth was increased by 19 000-fold [10]. An essential observation is that in all these studies vasculature arose from host tissues and grew into tumor implants. Tumor implants did not develop vasculature that then infiltrated or connected with host vessels. Later studies detailed the concepts of population dormancy, where tumors grew until the rate of peripheral cell growth and central cell necrosis was in balance [11, 12].

In 1971, Dr. Folkman published his discovery of a factor potentially responsible for angiogenesis [13]. His group assayed for angiogenesis using supernatant preparations of control (nonmalignant tissues) as well as Walker 256 carcinoma, mouse B16 melanoma, human neuroblastoma, Wilms tumor, and hepatoblastoma cells. They then prepared rats with a porous millitube placed into a dorsal air sac. The supernatants were injected into these tubes and then at 48 h the skin was harvested, histologic sections taken and scored for angiogenesis. In this seminal paper, Dr. Folkman's group definitively demonstrated that there was a factor that was probably composed of protein, MW 100, 000, and RNA that promoted neovascularization and had a mitogenic effect on capillary endothelium. In 1974, Dr. Folkman went on to publish results further confirming the angiogenic activity of the factor which he termed tissue angiogenesis factor (TAF). At that time, he also suggested that inhibitors of TAF, such as antibodies, would be antiangiogenic and could thereby be a potentially powerful cancer therapeutic [14].

Over the next decade, multiple other groups isolated other factors that were potentially angiogenic. However, these factors did not have mitogenic affects on vascular endothelial cells (ECs) and did not appear to be diffusible factors. A breakthrough unexpectedly arose through the independent, simultaneous efforts of multiple labs. In 1983, Dr. Dvorak's group published in *Science* an article on the identification of a factor they named vascular permeability factor (VPF). In their paper, they discussed that they found that intraperitoneal injection of hepatocarcinoma cells produced ascites. Analysis using the Miles assay revealed that a factor in the ascites fluid increased vascular permeability. Ascites' fluids from multiple tumor histologies in multiple species repeatedly demonstrated increased vascular permeability. They were able to isolate a causative protein and an IgG antibody against this protein, from immunized animals that blocked the effect of VPF. At this time, VPF had not been sequenced and it had not been tested for mitogenic activity [15, 16].

In 1989, Dr. Ferrara's group isolated a protein from a medium conditioned with bovine pituitary follicular cells. This protein had mitogenic effects directly on vascular ECs and was named VEGF [17]. At the same time, another group had identified a factor which was also mitogenic to ECs but was derived from a mouse pituitary cell line AtT20, named *vasculotropin* [18].

Plouet's group had sequenced vasculotropin [18], Ferrara's group had sequenced VEGF [17], and Senger and Keck's groups had sequenced VPF [15, 19]. It appeared that vasculotropin was a mouse ortholog of VEGF. And as fate would have it, VPF and VEGF were one and the same. Vasculotropin, VPF, and VEGF were all the same protein as TAF.

Research over the next decade, revealed that there were several splice variants and cleaved isoforms of VEGF with varying bioavailability and effects [16]. Furthermore, VEGF mRNA expression was shown to be highest in ischemic areas of tumors as found in research on glioblastoma multiforme [16, 20, 21]. The VEGF family has now come to include placental growth factor [22], VEGF-B [23], VEGF-C [24, 25], and VEGF-D [16, 26]. Also, the receptors for the VEGF family of proteins, VEGF1 [27] and VEGF2 [28–30] were later identified and isolated [16].

10.3 Anti-VEGF Cancer Therapeutics

10.3.1 Monoclonal Antibodies

10.3.1.1 Discovery and Development of Bevacizumab (Avastin®)

Following the initial identification and sequencing of VEGF, the sequence, structure, and function of splice variants and multiple isoforms has been studied. VEGF-A is encoded in a sequence comprised of eight exons with seven intervening introns [31, 32]. Splicing variations result in four isoforms, each of which has a signal sequence. The bioavailability of these isoforms is variable and is based on the presence or absence of extracellular matrix binding via heparin-binding domains. The cleavage of VEGF by plasmin and matrix metalloproteinases results in the release of diffusible, biologically active products: a possible mechanistic explanation to the angiogenic switch theory [33].

Preclinical research revealed multiple roles of VEGF including mitogenic affect on ECs lining arteries, veins, and lymphatics; increased vascular permeability; alteration of expression of surface molecules on ECs and thereby, modulation of leukocyte adhesion, prevention of EC apoptosis, and stimulation of angiogenesis in *in vivo* models [34, 35].

VEGF proteins are known to interact with two tyrosine kinase receptors: VEGFR1, VEGFR2. Whereas VEGF binding to VEGFR1 is of uncertain clinical significance, VEGF binding to VEGFR2 results in increased vascular permeability and promotes the growth and survival of ECs [35].

VEGF was discovered to be both, essential for physiologic angiogenesis as well as critical for tumor angiogenesis. Multiple studies revealed increased mRNA expression in a variety of malignant cells types (i.e., breast, ovarian, renal, gastrointestinal) and revealed a direct correlation with areas of tumoral hypoxia and increased VEGF levels [35].

In 1993, the therapeutic potential of VEGF blockade was first revealed in a publication from Dr. Ferrara's group. They injected nude mice with human tumor cells (A673 rhadbomyosarcoma, G55 glioblastoma multiforme, and SK-LMS-1 leiomyosarcoma) – all of which were known to express VEGF. The treatment groups received mouse anti-human monoclonal anti-VEGF antibody (A.4.6.1). The growth of all three tumor types was significantly impaired with reductions in tumor weights: 97% rhabdomyosarcoma, 80% G55, 70% SK-LMS-1. However, this could be explained by multiple theories including direct toxicity against tumor cells. To assess these possibilities, they performed *in vitro* assays and found that the survival of tumor cells was not directly affected by the presence of VEGF or the anti-VEGF antibody. However, histologic examination of tumors revealed decreased vascular density in the treatment groups. On the basis of these findings, they theorized that the antitumor effect of the anti-VEGF antibody was a result of an impairment in angiogenesis. This was the first study to show that inhibition of angiogenesis could result in a significant impairment in tumor growth [36]. Their lab went on to show similar tumor suppression in multiple other cells types, including metastatic colon cancer, prostate cancer, and breast cancer [37–39].

The discovery of VEGF and its functions as well as the realization of its therapeutic potential in preclinical animal models made it an ideal therapeutic to translate to clinical use. The pivotal step in this process came from Dr. Ferrara's lab in 1997 with the humanization of the mouse anti-human monoclonal antibody (A.4.6.1.). Bevacizumab does not bind VEGF-B or VEGF-C, has a terminal half-life of 17–21 days and does not elicit an antibody response in humans [35, 40]. The humanized antibody (rhuMab VEGF; bevacizumab, Avastin) was shown to be able to bind and neutralize all isoforms of human VEGF-A as well as biologically active fragments.

Prior to initiating clinical trials, an *in vivo* safety study was conducted in the primate *Macaca fascicularis*. The treatment groups received doses of bevacizumab up to 50 mg kg^{-1} with side effects including physeal dysplasia, suppression of female reproductive tract angiogenesis with resultant lower ovarian and uterine weights. These side effects appeared to halt with cessation of therapy and were in fact, reversible [35, 41].

10.3.1.2 **Clinical Trials**

Given the promising preclinical data, Genentech moved forward with bevacizumab and filed for an investigational new drug (IND) application in early 1997 and clinical trials began soon thereafter. We will begin by briefly discussing the phase I and phase II clinical trials and then discussing the approved clinical uses of bevacizumab in the context of the trials which garnered them approval for clinical use.

10.3.1.3 **Phase I and Phase II Trials**

Two phase I studies were published in the *Journal of Clinical Oncology* in 2001 that demonstrated that the addition of bevacizumab to accepted chemotherapy regimens not only did not add new, high-grade toxicities, but additionally did not compound known toxicities of those regimens. The first study by Margolin *et al.* enrolled 12 patients with solid tumors and treated them with doxorubicin, carboplatin, 5-FU with leucovorin and bevacizumab (3 mg kg^{-1}). Grade 3 toxicities were diarrhea (one patient), thrombocytopenia (two patients), and leucopenia (one patient) [42]. The second study by Gordon *et al.* enrolled 25 patients with metastatic solid organ cancer. These patients received bevacizumab (up to 10 mg kg^{-1}) with no grade III or IV adverse events specific to bevacizumab. Additionally, none of the study patients developed antibodies to bevacizumab [43].

Multiple phase II trials were conducted beginning in 1998, in which patients were treated with bevacizumab: renal cell cancer with progression on IL-2 [44], in combination with standard-of-care first-line therapy for metastatic colorectal cancer [45], stage IIIb/V non–small cell lung cancer (NSCLC) [46], recurrent breast cancer [47], hormone refractory prostate cancer [35, 48]. The results of the phase II trial with renal cancer and metastatic colorectal cancer revealed that bevacizumab significantly increased time to progression [44, 45].

10.3.2
Metastatic Colorectal Cancer

10.3.2.1 First-Line Therapy

In February of 2004, the US Food and Drug Administration (FDA) approved bevacizumab as a first-line agent in combination with irinotecan, 5-fluorouracil, and leucovorin (IFL) for patients with metastatic colorectal cancer. This approval was based on the promising results of a multicenter, international, phase-III-randomized control trial: 813 patients with untreated, metastatic colorectal cancer were enrolled in the study and randomized to IFL with bevacizumab ($5\ \mathrm{mg\ kg^{-1}}$, dosed every two weeks) or IFL alone. The primary end point was overall survival with secondary end points including progression-free survival, response rate, duration of response, safety and quality of life. Overall survival was 20.3 months in the IFL with bevacizumab group versus 15.6 months in the IFL with placebo group ($p < 0.001$). Progression-free survival was higher in the bevacizumab group 10.6 versus 6.2 months ($p < 0.001$) and the response rates were 44.8% versus 34.8% ($p = 0.004$). The median duration of response was 10.4 versus 7.1 months ($p = 0.001$) in the bevacizumab group versus the placebo group respectively. In this study, the only significant complications were manageable grade III hypertension [49].

10.3.2.2 Second-Line Therapy

In June 2006, the FDA approved use of bevacizumab in conjunction with 5-FU-based therapies in patients with metastatic colorectal cancer as a second-line therapy. This approval was based on a multicenter, international phase III trial that randomized 829 patients with metastatic colorectal cancer who had been treated with fluropyrimidine and irinotecan to receive either oxaliplatin, fluorouracil, and leucovorin (FOLFOX4) with bevacizumab or FOLOX4 alone or bevacizumab alone. The primary end point was overall survival with secondary end points of progression-free survival, response, and toxicity. The overall survival was 12.9 versus 10.8 versus 10.2 for FOLFOX4 with bevacizumab, FOLFOX4 alone, and bevacizumab alone respectively ($p = 0.0011$). Median progression-free survival was 7.3 versus 4.7 versus 2.7 for FOLFOX4 with bevacizumab, FOLFOX4 alone, and bevacizumab alone respectively ($p < 0.001$). Response rates were 22.7, 8.6, and 3.3% for FOLFOX4 with bevacizumab, FOLFOX4 alone, and bevacizumab alone respectively ($p < 0.001$). Toxicities associated with bevacizumab were hypertension, bleeding, and vomiting. This study demonstrated not only that the addition of bevacizumab to FOLFOX4 improved survival in patients with metastatic colorectal cancer, but also that bevacizumab alone was ineffective [50].

10.3.3
Metastatic Renal Cell Cancer (mRCC)

In July of 2009, the FDA granted approval for bevacizumab to be used in conjunction with interferon α (IFNα) for the treatment of metastatic renal cell cancer (mRCC). This approval was based on the results of the international, multicenter, randomized, double-blind trial B017705 (AVOREN). In that study, 649 patients with clear-cell mRCC with history of prior nephrectomy but without any prior systemic therapy were randomized to either bevacizumab with IFNα or IFN with placebo. The primary end point of the trial was overall survival. At final analysis, median overall survival was 22.9 versus 20.6 months in the bevacizumab + IFN versus IFN with placebo respectively. Progression-free survival was 10.4 versus 5.5 months and response rate was 21% versus 12% in the bevacizumab with IFN versus IFN with placebo respectively. On this basis, the authors concluded that bevacizumab with IFN should be standard-of-care first-line therapy for patients with mRCC as it improved both response rate and progression-free survival, and demonstrated a trend toward improved survival with minimal side effects [51].

10.3.4
Non–Small Cell Lung Cancer (NSCLC)

In October of 2006, the FDA approved use of bevacizumab in combination therapy with carboplatin and paclitaxel as first-line therapy in patients with unresectable, locally advanced, recurrent, or metastatic, nonsquamous, NSCLC. This approval was based on the result of the E4599 study which randomized 878 patients with metastatic or recurrent non–squamous cell, NSCLC to paclitaxel and carboplatin alone or paclitaxel and carboplatin with bevacizumab. The median survival was 12.2 months in the arm with bevacizumab versus 10.3 in the chemotherapy-alone group ($p = 0.003$). Median progression-free survival was 6.2 versus 4.5 months ($p < 0.001$). Response rates were 4.4% versus 0.7% respectively. There were 15 treatment-related deaths in the chemotherapy with bevacizumab group due to febrile neutropenia and pulmonary hemorrhage [52].

10.3.5
Recurrent Glioblastoma Multiforme

In May of 2009, the FDA approved use of bevacizumab for treatment of patients with glioblastoma multiforme who had progressed on prior therapy. This was based on the results of two studies: AVF3708g (BRAIN) and NCI 06-C-0064E. The AVF3708g study was a multicenter, randomized, noncomparative, parallel group study which included 167 patients with gliobastoma multiforme that had progressed on treatment with temozolomide and radiation therapy. The patients were randomized to either receive bevacizumab alone or in combination with irinotecan. According to the FDA analysis the objective response rate was 25.95 in the monotherapy arm with a median duration of response of 4.2 months and a 6-month progression-free survival rate of 36%. Grade 3–5 complications included retroperitoneal hemorrhage (1 patient), venous thromboemoblism (8%), and grade 3 hypertension (4.9%). According to Genentech, the combination-therapy group had an objective response rate of 37.8% with a median duration of response of 4.3 months and a 6-month progression-free survival rate of 50.3% (*http://www.fda.gov/ohrms/dockets/ac/09/briefing/2009-4427b1-01-FDA.pdf*) [53]

The NCI study (O6-C-0064E) was a single-arm, single-center trial of bevacizumab monotherapy in 56 patients with recurrent high-grade gliomas. The objective response rate according to the FDA was 19.6% with a median duration of response of 3.9 months. Treatment-related grade 3 or higher events included venous thromboembolism (12.7%), hypertension (3.6%), arterial thromboembolism (1 patient), and gastrointestinal perforation, as well as a wound healing complication. (*http://www.fda.gov/ohrms/dockets/ac/09/briefing/2009-4427b1-01-FDA.pdf*) [53].

10.3.6
Breast Cancer

In February of 2008, the US FDA approved use of bevacizumab as combination therapy with paclitaxel for untreated patients with metastatic, HER2 negative breast cancer. This approval was based on a single, open-label, randomized multicenter study (E2100) which randomized 722 patients with untreated, metastatic, HER2 negative breast cancer to receive paclitaxel alone or in combination with bevacizumab. The primary end point was progression-free survival. The objective response rate was 48.9% versus 22.2% ($p < 0.0001$) for the combination-therapy group versus paclitaxel alone. Importantly, the hazard ratio for progression-free survival was 0.48 (95% CI, 0.385–0.607, $p < 0.0001$) with a median progression-free survival of 11.3 versus 5.8 months respectively [54].

10.4
Other Anti-VEGF Approaches

10.4.1
Small Molecule Tyrosine Kinase Inhibitors

As antiangiogenic therapy both, theoretically and in early clinical trials demonstrated great promise, targets other than VEGF were sought. Among the most successful stories, was the discovery of tyrosine kinase inhibitors. Two of these, sorafenib and sunitinib, have become approved and part of the oncologic armamentarium.

10.4.1.1 Sorafenib (Nexavar®)

Discovery The MAPK cascade was one of the earliest intracellular pathways to be fully delineated and has been shown to play an integral role in the proliferation and prevention of apoptosis both, physiologically and pathologically. The cascade begins with activation of a-raf, b-raf, and c-raf (RAF1) and mutations in these isoforms are present in 30% of human tumors [55]. Ras, an upstream regular of RAF, is constitutively active in multiple human tumors including pancreatic (90%), colorectal (45%), hepatocellular (30%); NSCLCs (35%); melanomas (15%), and renal tumors (10%) [56]. Kasid *et al.* in 1989 worked on the inhibition of raf1 by specific antisense oligonucleotides. This resulted in significant inhibition of the growth of human lung, breast, and ovarian tumors in athymic mice. This was an early study showing the therapeutic potential of interventions focused on the Raf pathway [57].

In the early 1990s, a group of researchers at Bayer and Onyx collaborated to find inhibitors of the MAPK pathway [58, 59]. Over 200 000 compounds were tested for inhibitory activity. From these, sorafenib was found to be the most potent [60, 61]. *In vitro* studies demonstrated inhibition of Raf1 kinase, wild-type B-Raf, oncogenic b-raf, proangiogenic VEGFRs, platelet-derived growth factor (PDGF) receptor-β, and fibroblast growth factor (FGF) [62, 63]. Sorafenib inhibitory effects are via stabilization of the DGF (asparagine, glycine, phenylalanine) motif of Raf kinase, resulting in stabilization of the inactive conformation [64].

Many preclinical studies have demonstrated that sorafenib is able to inhibit tumor growth in human xenografts for tumor types including cololic, mammary, ovarian, pancreatic cancers; melanomas; thyroid, cholangiocarcinoma; multiple leukemia cell lines, and NSCLC. Tumor inhibition is likely to be multifactorial, but studies have shown that it is in part antiangiogenic as demonstrated by reductions in microvessel area, density, and in disruption of tumor vasculature [65].

Advanced Renal Cell Carcinoma In December of 2005, the US FDA approved use of sorafenib for the treatment of advanced renal cell carcinoma on the basis of a phase III, randomized, double-blind, placebo-controlled trial. Escudier *et al.* published in the New England Journal of Medicine in 2007, results from the target study group which randomly assigned 903 patients with inoperable or metastatic renal cell carcinoma who had one prior treatment, to receive either sorafenib or placebo to evaluate overall survival. Overall survival analysis in May of 2005 revealed a nonsignificant trend toward a reduced risk of death (hazard ratio 0.72 95% CI 0.54–0.94). Partial response rate was 10% versus 2% for sorafenib versus placebo ($p < 0.001$) However, median progression-free survival was significantly greater in the sorafenib group at 5.5 versus 2.8 months with a hazard ratio for disease progression of 0.44 (95% CI 0.35–0.55 $p < 0.01$). High-grade adverse events were hypertension and cardiac ischemia with lower grade events including diarrhea, rash, fatigue, and hand–foot skin reactions [66].

Unresectable Hepatocellular Carcinoma In November of 2007, the US FDA approved use of sorafenib for patients with advanced hepatocellular carcinoma based on the results of a phase III, multicenter, double-blind, placebo-controlled trial by the SHARP investigators. Llovet *et al.* published the results of this study which randomized 602 patients with advanced, untreated hepatocellular carcinoma to receive either sorafenib or placebo to evaluate overall survival and time to symptomatic progression. As secondary outcome measures they also evaluated time to radiologic progression and safety. There was an improvement in median overall survival in the sorafenib group of 10.7 versus 7.9 months (hazard ratio in the sorafenib group, 0.69; 95% CI, 0.55–0.87; $p < 0.001$) as well as a higher median time to radiologic progression (5.5 versus 2.8 months $p < 0.001$). However, there was no significant difference in time to symptomatic progression (4.1 versus 4.7, $p = 0.77$). Complications were similar to those experienced in the advanced renal cell carcinoma group and included diarrhea, weight loss, hand–foot skin reaction, and hypophosphatemia [67].

10.4.1.2 **Sunitinib (Sutent®)**

Discovery Researchers at Pfizer pharmaceutical sought to develop a cancer therapeutic that would inhibit both VEFGR2 and PDGFRβ. Preclinical and animal studies had shown that simultaneous inhibition was more efficacious than inhibiting either alone [68–69]. From their efforts, sunitinib was discovered and found to inhibit VEGFR2, PDGFRβ, KIT, and FLT3 [70].

Advanced Renal Cell Carcinoma In February 2007, the FDA granted regular approval of sunitinib for the treatment of advanced renal cell carcinoma on the basis of a phase III, multicenter, randomized trial which enrolled 750 patients with untreated mRCC. The patients were randomized to receive either sunitinib (50 mg PO once daily for four weeks) or IFNα (9 MU subcutaneously three times weekly) with a primary end point of progression-free survival and secondary end points of objective response rate, overall survival, patient-reported outcomes, and safety. Median progress-free survival was 11 months in the sunitinib group versus 5 months in the IFNα group (hazard ratio 0.42; 95% CI, 0.32–0.54; $p < 0.001$). Objective response rate was also higher in the sunitinib group at 31% versus 6% ($p < 0.001$). The only adverse event specific to sunitinib was diarrhea [71].

Gastrointestinal Stromal Tumor (GIST) In January of 2006, the US FDA approved use of sunitinib for the treatment of gastrointestinal stromal tumor (GIST) in patients who had disease progression on or who were intolerant to imatinib mesylate. This was based on the results of a randomized, double-blind placebo-controlled trial that randomized 270 patients to receive sunitinib and 105 patients to receive placebo. The median time to progression was longer in the sunitinib patients at 27.3 versus 6.4 weeks for placebo ($p < 0.0001$). The partial response rate was 6.8% and the progression-free survival was higher in the sunitinib group versus placebo at 24.1 versus 6 weeks ($p < 0.0001$) at interim analysis [72].

10.5
Non-VEGF Antiangiogenic Agents in Clinical Use

The field that Dr. Folkman helped define and develop has been in a state of rapid development over the past 20 years. In addition to the VEGF targeted therapies discussed above, the same approaches were applied to find other non-VEGF targets and to devise cancer therapeutics against them. Below is a description of the US FDA-approved agents which are in clinical use: cetuximab, erlotinib, temsirolimus, and everolimus (Table 10.1).

10.5.1 Monoclonal Antibodies

10.5.1.1 Cetuximab

Discovery VEGF and its receptor VEGFR2 have been shown to be suitable targets for antiangiogenic therapies. Additional growth factor receptors have been implicated in angiogenesis and targeting them may prove to be therapeutic. The epidermal growth factor receptor (EGFR), like VEFGR, is a tyrosine kinase receptor and is involved in regulation of cell proliferation, apoptosis, and angiogenesis. It has been implicated in pathologic oncologic tissues such as colorectal cancer and therefore, efforts were directed against inhibiting EGFR [73–76].

EGFR is known to be expressed on multiple human tumor types including those of the prostate, head and neck, bladder, ovary, and breast. Preclinical animal studies showed significant antitumor activity of humanized mouse monoclonal antibody (C255) against EGFR for treatment of human epithelial carcinoma xenografts (A431) in aythmic nude mice [77]. Humanized, mouse monoclonal antibody against endothelial growth factors (EGFs) in combination with irinotecan was also shown to have significant antitumor effects against human colorectal cancer xenografts in athymic mice [78].

Metastatic Colorectal Cancer In February of 2004, the US FDA approved use of cetuximab in combination with irinotecan as second-line therapy in patients with metastatic colon cancer who are either refractory to or intolerant of irinotecan therapy alone. Cunningham *et al.* [79] published the results of a multicenter, randomized, controlled trial where 329 patients who had progressive disease during or within three months after treatment with irinotecan were randomized to receive cetuximab and irinotecan or cetuximab alone. If patients progressed on cetuximab alone, they could crossover into the combination-therapy arm. Patients were evaluated for response, tumor progression, survival, and side effects. The combination-therapy group's response rate (22.9% versus 10.8%, $p = 0.007$), median time to progression (4.1 versus 1.5 months, $p < 0.001$) were significantly better than the monotherapy group. However, median survival was not statistically significant at 8.6 versus 6.9 months ($p = 0.48$) respectively. Toxicity in the combination group was more frequent but similar to that seen with irinotecan alone. That being said neutropenia, diarrhea, and dyspnea were more common in the combination-therapy group and the three most common cetuximab monotherapy complications were dyspnea (13%) asthenia (10.4%), and abdominal pain (5.2%) or acnelike rash (5.2%).

As cetuximab entered the clinical arena and data matured, researchers attempted to define patient subgroups that would benefit and more importantly, identify those subgroups that would not benefit from therapy. Karapetis *et al.* [80] published the results of a study that addressed this very question. They analyzed the tissue from tumors of 394 patients with colorectal cancer who had been randomized to receive cetuximab plus best supportive care or cetuximab alone. Overall, 42.3% of tumors had K-ras mutations. In the best supportive care group, there were no differences in survival with relation to K-ras mutation status. Importantly, the effectiveness of cetuximab was significantly affected by K-ras mutation status ($p < 0.001$). In tumors with the wild-type K-ras gene, cetuximab improved overall survival compared to best supportive care alone (9.5 vs 4.8 months, $p < 0.001$). However in patients with mutated K-ras tumors, there was no difference in overall survival between the cetuximab with best supportive care and best supportive care alone groups. On the basis of this study and multiple other retrospective trials, in July of 2009, the FDA changed the indications for usage of cetuximab and did not recommend it that it be used in patient with known K-ras mutations in codon 12 or 13.

Squamous Cell Carcinoma of the Head and Neck In March of 2006, the US FDA approved use of cetuximab as combination therapy with radiation for patients with locally or regionally advanced squamous cell carcinoma of the hand and neck. Bonner *et al.* [81] published the results of a multinational, randomized study comparing high-dose radiotherapy alone or in combination with cetuximab (400 mg m^{-2} followed by 25 mg m^{-2} weekly) for patients with locally or regionally advanced squamous cell carcinoma of the head and neck. The primary end point was duration of control of locoregional disease with secondary end points of overall survival, progression-free survival, response rate, and safety. Combination therapy was superior to radiotherapy alone with regard to median duration of locoregional control (24.4 versus 14.9 months, $p = 0.005$), overall survival (49.0 versus 29.3 months, $p = 0.03$) and progression-free survival (hazard ratio 0.70, $p = 0.006$). Grade 3 toxicities that were significantly higher in the combination-therapy group were acneiform rash and infusion reactions.

10.5.2 Small Molecule Tyrosine Kinase Inhibitors (smTKI)

10.5.2.1 Erlotinib

Discovery As discussed in the discovery of cetuximab, the EGF receptor EGFR is present on a variety of human solid tumors. Like VEGFR, EGFR is a tyrosine kinase receptor which on binding to a ligand, dimerizes, autophosphorylates, and initiates an intracellular cascade predominantly involved with the promotion of cellular proliferation and inhibition of apoptosis [74]. The ERBB family of receptors includes EGFR (ERBB1, HER1), HER2 (ERBB2), HER3 (ERBB3), and HER4 (ERBB4) [82]. Inhibition of this receptor by erlotinib was shown to inhibit tumor growth in multiple tumor types including, prostate, breast, ovarian, colon, small cell lung, and non-small cell lung cancers. Preclinical studies also demonstrated that combination therapy with other cytotoxic agents or radiotherapy can have improved tumor inhibition and even improve survival [83]. From these efforts, erlotinib was discovered and found to have significant antitumor effects. An alternate smTKI, gefitinib (Iressa), had been approved by the US FDA in 2003 but was then essentially removed from the market in 2005, for failure to show benefit in matured studies.

Locally Advanced or Metastatic Non–Small Cell Lung Cancer In November of 2004, the US FDA approved use of erlotinib for monotherapy in patients with locally advanced or metastatic NSCLC. This was based on the results of a double-blinded, multicenter, multinational, randomized trial which randomized 731 patients 2 : 1 to receive either erlotinib or placebo. The primary objective was overall survival with secondary objectives of progression-free survival, response rate, response duration, and quality of life. The erlotinib group had a longer median survival at 6.67 months compared to placebo at 4.7 months. Erlotinib also improves progression-free survival (9.9 versus 7.7 weeks) and had a response rate of 8.9% with a median duration of response of 34.3 weeks. Rash and diarrhea were the most common low- and high-grade adverse events [68].

Pancreatic Cancer In November of 2005, erlotinib was approved as combination therapy with gemcitabine in patients with locally advanced, unresectable, or metastatic pancreatic cancer. This was based on a single, multicenter, double-blinded, placebo-controlled, randomized, phase III trial of erlotinib (100/150 mg per day orally) with gemcitabine versus placebo with gemcitabine as first-line chemotherapy for patients with locally advanced or metastatic pancreatic carcinoma in 569 patients. The primary end point of the study was overall survival. Overall, the combination-therapy group with erlotinib faired better. The overall survival ratio was better with a corresponding improvement in hazard ratio (0.82 95% CI, 0.69–0.99; $p = 0.038$). One-year survival was improved at 23% versus 17% ($p = 0.023$). Progression-free survival was improved with a corresponding hazard ratio of 0.77 (95% CI, 0.64–0.92; $p = 0.004$). Response rates were not statistically significant and there was no greater incidence of high-grade adverse events [84].

10.5.3
Inhibitors of Mammalian Target of Rapamycin (mTOR)

10.5.3.1 Discovery

The P13K/AKT/mTOR is essential to cancer cell metabolism. mTOR, the downstream effector of P13K activation, controls the balance between catabolism and anabolism and can thereby regulate tumor growth. Additionally, mTOR's downstream targets regulate apoptic cell death. In relation to angiogenesis, mTOR affects tumor growth via controlling the maintenance of ECs and pericytes. A multitude of vascular growth factors such as VEGF activate this pathway. Hypoxia, thought to be a general stimulator of tumor angiogenesis, activates hypoxia inducible factors (HIFs) which increase expression of factors that promote angiogenesis. mTOR activation also promotes increases in transcription of HIF1α [85]. Given the potential significance of this pathway in cancer, efforts were taken to find inhibitors. As is often the case in science, an inhibitor already existed and was in clinical use but its potential in cancer was unrealized. Rapamycin has been developed from a naturally occurring product, both as an antifungal and as an immunosuppressant. Temsirolimus and Everolimus are derivatives of rapamycin which inhibit mTOR.

10.5.3.2 Advanced Renal Cell Carcinoma

Temsirolimus In May of 2007, temsirolimus was approved for the treatment of patients with mRCC. Hudes *et al.* [86] published the results of a multicenter, phase III trial of 626 patients with untreated, poor-prognosis renal cell carcinoma randomized to receive either temsirolimus (25 mg IV weekly) or 3 MU of IFNα subcutaneously three times weekly or combination therapy with 15 mg of temsirolimus IV weekly and 6 MU of IFNα three times weekly. The primary end point was survival. The temsirolimus-alone group had a better overall survival (hazard ratio for death, 0.73; 95% CI, 0.58–0.92; $p = 0.008$) and progression-free survival (5.5 versus 3.1 months $p < 0.001$) compared to the IFNα-alone group. However, the addition of temosorilimus to IFNα did not have an improved survival compared to IFNα alone. Rash, peripheral edema, hyperglycemia, and hyperlipidemia were more common in the temsorilimus group.

Everolimus In March of 2009, everolimus was approved for patients with advanced renal cell carcinoma who had failed therapy with either sunitinib or sorafenib based on a phase III, randomized, double-blinded trial. Motzer *et al.* in 2008 reported on a study which randomized 410 patients with mRCC to receive either everolimus (10 mg once daily) or placebo with a primary end point of progression-free survival. The trial was terminated early due to a significant benefit seen with regard to efficacy with 65% progression on placebo compared to 37% on everolimus with a hazard ratio of 0.3 (95% CI 0.22–0.40, $p < 0.0001$). The primary end point of progression-free survival was also better in the everolimus group compared to placebo 4 versus 1.9 months. Stomatitis, rash, and fatigue were more common in the everolimus group with grade 3 pneumonia being the only common high-grade adverse event.

10.6 Future Direction of Antiangiogenesis Therapies

Dr. Judah Folkman's vision of a new field of cancer therapeutics has become a reality. What began with the discovery of VEGF and later, the discovery of the potential antitumor effect of its inhibition is now an arena in which tens of thousands of agents are being tested against thousands of targets. The hope is to find better targets and more powerful, targeted therapies. In the midst of this, the agents described in this chapter are being used in clinical practice and data is maturing. The true impact of these agents is yet to be discovered in the settings for which they are currently approved, new diseases for which they may be therapeutic, and new combinations of therapies using known radiochemotherapies. As in its inception, the growth of therapeutic antiangiogenesis will rely on a combination of good fortune and the relentless work of researchers worldwide.

References

1. Ide, A.G., Baker, N.H., and Warren, S.L. (1939) Vascularization of the Brown Pearce rabbit epithelioma transplant as seen in the transparent ear chamber. *Am. J. Roentgenol.*, **42**, 891–899.
2. (a) Green, H.S.N. (1941) Heterologous transplantation of mammalian tumors: I. The transfer of rabbit tumors to alien species. *J. Exp. Med.*, **73**, 461–474; (b) Orlandini, M., Marconcini, L., Ferruzzi, R., and Oliviero, S. (1996) Identification of a c-fos-induced gene that is related to the PDGF/VEGF growth factor family. *Proc. Natl. Acad. Sci. U.S.A.*, **93**, 11675–11680.
3. Algire, G.H., and Chalkley, H.W. (1945) Vascular reactions of normal and malignant tissue in vivo. *J. Natl. Cancer Inst.*, **6**, 73–85.
4. Greenblatt, M. and Shubi, P. (1968) Tumor angiogenesis: transfilter diffusion studies in the hamster by the transparent chamber technique. *J. Natl. Cancer Inst.*, **41**, 111–124.
5. Ehrman, R.L. and Knoth, M. (1968) Choriocarcinoma: transfilter stimulation of vasoproliferation in the hamster cheek pouch studied by light and electron microscopy. *J. Natl. Cancer Inst.*, **41**, 1329–1341.
6. Gimbrone, M.A. Jr., Cotran, R.S., and Folkman, J. (1974) Tumor growth and neovascularization: an experimental model using rabbit cornea. *J. Natl. Cancer Inst.*, **52**, 413–427.
7. Knighton, D., Ausprunk, D., Tapper, D., and Folkman, J. (1977) Avascular and vascular phases of tumour growth in the chick embryo. *Br. J. Cancer*, **35**, 347–356.
8. Gimbrone, M.A., Leapman, S.B., Cotran, R.S., and Folkman, J. (1972) Tumor dormancy in vivo by prevention of neovascularization. *J. Exp. Med.*, **136**, 261–276.
9. Gimbrone, M.A., Leapman, S., Cotran, R.S., and Folkman, J. (1973) Tumor angiogenesis: iris neovascularization at a distance from experimental intraocular tumors. *J. Natl. Cancer. Inst.*, **50**, 219–228.
10. Brem, S., Brem, H., Folkman, J., Finkelstein, D., and Patz, G. (1976) Prolonged tumor dormancy by prevention of neovascularization in the vitreous. *Cancer Res.*, **36**, 2807–2812.
11. Folkman, J. and Hochberg, M. (1983) Self-regulation of growth in three dimensions. *J. Exp. Med.*, **138**, 745–753.
12. Ribatti, D. (2008) Judah Folkman, a pioneer in the study of angiogenesis. *Angiogenesis*, **11**, 3–10.
13. Folkman, J., Merler, E., Abernathy, C., and Williams, G. (1971) Isolation of a tumor factor responsible for angiogenesis. *G. J. Exp. Med.*, **133**, 275–288.
14. Folkman, J. (1974) Tumor angiogenesis factor. *Cancer Res.*, **34**, 2109–2113.
15. Senger, D.R., Galli, S.J., Dvorak, A.M., Perruzzi, C.A., Harvey, V.S., and Dvorak. H.F. (1983) Tumor cells secrete a vascular permeability factor that promotes accumulation of ascites fluid. *Science*, **219**, 983–985.
16. Ferrara, N. (2002) VEGF and the quest for tumour angiogenesis factors. *Nat. Rev. Cancer*, **2** (10), 795–803.
17. Ferrara, N. and Henzel, W.J. (1989) Pituitary follicular cells secrete a novel heparin-binding growth factor specific for vascular endothelial cells. *Biochem. Biophys. Res. Commun.*, **161**, 851–858.
18. Plouet, J., Schilling, J., and Gospodarowicz, D. (1989) Isolation and characterization of a newly identified endothelial cell mitogen produced by AtT20 cells. *EMBO J.*, **8**, 3801–3808.
19. Keck, P.J. *et al.* (1989) Vascular permeability factor, an endothelial cell mitogen related to PDG. *Science*, **246**, 1309–1312.
20. Shweiki, D., Itin, A., Soffer, D., and Keshet, E. (1992) Vascular endothelial growth factor induced by hypoxia may

mediate hypoxia-initiated angiogenesis. *Nature*, **359**, 843–845.

21. Plate, K.H., Breier, G., Weich, H.A., and Risau, W. (1992) Vascular endothelial growth factor is a potential tumour angiogenesis factor in human gliomas in vivo. *Nature*, **359**, 845–848.
22. Maglione, D., Guerriero, V., Viglietto, G., Delli-Bovi, P., and Persico, M.G. (1991) Isolation of a human placenta cDNA coding for a protein related to the vascular permeability factor. *Proc. Natl. Acad. Sci. U.S.A.*, **88**, 9267–9271.
23. Olofsson, B. *et al.* (1996) Vascular endothelial growth factor B, a novel growth factor for endothelial cells. *Proc. Natl. Acad. Sci. U.S.A.*, **93**, 2576–2581.
24. Joukov, V. *et al.* (1996) A novel vascular endothelial growth factor, VEGF-C, is a ligand for the Flt4 (VEGFR-3) and KDR (VEGFR-2) receptor tyrosine kinases. *EMBO J.*, **15**, 1751.
25. Lee, J. *et al.* (1996) Vascular endothelial growth factor-related protein: a ligand and specific activator of the tyrosine kinase receptor Flt4. *Proc. Natl. Acad. Sci. U.S.A.*, **93**, 1988–1992.
26. Orlandini, M., Marconcini, L., Ferruzzi, R., and Oliviero, S. (1996) Identification of a c-fos-induced gene that is related to the platelet-derived growth factor/vascular endothelial factor family. *Proc. Natl. Acad. Sci. U.S.A.*, **93**, 11675–11680.
27. de Vries, C. *et al.* (1992) The FMS-like tyrosine kinase, a receptor for vascular endothelial growth factor. *Science*, **255**, 989–991.
28. Terman, B.I. *et al.* (1992) Identification of the KDR tyrosine kinase as a receptor for vascular endothelial cell growth factor. *Biochem. Biophys. Res. Commun.*, **187**, 1579–1586.
29. Millauer, B. *et al.* (1993) High affinity VEGF binding and developmental expression suggest Flk-1 as a major regulator of vasculogenesis and angiogenesis. *Cell*, **72**, 835–846.
30. Quinn, T.P., Peters, K.G., De Vries, C., Ferrara, N., and Williams, L.T. (1993) Fetal liver kinase 1 is a receptor for vascular endothelial growth factor and is selectively expressed in vascular endothelium. *Proc. Natl. Acad. Sci. U.S.A.*, **90**, 7533–7537.
31. Houck, K.A. *et al.* (1991) The vascular endothelial growth factor family: identification of a fourth molecular species and characterization of alternative splicing of RNA. *Mol. Endocrinol.*, **5**, 1806–1814.
32. Tischer, E. *et al.* (1991) The human gene for vascular endothelial growth factor. Multiple protein forms are encoded through alternative exon splicing. *J. Biol. Chem.*, **266**, 11947–11954.
33. Bergers, G. *et al.* (2000) Matrix metalloproteinase-9 triggers the angiogenic switch during carcinogenesis. *Nat. Cell Biol.*, **2**, 737–744.
34. Ferrara, N. and Davis-Smyth, T. (1997) The biology of vascular endothelial growth factor. *Endocrin. Rev.*, **18**, 4–25.
35. Ferrara, N. *et al.* (2004) Discovery and development of bevacizumab, an anti-VEGF antibody for treating cancer. *Nat. Rev. Drug Discovery*, **3**, 391–400.
36. Kim, K.J. *et al.* (1993) Inhibition of vascular endothelial growth factor-induced angiogenesis suppresses tumor growth in vivo. *Nature*, **362**, 841–844.
37. Warren, R.S., Yuan, H., Matli, M.R., Gillett, N.A., and Ferrara, N. (1995) Regulation by vascular endothelial growth factor of human colon cancer tumorigenesis in a mouse model of experimental liver metastasis. *J. Clin. Invest.*, **95** (4), 1789–1797.
38. Borgstrom, P., Bourdon, M.A., Hillan, K.J., Sriramarao, P., and Ferrara, N. (1998) Neutralizing anti-vascular endothelial growth factor antibody completely inhibits angiogenesis and growth of human prostate carcinoma micro tumors in vivo. *Prostate*, **35**, 1–10.
39. Borgstrom, P., Gold, D.P., Hillan, K.J., and Ferrara, N. (1999) Importance of VEGF for breast cancer angiogenesis in vivo: implications from intravital microscopy of combination treatments with an anti-VEGF neutralizing monoclonal antibody and doxorubicin. *Anticancer Res.*, **19**, 4203–4214.
40. Presta, L.G. *et al.* (1997) Humanization of an anti-VEGF monoclonal antibody for the therapy of solid tumors

and other disorders. *Cancer Res.*, **57**, 4593–4599.

41. Ryan, A.M. *et al.* (1999) Preclinical safety evaluation of rhuMAbVEGF, an antiangiogenic humanized monoclonal antibody. *Toxicol. Pathol.*, **27**, 78–86.
42. Margolin, K. *et al.* (2001) Phase Ib trial of intravenous recombinant monoclonal antibody to vascular endothelial growth factor in combination with chemotherapy in patients with advanced cancer: pharmacologic and long-term safety data. *J. Clin. Oncol.*, **19**, 851–856.
43. Gordon, M.S. *et al.* (2001) Phase I safety and pharmacokinetic study of recombinant human anti-vascular endothelial growth factor in patients with advanced cancer. *J. Clin. Oncol.*, **19**, 843–850.
44. Yang, J.C. *et al.* (2003) A randomized trial of bevacizumab, an anti- VEGF antibody, for metastatic renal cancer. *N. Engl. J. Med.*, **349**, 427–434.
45. Kabbinavar, F. *et al.* (2003) Phase II, Randomized Trial Comparing Bevacizumab Plus Fluorouracil (FU)/Leucovorin (LV) with FU/LV Alone in Patients with Metastatic Colorectal Cancer.
46. DeVore, R. *et al.* (2000) A randomized phase II trial comparing rhumab VEGF (recombinant humanized monoclonal antibody to vascular endothelial cell growth factor) plus carboplatin/paclitaxel (CP) to CP alone in patients with stage IIIB/IV NSCLC. *Proc. Am. Soc. Clin. Oncol.*, **A1896**.
47. Cobleigh, M.A. *et al.* (2003) A phase I/II dose-escalation trial of bevacizumab in previously treated metastatic breast cancer. *Semin. Oncol.*, **30**, 117–124.
48. Reese, D.M. *et al.* (2001) A Phase II trial of humanized antivascular endothelial growth factor antibody for the treatment of androgen-independent prostate cancer. *Prostate J.*, **3** (2), 65–70.
49. (a) Hurwitz, H. *et al.* (2004) Bevacizumab plus irinotecan, fluorouracil, and leucovorin for metastatic colorectal cancer. *N. Engl. J. Med.*, **350** (23); (b)Kabbinavar, F. *et al.* (2003) Phase II, Randomized Trial Comparing Bevacizumab Plus Fluorouracil (Fu)/Leucovorin (LV) with Fu/LV Alone in Patients with Metastatic Colorectal Cancer. *J. Clin. Oncol.*, **21**, 60–65.
50. Giantonio, B.J. *et al.* (2007) Bevacizumab in combination with oxaliplatin, fluorouracil, and leucovorin (FOLFOX4) for previously treated metastatic colorectal cancer: results from the Eastern cooperative oncology group study E 3200. *J. Clin. Oncol.*, **25** (12) 1539–1544.
51. (a) Escudier, B.J. *et al.* (2009) Final results of the phase III, randomized, double-blind AVOREN trial of first-line bevacizumab (BEV) + interferon-alpha 2a (IFN) in metastatic renal cell carcinoma (mRCC). *J. Clin. Oncol.*, **27** (15S); 5020. (b) Warren, R.S. *et al.* (1995) Regulation by vascular endothelial growth factor of human colon cancer tumorigenesis in a mouse model of experimental liver metastasis. *J. Clin. Invest.*, **95**, 1789–1797.
52. Sandler, A. *et al.* (2006) Paclitaxel-carboplatin alone or with bevacizumab for non-small-cell lung cancer. *N. Engl. J. Med.* **355** (24), 2542–2550.
53. Bredel, M. (2009) Keynote comment: translating biological insights into clinical endpoints in neuro-oncology. *Lancet*, **10**, 923–924.
54. Gray, R. *et al.* (2009) Independent review of E2100: a phase III trial of bevacizumab plus paclitaxel versus paclitaxel in women with metastatic breast cancer. *J. Clin. Oncol.*, **27**, 4966–4972.
55. Kolch, W., Kotwaliwale, A., Vass, K., and Janosch, P. (2002) The role of Raf kinases in malignant transformation *Expert Rev. Mol. Med.*, **4**, 1–18.
56. Downward, J. (2003) Targeting RAS signalling pathways in cancer therapy. *Nat. Rev. Cancer*, **3**, 11–22.
57. Kasid, U. and Dritschilo, A. (2003) RAF antisense oligonucleotide as a tumor radiosensitizer. *Oncogene*, **22**, 5876–5884.
58. Riedl, B. *et al.* (2001) Potent Raf kinase inhibitors from the diphenylurea class: structure activity relationships. *Clin. Cancer Res.*, **20**, 83a.
59. Smith, R.A. *et al.* (2001) Discovery of heterocyclic ureas as a new class of raf kinase inhibitors: identification of a second generation lead by a combinatorial

chemistry approach. *Bioorg. Med. Chem. Lett.*, **11**, 2775–2778.

60. Lowinger, T.B., Riedl, B., Dumas, J., and Smith, R.A. (2002) Design and discovery of small molecules targeting raf-1 kinase. *Curr. Pharm. Des.*, **8**, 2269–2278.
61. Wilhelm, S. *et al.* (2001) BAY 43-9006, a novel Raf-1 kinase inhibitor (RKI) blocks the Raf/MEK/ERK pathway in tumor cells. *Proc. Am. Assoc. Cancer Res.*, **42**, 923.
62. Wilhelm, S.M. *et al.* (2004) BAY 43-9006 exhibits broad spectrum oral anti-tumor activity and targets the Raf/ MEK/ERK pathway and receptor tyrosine kinases involved in tumor progression and angiogenesis. *Cancer Res.*, **64**, 7099–7109.
63. Carlomagno, F. *et al.* (2006) BAY 43-9006 inhibition of oncogenic RET mutants. *J. Natl. Cancer Inst.*, **98**, 326–334.
64. Wan, P.T. *et al.* (2004) Mechanism of activation of the RAF-ERK signaling pathway by oncogenic mutations of B-RAF. *Cell*, **116**, 855–867.
65. Wilhelm, S., Carter, C., Lynch, M., Lowinger, T., Dumas, J., Smith, R.A., Schwartz, B., Simantov, R., and Kelley, S. (2006) Discovery and development of sorafenib: a multikinase inhibitor for treating cancer. *Nat. Rev. Drug Discovery*, **5**, 835–844.
66. Escudier, B., Eisen, T., Stadler, W.M., Szczylik, C., Oudard, S., Siebels, M., Negrier, S., Chevreau, C., Solska, E., Desai, A.A., Rolland, F., Demkow, T., Hutson, T.E., Gore, M., Freeman, S., Schwartz, B., Shan, M., Simantov, R., and Bukowski, R.M., TARGET Study Group (2007) Sorafenib in advanced clear-cell renal-cell carcinoma. *N Engl J Med.*, **356** (2), 125–134.
67. Llovet, J.M., Ricci, S., Mazzaferro, V., Hilgard, P., Gane, E., Blanc, J.F., de Oliveira, A.C., Santoro, A., Raoul, J.L., Forner, A., Schwartz, M., Porta, C., Zeuzem, S., Bolondi, L., Greten, T.F., Galle, P.R., Seitz, J.F., Borbath, I., Häussinger, D., Giannaris, T., Shan, M., Moscovici, M., Voliotis, D., and Bruix, J., SHARP Investigators Study Group (2008) Sorafenib in advanced hepatocellular carcinoma. *N. Engl. J. Med.*, **359** (4), 378–390.
68. Cohen, M.H. *et al.* (2005) FDA drug approval summary: erlotinib (Tarceva®) tablets. *Oncologist*, **10**, 461–466.
69. Ikezoe, T. *et al.* (2006) The antitumor effect of sunitinib (formerly SU11248) against a variety of human hematologic malignancies: enhancement of growth inhibition via inhibition of mammalian target of rapamycin signalling. *Mol. Cancer Ther.*, **5**, 2522–2530.
70. Bergers, G. *et al.* (2003) Benefits of targeting both pericytes and endothelial cells in the tumor vasculature with kinase inhibitors. *J. Clin. Invest.*, **11** (9) 1287.
71. Motzer, R.J. *et al.* (2007) Sunitinib versus interferon alfa in metastatic renal-cell carcinoma. *N. Engl. J. Med.*, **356** (2), 115–124.
72. Goodman, V.L. *et al.* (2007) Approval summary: sunitinib for the treatment of imatinib refractory or intolerant gastrointestinal stromal tumors and advanced renal cell carcinoma. *Clin. Cancer Res.*, **13** (5) 1367.
73. Baselga, J. (2001) The EGFR as a target for anticancer therapy – focus on cetuximab. *Eur. J. Cancer*, **37**, S16–S22.
74. Dancey, J. and Sausville, E.A. (2003) Issues and progress with protein kinase inhibitors for the treatment of cancer. *Nat. Rev. Drug Discovery*, **2**, 296–313.
75. Salomon, D.S. *et al.* (1995) Epidermal growth factor-related peptides and their receptors in human malignancies. *Crit. Rev. Oncol. Hematol.*, **19**, 183–232.
76. Graham, J. (2004) Cetuximab. *Nat. Rev. Drug Discovery*, **3**, 549–550.
77. Goldstein, N.I. *et al.* (1995) Biological efficacy of a chimeric antibody to the epidermal growth factor receptor in a human tumor xenograft model. *Clin. Cancer Res.*, **1**, 1311–1318.
78. Prewett, M.C. *et al.* (2002) Enhanced antitumor activity of antiepidermal growth factor receptor monoclonal antibody IMC-C225 in combination with irinotecan (CPT-11) against human colorectal tumor xenografts. *Clin. Cancer Res.*, **8**, 994–1003.

79. Cunningham, D. *et al.* (2004) Cetuximab monotherapy and cetuximab plus irinotecan in irinotecan-refractory metastatic colorectal cancer, *N. Engl. J. Med*, **351** (4), 337–345.
80. Karapetis, C.S., Khambata-Ford, S., Jonker, D.J., O'Callaghan, C.J., Tu, D., Tebbutt, N.C., Simes, R.J., Chalchal, H., Shapiro, J.D., Robitaille, S., Price, T.J., Shepherd, L., Au, H.J., Langer, C., Moore, M.J., and Zalcberg, J.R. (2008) K-ras mutations and benefit from cetuximab in advanced colorectal cancer. *N. Engl. J. Med.*, **359** (17), 1757–1765.
81. Bonner, J.A. *et al.* (2006) Radiotherapy plus cetuximab for squamous cell carcinoma of the head and neck. *N. Engl. J. Med* **354** (6), 567–578.
82. Yarden, Y. and Sliwkowski, M.X. (2001) Untangling the ErbB signaling network. *Nat. Rev. Mol. Cell Biol.*, **2**, 127–137.
83. Baselga, J. (2002) Why the epidermal growth factor receptor? The rationale for cancer therapy. *Oncologist*, **7** (S4), 2–8.
84. Moore, M.J. *et al.* (2007) Erlotinib plus gemcitabine compared with gemcitabine alone in patients with advanced pancreatic cancer: a phase III trial of the National Cancer Institute of Canada Clinical Trials Group. *J. Clin. Oncol.*, **25** (15).
85. Faivre, S. *et al.* (2006) Current development of mTOR inhibitors as anticancer agents. *Nat. Rev. Drug Discovery*, **5**.
86. Hudes, G. *et al.* (2007) Temsirolimus, inteferon alfa, or both for advanced renal-cell carcinoma, *N. Engl. J. Med.*, **356**, 22.

11
Combination of Antiangiogenic Therapy with Other Anticancer Therapies

Raffaele Longo, Francesco Torino, and Giampietro Gasparini

11.1
Introduction

Angiogenesis is a necessary process allowing tumors to grow beyond 1–2 mm^3 in size and it facilitates metastasis because the tissue oxygen diffusion limit is 100–200 μm, which corresponds to three to five cell layers around a blood vessel [1, 2]. This dynamic process is induced by complex molecular mechanisms and/or by local alterations such as hypoxia, glucose deprivation, oxidative, and mechanical stresses [1, 2] and is tightly regulated by pro- and anti-angiogenesis growth factors (i.e., endogenous inhibitors) [1–3]. Proangiogenic growth factors, such as vascular endothelial growth factor (VEGF), fibroblast growth factors (FGFs), and platelet-derived growth factor (PDGF) are released into the microenvironment by malignant, inflammatory, and other stromal cells in response to various stimuli. Activated ECs, as well as local stromal cells and circulating endothelial progenitors (CEPs), secrete several enzymes including matrix metalloproteases (MMPs) that break down the extracellular matrix and allow ECs to invade surrounding tissues, proliferate, and migrate [1–5]. Plasmin generated by the urokinase plasminogen activator (uPA), regulated by the urokinase plasminogen activator-receptor (uPA-r), and the tissue-type plasminogen activator inhibitor-1 (TTPAI-1), is a key mediator of these processes [5]. The migration of both ECs and CEPs is regulated by adhesion molecules such as integrins $\alpha_v\beta_3$, and $\alpha_v\beta_5$ [6].

VEGF expression is regulated by the von-Hippel Lindau (VHL) gene and VEGF in turn stimulates several processes: (i) invasion of ECs by induction of MMPs, uPA, uPA-r, and TTPAI-1; (ii) migration of ECs by activation of p38 MAP kinase, nitric oxide, and focal adhesion kinase (FAK); (iii) survival of ECs by induction of phosphatidylinositol 3-kinase/protein kinase B (PI3K/PKB) pathway, inhibition of caspases, and upregulation of the inhibitors of the apoptosis family, including survivin and Bcl2; (iv) permeability of ECs by vesicovascular organelles, endothelial fenestrations, and opening of junctions between adjacent ECs; and finally, (v) proliferation of ECs by activation of mitogen-activated protein kinases, such as extracellular signal-regulated kinases Erk1/2, JNK/SAPK (stress-activated protein kinase), and protein kinase C (PKC) [7, 8].

Tumor Angiogenesis – From Molecular Mechanisms to Targeted Therapy. Edited by Francis S. Markland, Stephen Swenson, and Radu Minea
Copyright © 2010 WILEY-VCH Verlag GmbH & Co. KGaA, Weinheim
ISBN: 978-3-527-32091-2

Tumor endothelium loses the normal organized structure of the vascular network and it is phenotypically different from normal vessels [1–3]. The structural abnormalities of the vasculature alter permeability, tumor perfusion, and induce enhanced interstitial fluid pressure [9]. Tumor ECs divide up to 30–40 times more frequently than normal ECs and preferentially overexpress the following cell-surface molecules: integrin $\alpha_V\beta_3$, E-selectin, endoglin, endosialin, and vascular endothelial growth factor receptors (VEGFRs) all stimulating endothelium adhesion and migration [2, 3]. A mutual stimulation occurs between the stroma and tumor parenchyma, which sustains malignant growth, progression, and metastasis [3]. Therefore, the interaction between the tumor cell and the vascular system should be considered as a functional unit with regard to tumor growth.

11.2 Antiangiogenic Therapy and Immunomodulation

Several clinical studies have evaluated the combination of antiangiogenic compounds, such as anti-VEGF monoclonal antibodies and/or VEGF tyrosine kinase inhibitors (TKIs), with interferon (IFN) in renal cell carcinoma (RCC) and melanoma.

A randomized phase III trial has evaluated, in 649 patients with previously untreated metastatic RCC, the combination of IFNα-2a (9 MIU subcutaneously (sc) three times weekly for a maximum of 52 weeks) and bevacizumab (10 mg kg^{-1} every two weeks until disease progression) ($n = 327$) versus IFNα-2a and placebo ($n = 322$) (*BO17705E Trial*). Median duration of progression-free survival (PFS) was significantly longer in the experimental arm than it was in the control group (10.2 vs. 5.4 months; hazard ratio: 0.63, 95% CI: 0.52–0.75; $p = 0.0001$, irrespective of Memorial Sloan-Kettering Cancer Center (MSKCC) risk category or reduced-doses of IFN α-2a. Although the data on overall survival (OS) were not mature, median OS has not yet been reached the bevacizumab + IFNα-2a group versus 19.8 months in the control. Overall, 214 (70%) patients obtained tumor shrinkage in the experimental arm, compared with 112 (39%) of the control group. The median duration of response and stable disease (SD) were longer and time of induction of response was shorter in the bevacizumab + IFNα-2a group. The most common grade 3 or worse toxicity reported in the experimental group was fatigue (12% vs. 8%) and asthenia (10% vs. 7%). Grades 3 and 4 adverse events in patients receiving bevacizumab included four gastrointestinal perforations (1%) and 10 thromboembolic events (3%). Seven (2%) patients with hypertension of any severity discontinued treatment due to this event and 16 (5%) patients discontinued due to proteinuria of any severity [10].

Another randomized phase III trial has evaluated bevacizumab (10 mg kg^{-1} intravenously every two weeks) plus IFNα (9 MIU sc three times weekly) versus the same dose and schedule of IFN monotherapy in 732 patients with previously untreated, clear cell mRCC (*CALGB 90206 Trial*) [11].

The median PFS was 8.5 months in patients receiving the combined treatment (95% CI, 7.5–9.7 months) versus 5.2 months (95% CI, 3.1–5.6 months) in patients receiving only IFN ($p < 0.0001$). The adjusted hazard ratio was 0.71 (95% CI, 0.61–0.83; $p < 0.0001$). Bevacizumab plus IFNα induced a higher overall response rate (ORR) as compared to IFNα (25.5% (95% CI, 20.9–30.6%) versus 13.1% (95% CI, 9.5–17.3%); $p < 0.0001$). Overall toxicity was greater for bevacizumab plus IFNα, including a significant higher grade 3 hypertension (9% vs. 0%), anorexia (17% vs. 8%), fatigue (35% vs. 28%), and proteinuria (13% vs. 0%) [11].

Based on the above results, the US FDA approved this schedule for first-line therapy of advanced RCC.

A phase I study was undertaken to determine the maximum tolerated dose (MTD), and pharmacokinetics of sorafenib, a Raf and multiple VEGFR TKI, plus IFNα in advanced RCC or melanoma patients. Dynamic contrast-enhanced ultrasonography (DCE-US) was used to monitor the effects of sorafenib/IFNα on tumor vasculature. Patients received 28-day cycles of continuous, oral sorafenib twice daily and sc IFNα thrice weekly: sorafenib 200 mg twice daily plus IFNα 6 MIU thrice weekly (cohort 1); and sorafenib 400 mg twice daily plus IFNα 6 MIU thrice weekly (cohort 2); or plus IFNα 9 MIU thrice weekly (cohort 3). Thirteen patients received at least one dose of sorafenib plus IFNα (12 RCC; one melanoma). The MTD was not reached (only one dose-limiting toxicity (DLT) occurred (grade 3 asthenia)). Most frequently reported drug-related adverse events were grade 2 or less in severity, including fatigue, diarrhea, nausea, alopecia, and hand–foot skin reaction. One (7.7%) RCC patient achieved partial response (PR) and eight (61.5%) had SD (including the melanoma patient). Good responders assessed by DCE-US had increased PFS and OS, relative to poor responders. IFNα had no effect on the pharmacokinetics of sorafenib. There were no significant changes in absolute values of lymphocytes, levels of proangiogenic cytokines, or inhibition of phosphorylated Erk in T cells or natural killer cells, with combination therapy [12].

Gollob *et al.* evaluated the combination of sorafenib plus IFNα-2b in 40 patients with metastatic RCC as first- or second-line therapy in a phase II study. Treatment consisted of eight-week cycles of sorafenib 400 mg orally bid plus IFNα-2b 10 MIU sc three times a week followed by a two-week break. The response rate (RR) was 33% (95% CI, 19–49%; 13 of 40 patients), including 28% partial ($n = 11$) and 5% complete responses ($n = 2$). Responses were seen both in treatment-naïve and in interleukin-2 (IL-2) pretreated patients within the first two cycles. The median duration of response was 12 months (95% CI, 6–13 months). Forty-five percent of patients had SD for at least one cycle, and 12% had progressive disease (PD). Responses occurred quickly, with the majority of tumor regressions observed within the first cycle. Although the most common sites of response were lungs and nodes, dramatic responses were also seen in patients with bulky disease in pleura, liver, and pancreas. With a median follow-up of 14 months, median PFS was 10 months (95% CI, 8–18 months), and median OS had not yet been reached. Fatigue, anorexia, anemia, diarrhea, hypophosphatemia, rash, nausea, and weight loss were the most common toxicities. However, dose reductions were required in 65% of patients [13].

A second phase II trial (*SWOG 0412 Trial*) evaluated the activity of combined treatment with IFNα-2b (10×10^6 IU sc three times weekly) and sorafenib (400 mg orally bid) in 62 patients with advanced RCC as first-line therapy. Twelve (19%) of the 62 assessable patients achieved a confirmed PR. An additional 31 (50%) had an unconfirmed PR or SD as best response. The median duration of confirmed responses was 8 months (range: 2–20+ months). The median PFS was 7 months (95% CI, 4–11 months). The 6- and 12-month PFS rates were 53% (95% CI, 40–66%) and 37% (95% CI, 25–50%) respectively. The most common adverse events were fatigue, anorexia, anemia, diarrhea, nausea, rigors/chills, leukopenia, fever, and transaminases elevation. *VHL* gene mutations were detected in four (22%) of the 18 available archival tumor specimens. The four mutations included two truncating mutations (one nonsense in exon 3 and one frameshift in exon 2) and two missense mutations (one each in exons 1 and 3). Immunohistochemistry analysis of p-MAPK, p-p38, and p-AKT expression in archival tumor tissue was available for 22 eligible patients (23 for p-MAPK). p-MAPK expression was undetectable in all but one tumor, whereas p-p38 expression was elevated in all tumors, precluding analysis of these markers for clinical outcome predictive value. p-AKT was overexpressed (>20% staining) in 18 (82%) of 22 patients. Two of three patients with 0% p-AKT staining had a PR. Median p-AKT expression was nonsignificantly higher (95%) among 16 nonresponders than among the six responders (53%). Although the small sample size of this study prevents formal statistical analysis, these findings are not inconsistent with a report that has suggested that certain *VHL* mutations (those that truncate or shift the reading frame) may be associated with longer PFS in patients receiving VEGF-targeted therapy [14].

The combination of bevacizumab and pegylated (or polyethyleneglycolated i.e., PEG) IFNα-2b has been tested in patients with metastatic or unresectable carcinoid tumors. Forty-four patients in therapy with octreotide were randomly assigned to 18 weeks of treatment with bevacizumab or PEG-IFNα-2b. At disease progression or after 18 weeks (whichever occurred earlier), the patients received bevacizumab plus PEG-IFNα-2b until progression. Functional computer tomography (CT) scans were performed to measure the effects on tumor blood flow. In the bevacizumab arm, four patients (18%) achieved confirmed PR, 17 patients (77%) had SD, and one patient (5%) had a PD. In the PEG-IFNα-2b arm, 15 patients (68%) had SD and six patients (27%) a PD. PFS rates after 18 weeks of monotherapy were 95% in bevacizumab versus 68% on the PEG-IFNα-2b arm. The overall median PFS for all 44 patients was 63 weeks. When compared with paired baseline measurements on functional CT scans, a decrease of 49% ($p < 0.01$) and 28% ($p < 0.01$) in tumor blood flow at day 2 and week 18, respectively, was observed among patients treated with bevacizumab. No significant changes in tumor blood flow were observed following PEG-IFN. PEG-IFNα-2b treatment was associated with decreased plasma basic fibroblast growth factor (bFGF; $p = 0.04$) and increased IL-18 ($p < 0.01$). No significant changes in bFGF or IL-18 were observed following treatment with bevacizumab [15].

Despite these encouraging results, several important challenges include the following: the lack of validated predictive markers for the selection of the patients who are most likely to achieve benefit of antiangiogenic treatments. Second, for agents such as bevacizumab and anti-VEGF TKIs that are preferentially cytostatic drugs, it is very difficult to rigorously define the DLT because the (graphical) relationship between toxicity and drug dosage is generally quite flat. In addition, the optimal biological dose of these compounds is probably below the MTD and could spare patients from unnecessary toxicity. Third, combining anti-VEGF drugs with immunomodulant agents, such as IFNs, could lead to unexpected pharmacokinetic and/or pharmacological interactions and sometimes, to an increased toxicity. As reported in the studies of RCC, this latter aspect, particularly the adverse events commonly related to IFN, could limit further development of these combinations. Future, well-designed, prospective and randomized phase III studies should answer these concerns.

11.3 Anti-VEGF and Anti-EGFR/HER-2 Combinations

Robust experimental evidence has shown that both EGFR and HER-2 pathways and VEGF-dependent angiogenesis are functionally linked [16]. VEGF signaling is upregulated by EGFR expression and, conversely, VEGF upregulation independent of EGFR signaling seems to contribute to the resistance to EGFR inhibition [17]. Combined inhibition of EGFR and VEGF signaling interferes with a molecular feedback loop responsible for acquired resistance to anti-EGFR agents and promotes apoptosis [16]. To this end, inhibition of both these pathways could improve antitumor efficacy and overcome resistance to EGFR therapy.

There are two main strategies pursued in preclinical and clinical studies, the first being the combination of different drugs that selectively inhibit VEGF and EGFR/HER-2 pathways, such as monoclonal antibodies and small TKIs. The second strategy is based on the single, multitargeted small-molecule inhibitors that simultaneously block VEGF and EGFR/HER-2 receptors, such as ZD6474 (vandetanib) and AEE 788.

Two phase II studies have evaluated the combination of bevacizumab and erlotinib in untreated patients with metastatic RCC [17, 18].

In the first study, the RR was of 25% with 61% of SD. The median OS and one-year PFS were 11 months and 43% respectively. Treatment was generally well tolerated and only two patients discontinued treatment due to skin toxicity [18].

The second study compared the same combination versus bevacizumab alone in 104 untreated metastatic RCC patients. The median PFS was 9.9 months in the experimental arm versus 8.5 months ($p = 0.58$) and the RR was 14 and 13% respectively. Median OS was similar in the two groups. The most common grade 3/4 toxicity included hypertension, rash, proteinuria, diarrhea, and hemorrhage [19].

A small phase II study performed in heavily pretreated patients with metastatic breast cancer showed that the combination of bevacizumab and erlotinib was safe and active [20].

In patients with HER-2 positive refractory breast cancer, the combination of bevacizumab and trastuzumab induced a RR of 40% without relevant toxicity [21].

A randomized phase II (*BOND-2 Trial*) study evaluated the activity and safety of bevacizumab and cetuximab in patients with irinotecan refractory colorectal cancer. Patients in arm A (43) received irinotecan at the same dose and schedule as last received before study entry, plus cetuximab 400 mg m^{-2} loading dose, then weekly cetuximab 250 mg m^{-2}, plus bevacizumab 5 mg kg^{-1} administered every other week. Patients in arm B (40) received the same schedules of cetuximab and bevacizumab as those in arm A without irinotecan. Toxicities were as would have been expected from the single agents. Time to progression (TTP) and RR were 7.3 months/37% and 4.9 months/20% for arms A and B respectively. The OS was 14.5 months (arm A) and 11.4 months (arm B) [22].

An interim analysis of toxicity performed in 755 patients with metastatic colorectal cancer randomly assigned between treatment with capecitabine, oxaliplatin, and bevacizumab $\pm$ cetuximab (*CAIRO-2 Trial*), documented that the incidence of overall grade 3/4 toxicity was significantly higher in the cetuximab-containing arm as compared with the control arm (81% vs. 72%, $p = 0.03$). This difference was mainly related to cetuximab-related skin toxicity. The addition of cetuximab did not result in an increase of gastrointestinal toxicity or treatment-related mortality. Until now, there are no available data regarding the efficacy of this combination [23, 24].

Panitumumab, a fully human antibody targeting the EGFR, was recently evaluated in a randomized phase III trial (*PACCE Trial*) in combination with bevacizumab and chemotherapy (oxaliplatin- and irinotecan-based) as first-line treatment in 823 metastatic colorectal cancer patients. Panitumumab was discontinued after a planned interim analysis that showed worse efficacy in the experimental arm. In the final analysis, median PFS and OS were 10/19.4 and 11.4/24.5 months for the panitumumab and control arm respectively. Grade 3/4 adverse events in the oxaliplatin cohort (panitumumab vs. control) included skin toxicity (36% vs. 1%), diarrhea (24% vs. 13%), infections (19% vs. 10%), and pulmonary embolism (6% vs. 4%). Increased toxicity without evidence of improved efficacy was observed in the panitumumab arm of the irinotecan cohort. KRAS (Kirsten rat sarcoma viral oncogene homolog) analyses showed adverse outcomes for the panitumumab arm in both, wild-type and mutant groups [25].

A phase I/II trial evaluated the combination of erlotinib (150 mg per day orally) and bevacizumab (15 mg kg^{-1} every three weeks) in second-line patients with advanced, nonsquamous, non-small-cell lung cancer (NSCLC) with interesting results (RR: 14.3%, PFS: 6.2 months, and median OS 12.6 months) [26].

A multicenter, randomized phase II trial evaluated the safety of combining bevacizumab with either chemotherapy (docetaxel or pemetrexed) or erlotinib in 120 patients with NSCLC progressed during or after one platinum-based regimen. No unexpected adverse events were noted. Fewer patients (13%) in the bevacizumab–erlotinib arm discontinued treatment as a result of adverse events than in the chemotherapy alone (24%) or bevacizumab–chemotherapy (28%) arms. The incidence of grade 5 hemorrhage in patients receiving bevacizumab was 5.1%. Although not statistically significant, the risk of PD or death was 0.66 (95% CI, 0.38–1.16) in patients treated with bevacizumab–chemotherapy versus 0.72 (95% CI, 0.42–1.23) among patients treated with bevacizumab–erlotinib. One-year survival rate was 57.4% for bevacizumab–erlotinib and 53.8% for bevacizumab–chemotherapy compared with 33.1% for chemotherapy alone [27].

A number of studies are testing the combination of bevacizumab with anti-EGFR agents, such as cetuximab and/or erlotinib, alone or plus chemotherapy, in pancreatic cancer.

A recent study evaluated the combination of sunitinib with gefitinib, an EGF-receptor inhibitor in metastatic RCC with clear cell component. The phase I part of the study was conducted to determine the MTD of sunitinib in combination with gefitinib and subsequently, patients were enrolled in the phase II part to further evaluate safety and antitumor activity. Forty-two patients were enrolled, 11 in phase I and 31 in phase II. Twenty-eight patients (67%) had prior cytokine therapy, 11 (26%) had no prior cytokine therapy, and 3 (7%) received prior vaccine therapy. Two DLTs were observed at the 50-mg dose level (DL) of sunitinib (grade 3 fatigue and grade 2 ejection fraction decline), and 37.5 mg on schedule 4/2 in combination with gefitinib 250 mg daily was determined to be the MTD. Eleven patients (30%) achieved a PR and 15 (42%) an SD. Median duration of response and PFS were 9.2 and 11.2 months respectively. The most common grade 3 treatment-related adverse events observed in phase I were diarrhea and nausea ($n = 2$), and for phase II were diarrhea (10%) and gastrointestinal hemorrhage (6%). Two phase II patients were withdrawn from the study due to ejection fraction decline and cardiac arrhythmia, both of which were reversible after discontinuation of the treatment [28].

Sunitinib is also being combined with erlotinib in a phase II trial of patients with metastatic RCC that began recruitment as of 11 November 2006, and is still accruing patients at this time.

Vandetanib (Zactima; ZD6474) is an oral TKI that selectively targets VEGFR-, EGFR-, and RET-dependent signaling [29]. Phase I evaluation showed that vandetanib monotherapy was generally well tolerated at doses of ≤300 mg per day, and its half-life of approximately 120 h supports a once-daily dosing [30, 31]. In patients with previously treated NSCLC, including those with squamous histology and stable brain metastases, PFS prolongation was observed with vandetanib 100 mg per day plus docetaxel versus docetaxel alone [32].

A recent, randomized phase II study (*6474IL/0007 Trial*) compared the activity of vandetanib monotherapy versus vandetanib with paclitaxel and carboplatin (VPC) versus paclitaxel and carboplatin (PC) in previously untreated patients with NSCLC. The risk of progression was reduced in patients receiving VPC ($n = 56$) versus PC ($n = 52$; hazard ratio = 0.76, one-sided $p = 0.098$); median PFS was 24 weeks (VPC) and 23 weeks (PC). The vandetanib monotherapy arm ($n = 73$) was discontinued after a planned interim PFS analysis met the criterion for discontinuation (hazard ratio > 1.33 vs. PC). OS was not significantly different between patients receiving VPC or PC. Increased rates of hypertension (32% vs. 4%), rash (50% vs. 25%), and diarrhea (52% vs. 33%) occurred with VPC versus PC. No hemorrhage events requiring intervention were observed [33].

11.4
Miscellaneous Anti-VEGF Combinations

Ongoing exploratory clinical trials are also evaluating combinations of bevacizumab with other targeted agents, including VEGFR TKIs and mTOR (mammalian target of rapamycin) inhibitors.

Combinatorial strategies using signal inhibitory agents with related targets have the potential for induced biochemical and clinical synergism, with the expectation that therapeutic interruption of pathways in series (vertical inhibition) may be successful using lower doses of complementary agents that intersect the pathway at multiple sites.

Recently, a phase I/II study tested the combination of bevacizumab and sorafenib in 31 patients with advanced solid tumors [34]. A phase I dose-escalation trial of sorafenib and bevacizumab was initiated at below-recommended single-agent doses because of possible overlapping toxicity: sorafenib 200 mg orally twice daily and bevacizumab intravenously at 5 mg kg^{-1} (DL1) or 10 mg kg^{-1}(DL2) every two weeks. DLT in DL2 was grade 3 proteinuria and thrombocytopenia. Adverse events included hypertension, hand–foot syndrome (HFS), diarrhea, transaminitis, and fatigue. PRs were seen in 6 (43%) of 13 patients with ovarian cancer (response duration range, 4 to 22+ months) and 1 of 3 patients with RCC (response duration, 14 months). Clinical benefit ≥ 4 months (median, 6 months; range, 4–22+ months) was reported in 22 (59%) of 37 assessable patients. The majority (74%) required sorafenib dose reduction to 200 mg per day at a median of four cycles (range, 1–12 cycles) [34].

This study suggests a potential clinical benefit of this combination, albeit with therapy-limiting toxicities. The combination of the low-dose sorafenib and bevacizumab was at least supra-additive in adverse events with a greater intensity, rapidity, and frequency of side-effects. A phase II study is evaluating this combination that holds promise for clinical benefit in multiple tumor types, especially ovarian cancer.

The safety and MTD of sunitinib in combination with bevacizumab was assessed in a Phase I trial. Cohorts of three to six patients with metastatic RCC received escalating doses of sunitinib (DLs: 25, 37.5, and 50 mg po) daily for four weeks followed by two weeks off with fixed-dose bevacizumab (10 mg kg^{-1} IV) every two weeks continuously. Two DLTs of G4 hemorrhage occurred (1 cohort 2, 1 cohort 3). The MTD was sunitinib 50 mg plus bevacizumab 10 mg kg^{-1}. However, with chronic treatment, additional toxicities were observed, including G3/4 HTN (14/25 overall, 9/12 cohort 3), G3/4 proteinuria (6/25 overall, 4/12 cohort 3),

G3 thrombocytopenia (6/12 cohort 3), G3 reversible posterior leukoencephalpathy syndrome (RPLS) (2/12 cohort 3), and G3 microangiopathic hemolytic anemia (2/12 cohort 3). One pt in cohort 2 died of a myocardial infarction. Other toxicities were reversible. In total, 10/25 (40%) patients required sunitinib dose reductions (2/6 cohort 2, 8/12 cohort 3) and 11/25 (44%) patients were withdrawn for toxicity (2/7 cohort 1, 2/6 cohort 2, 7/12 cohort 3). The median number of treatment cycles received was seven for cohort 1 (range 1–14), three for cohort 2 (range 1–9), and four for cohort 3 (range 1–8). Of 25 patients, 13 (52%) had a confirmed objective response (1 complete response and 12 PRs) [35].

In a multicenter phase II trial, the combination of bevacizumab and everolimus (RAD001), an mTOR inhibitor, was evaluated in 59 patients with metastatic RCC, previously treated (29, Group B) or not (30, Group A) with sorafenib or sunitinib. All patients received bevacizumab (10 mg kg^{-1} IV every two weeks) and RAD001 (10 mg orally daily). 90% of patients who completed eight weeks of treatment had objective response (21%) or minor response/SD (69%). Efficacy was better in Group A. Grade 3/4 proteinuria occurred in 10 patients (19%); other grade 3/4 toxicity was uncommon (fatigue 9%, stomatitis 8%). Grade 1/2 toxicities were common including fatigue (68%), skin rash/pruritus (55%), mucositis/stomatitis (49%), hyperlipidemia (45%), nausea (40%), and hypertension (25%) [36].

A phase II, single-arm, open-label, multicenter study evaluated the safety and efficacy of sunitinib in 61 patients with bevacizumab-refractory metastatic RCC. Fourteen patients (23.0%; 95% CI, 13.2–35.5%) experienced a PR, with a median response duration of 44.1 weeks (95% CI, 25.0–102.7 weeks). Thirty-six patients (59%) had an SD and five patients (8%) a PD. Median PFS and OS were 30.4 weeks (95% CI, 18.3–36.7 weeks) and 47.1 weeks (95% CI, 36.9–79.4 weeks) respectively. Overall, 41 patients (67%) discontinued treatment during the study: 33 (54%) because of PD and seven (12%) due to adverse events; one patient withdrew consent. The reported treatment-related toxicity was fatigue, diarrhea, and nausea, of mild or moderate intensity (grade 1 or 2). The reported treatment-related grade 3 adverse events were fatigue (34%), hypertension (18%), and HFS (10%). Significant changes in the mean plasma levels of VEGF-A, sVEGFR-3, and PlGF were observed at the first cycle of sunitinib therapy. After 28 days, mean plasma VEGF-A levels increased 2.8-fold (range: 0.4- to 13.6-fold) greater than baseline and PlGF levels increased 3.9-fold (range: 0.8- to 20.4-fold). In contrast, mean sVEGFR-3 and VEGF-C levels decreased by 37.6 and 22.7% respectively. VEGF-A and sVEGFR-3 returned to near-baseline levels after the two-week off-treatment period with similar changes in subsequent cycles [37].

Patients with baseline sVEGFR-3 and VEGF-C levels lower than the median baseline values (sVEGFR-3, 47 000 pg ml^{-1}; VEGF-C, 722.1 pg ml^{-1}) had longer PFS than did patients with levels greater than the median. In contrast, there was no correlation between baseline plasma VEGF-A or PlGF levels and PFS (data not shown). Similarly, baseline levels of sVEGFR-3 and VEGF-C were significantly lower in patients with PR compared with patients with SD or PD, whereas baseline levels of VEGF-A and PlGF showed no significant correlation with the RR (data not shown). Nevertheless, it is substantial, and strongly suggests that sunitinib can overcome resistance mechanisms that develop in response to exposure to VEGF-binding agents. Whether this depends on its inhibitory properties against VEGFR or against PDGFR and other TKs remains to be determined [37].

In a retrospective report, 30 patients with metastatic RCC received Sunitinib or Sorafenib following prior antiangiogenic therapy. Thirteen of 16 patients who received Sunitinib and 10 of 14 patients treated with Sorafenib had some tumor shrinkage. The median TTP for the entire cohort was 10.4 months. Prior response to an antiangiogenic agent (AIA) did not predict subsequent clinical benefit to either sunitinib or sorafenib [38].

In another retrospective study, in 90 patients with metastatic RCC receiving sunitinib or sorafenib in sequence (68 patients received sorafenib first, 22 received sunitinib first), PR or SD was observed as best response in the second treatment in both sequences. Only six patients had PD with both drugs [39]. These results suggest the lack of cross-resistance between these two drugs and support their sequential use in metastatic RCC.

Another report also supports this hypothesis. Twenty-three patients received sorafenib followed by sunitinib (Group A), and 14 patients received sunitinib followed by sorafenib (Group B). Three patients (13%) in Group A and two (14%) in Group B switched due to toxicity, while all remaining patients switched due to PD. For patients in Group A, median duration of SD after starting sunitinib was 32 weeks (range 6–37 weeks); 9% could not be evaluated for response to treatment; 34% had PD; and four patients (17%) had a PR. For patients in Group B, median duration of SD after starting sorafenib was eight weeks (range 4–10 weeks); 7% were not evaluated for response to treatment; 50% had PD; and one patient (7%) had no evidence of disease after tumor resection. The median duration of disease control in Groups A and B was 42 weeks (range: 10–113 weeks) and 30.5 weeks (range: 4–56 weeks) respectively [40].

A multicenter, open-label, phase II study tested the activity of Axitinib (AG-013736), a potent inhibitor of VEGFRs, in 42 evaluated patients with metastatic RCC progressing after therapy with sorafenib or sunitinib. PR was observed in 6 patients (14%; 95% CI: 5–29%) and SD in 15 patients (36%). Twelve patients (29%) experienced PD, and 9 patients (21%) withdrew due to adverse events. Overall, 57% of patients experienced some degree of tumor regression. With a median follow-up of 5.3 months, the median PFS was not reached. Preliminary analysis indicates overall median PFS > 7.1 months. Treatment-related grade 3/4 toxicity included hypertension (16%), fatigue (14%), and HFS (14%) [41].

A phase III, randomized, double-blind, placebo-controlled trial evaluated everolimus, at 10 mg once daily ($n = 272$), versus placebo ($n = 138$) in patients with metastatic RCC progressed on sunitinib, sorafenib, or both (*RECORD-1 Trial*). The primary endpoint was PFS, assessed by a blinded, independent central review. The results of the second interim analysis suggests a significant difference in efficacy between the two arms after 191 progression events (101 (37%) events in the everolimus group, 90 (65%) in the placebo group; hazard ratio 0.30, 95%

CI 0.22–0.40, $p < 0.0001$; median PFS 4.0 vs. 1.9 months). The most commonly reported toxicity included stomatitis (40% in the everclimus group vs. 8% in the placebo group), rash (25% vs. 4%), and fatigue (20% vs. 16%). Pneumonitis (any grade) was detected in 22 (8%) patients in the everolimus group, of whom eight had pneumonitis of grade 3 severity [42].

11.5 Antiangiogenic Therapy and Radiation Treatment

11.5.1 Biological Aspects

Ionizing radiation is effective as a primary curative cancer treatment modality in many tumors and over half of all cancer patients receive radiation therapy during their course of treatment [43]. The combination of cytotoxic chemotherapeutic agents with radiation may obtain better local disease control and survival advantage than radiotherapy (RT) alone [43, 44]. Unfortunately, following RT or chemoradiotherapy (CRT), a substantial proportion of patients fail to achieve long-term local or distant tumor control. Integration of targeted agents into RT or CRT may offer greater improvements in therapeutic activity.

Over the last few years it has been acknowledged that radiation treatment does not only affect malignant cells within the target volume, but also other cell populations of tumor microenvironment, including proliferating endothelial cells leading to angiogenesis [45]. Moreover, tumor vascularization has been progressively recognized as a key factor influencing the activity ofRT. As a consequence the combination of AIAs with radiation therapy has become an active field of research [46].

The mechanisms of interaction between tumor vasculature and ionizing radiation are complex and not completely understood. This is a major hurdle to the rational combination of AIAs to radiation therapy.

The response of any cell type to radiation is strongly dependent on oxygen concentration [47, 48] and because AIAs have been shown to improve the oxygenation status of tumors [49–57], the potential for a greater than additive effect clearly exists when AIAs and radiation are combined [58].

In particular, hypoxia is a typical status of cancer cells, has a key role in tumor growth and progression, and influences the activity of cytotoxic agents and RT. A two- to three-fold higher radiation dose is required to kill hypoxic cells compared with well-oxygenated cells (oxygen-enhanced effect) [59, 60]. Hypoxia leads to radiation resistance because of lack of oxygen to facilitate DNA damage by radiation-induced free radicals. Hypoxic conditions also create a microenvironment in which tumor cells become less angiogenesis-dependent, more apoptosis resistant, more capable of existing under hypoxic conditions, and more malignant [61–63] because of the development of genomic instability and mutant genotypes regulating apoptosis/survival signaling pathways [64]. The degree of intratumoral hypoxia is positively correlated with the expression of the transcription factor hypoxia-inducible factor (HIF)-1 [65]. Under hypoxic conditions, HIF-1 is responsible for transactivation of several target genes including VEGF, leading to increased angiogenesis and vascular permeability in tumors [66] and to more aggressive tumor phenotype [67]. In addition, HIF-1 may increase radioresistance of solid tumors, independently of the tumor oxygenation status [68].

Evidence exists that radiation therapy can itself induce intensification of neoangiogenic processes, contributing to radioresistance [69]. Direct upregulation of VEGF after irradiation of various cancer cell lines has been reported [70] and radiation exposure of endothelial cells is responsible for production of several proangiogenic cytokines, including TGF-β, bFGF, PDGF, IFN-γ, TNF-α, IL-4/5, and VEGF [47, 71]. Sonveaux *et al.* [72] reported that RT activates the nitric oxide pathways in endothelial cells leading to their migration and sprouting, the first step of angiogenesis. Garcia-Barros *et al.* [73] demonstrated that tumor response to irradiation was mainly dependent on microvascular damage. The increased tumor cell proliferation that is often recorded after radiation may be the result of increased proliferation in the tumor stem cell compartment [74].

It has been postulated that exposure to AIAs, by destroying the vasculature, would severely compromise the delivery of oxygen and therapeutics to the solid tumor and thus produce hypoxia that would render many chemotherapeutics, as well as radiation, less effective [52, 75–79]. However, preclinical experiments demonstrated that this paradox is not true. Beginning with the seminal work of Teicher, several preclinical and clinical studies have shown that antiangiogenic therapy improves the outcome of cytotoxic therapies, including RT [80, 81]. In addition, Jain hypothesized that the appropriate application of AIAs can improve both, the structure and the function of tumor vessels, the abnormal tumor vasculature (tumor vessels normalization), resulting in more efficient delivery of drugs and oxygen to the targeted cancer cells [82]. Increased penetration of drugs throughout the tumor would enhance the outcome of chemotherapy, whereas the ensuing increased level of oxygen would enhance the efficacy of radiation therapy and many chemotherapeutic agents. However, sustained or aggressive antiangiogenic regimens may eventually prune away these vessels, resulting in a vasculature that is both, resistant to further treatment and inadequate for delivery of drugs or oxygen [82].

The studies that investigated the time dependency of this AIA-induced improvement in tumor oxygenation reported that the window of opportunity to exploit this possibility is short [82–84]. Indeed, when combined with radiation, the only time synergy was observed was when the radiation was administered at the time of maximal reduction of tumor hypoxia [85]. Irradiating immediately before or after this period only resulted in an additive response to the AIAs and radiation treatment, although hypoxia was still significantly reduced at these times. This suggests that changes in oxygenation may not be the only factor involved and unless it is possible to accurately predict the window of opportunity for each drug and tumor type, the potential for exploiting oxygenation modification by AIAs remains minimal [59].

11.5.2
Combination of Angiogenesis Inhibition with Radiotherapy: Preclinical Evidence

Numerous preclinical studies have investigated the potential of combining AIAs with radiation [46, 85]. The radiation treatments have involved both single and fractionated schedules. In single radiation treatments, the total doses given was highly variable; in the fractionated studies, relevant differences regard both, the number of fractions given and the time over which the doses were delivered.

The AIAs evaluated include nonspecific as well as targeted molecules, in particular anti-VEGF or VEGFR antibody or TKIs. The drug doses and treatment times used were heterogeneous. The AIAs were administered during the radiation treatment [86–91], before [50, 78, 80, 92, 93] or after radiation [76, 94, 95], or in combination with irradiation [52, 54, 69, 71, 77, 83, 95–105]. In addition, different parameters of tumor response were used, and only two studies evaluated a more important endpoint; that is, TCD50 as a measure of tumor cure [96, 98]. However, the majority of experimental combination consistently resulted in improved tumor growth delay, with synergistic or at least additive effects [46].

According to some preclinical evidence, tumor vasculature normalization obtained by anti-VEGF agents allow a "therapeutic window" whereby improving oxygenation for RT efficacy [82] and antiangiogenics, may sensitize the vasculature to radiation damage [90] to deliver AIAs before or concomitantly with radiation. This might be the most advantageous scheduling option. Accordingly, Gorski showed that the concurrent administration of angiostatin and RT obtained the best efficacy in the treatment of Lewis lung carcinoma [71]. Winkler *et al.* [82] demonstrated that a delay in RT following VEGFR2 blockade with the monoclonal antibody DC101 was associated with improved tumor response, as compared to concurrent delivery in a human glioblastoma xenograft.

Cediranib (AZD2171) sensitized a human tumor xenograft to radiation by increasing the proportion of functional vessels. In addition, when the drug was administered concomitantly with radiation, hypoxia was increased together with the antivascular effect [106, 107].

The rationale for using AIAs after RT is based on the damage caused by a single or fractionated dose of radiation [108, 109] and on the hypothesis that irradiated endothelial cells might be more sensitive to vasculature-targeting agents [110]. Furthermore, if AIAs disrupt tumor vasculature and the recovery of abnormal tumor vessels after transient vessel normalization, tumor perfusion and oxygenation and radiation efficacy are reduced [111]. Murata *et al.* demonstrated that TNP-470 administered concomitantly with radiation reduced curability of tumors, suggesting that the drug should be administered after radiation. Similarly, the VEGFR antagonist vatalanib (PTK787/ZK222584) enhanced radiation response only when administered after RT in a human squamous cell carcinoma xenograft [103]. Furthermore, higher activity of vandetanib (ZD6474), a VEGFR- and EGFR-TKI, was obtained when administered immediately after fractioned RT [112].

Human tumor xenografts recurring after RT seem more sensitive to anti-VEGFR-2 treatment than treatment-naïve tumors [113] and VEGFR-targeting agents appear more active against irradiated tumor vasculature [103, 107, 108].

When trimodality treatment (AIAs, cytotoxic drugs, andRT) had been evaluated in some preclinical experiments, a more than additive effect has been demonstrated [114, 115].

11.5.3
Antiangiogenic Therapy with Radiation or Chemoradiation Therapy: Clinical Trials

A phase I trial explored the MTD and DLT of escalating dose bevacizumab (2.5–10 mg kg^{-1}), added to fluorouracil, hydroxyurea, and RT in 43 patients with recurrent, previously radiated, or treatment-naive poor prognosis head and neck cancer. Results indicated that bevacizumab can be integrated with 5-fluoruracil-hydroxyurea (FHX) CRT at a dose of 10 mg mq^{-1} every two weeks with decreased chemotherapy doses because of neutropenia. Bevacizumab might be considered responsible for fistula formation or tissue necrosis [116]. In fact, five patients (11.6%) developed fistulas, with histologically documented radionecrosis or residual tumor and an additional four patients experienced ulceration and wound healing in the treatment field.

Willet *et al.* designed a phase I study in which patients affected by primary locally advanced rectal adenocarcinoma (LARC) were treated with bevacizumab every two weeks, followed by bevacizumab with 5-fluorouracil continuous infusion and external beam radiation to the pelvis, and surgery seven weeks after treatment completion. The study included the analysis of several noninvasive and invasive surrogates for response to the combined modality therapy. The authors concluded that a single infusion of the VEGF-specific antibody bevacizumab decreases tumor perfusion, vascular volume, microvascular density, interstitial fluid pressure and the number of viable, circulating endothelial and progenitor cells, and increases the fraction of vessels with pericyte coverage in patients with rectal carcinoma. Thus, the combination of bevacizumab with CRT was potentially effective [117]. A subsequent study demonstrated increased toxicity when the dose of bevacizumab was escalated up to 10 mg kg^{-1}. Pathologic evaluation of the surgical specimens for staging following completion of all therapy in the patients receiving bevacizumab at the dose of 10 mg kg^{-1}, showed two complete pathologic responses (pCR), as compared to no pCR in the 5-mg kg^{-1} bevacizumab group. Interestingly, the complete responses were seen in the two patients experiencing DLT: pulmonary embolism and ileal perforation respectively. Histological evaluation of the surgical specimens of the three other patients on 10 mg kg^{-1} BV showed ulceration and fibrosis with residual microscopic disease (consistent with Manard grade 3–4 features) [118].

Following these results, two phase II trials exploring the safety and activity of bevacizumab in combination with capecitabine and standard RT as neoadjuvant treatment of LARC have been designed. One of the studies is completed. In this study 42/43 patients enrolled who were affected by LARC, received 5 mg kg^{-1} IV bevacizumab, every two weeks for four courses, started two weeks before CRT. Capecitabine (825 mg mq^{-1} bid) was taken throughout RT period, including weekend rest. RT was delivered with a three-field technique (50.4 Gy over 5.5 weeks with a concomitant boost approach). Surgery, including total mesorectal excision, was performed 6–8 weeks after RT. Of the 43 patients enrolled, 40 were evaluated for response and 42 for toxicity. One patient had unrecognized metastatic disease and died from PD; two patients refused surgery. Six patients obtained a pCR (intent-to-treat analysis: 14.9%). Twenty-one patients were downstaged (48.8%), 11 of whom had only focal residual disease (25.6%). Downsizing was obtained in three patients (6.9%), while in nine patients (20.9%) disease remained stable. Three patients (6.97%) showed distant metastases at restaging after CRT. Two underwent surgery, one with concomitant liver metastasectomy. Forty patients (93%) were alive at 16.7 months of median follow-up (range 14–780 days). Sphincter-sparing surgery was achieved in 31 patients (72.1%). During CRT, no G4 toxicity was reported. The G3 toxicity included: diarrhea (three patients), neutropenia (two patients), and hypokalemia (two patients). The most frequent side-effects were: G1/2 diarrhea (17 pts; 39.5%), G1/2 proctitis (11 pts; 25.6%), and G1 rectal bleeding (9 pts; 21%). HFS was mild and transient (only 1 pt with G2 HFS). Bevacizumab-related toxicities were hypertension in five patients (11.6%; G1 = 2 patients; G2 = 3 patients), proteinuria in one patient (2.3%; G2), and epistaxis in one patient (2.3%; G2). No thromboembolic disease, perforation or major hemorrhagic events were reported. One patient had bowel perforation and died; two patients presented dehiscence of anastomosis; one patient presented intestinal occlusion; and two patients experienced myocardial ischemia resolved with medical treatment [119].

In the other study by Crane *et al.*, 25 patients affected by LARC underwent neoadjuvant RT (50.4 Gy in 28 fractions over 5.5 weeks using a posterior three-field technique) with capecitabine (900 mg m^{-2} PO twice daily only on days of radiation) and bevacizumab (5 mg kg^{-1}IV every two weeks for three doses). Surgical resection followed a median of about seven weeks after the completion of CRT. Eight of 25 patients (32%) had pCR, and 24% had <10% viable tumor cells in the specimen. Overall sphincter preservation rate was 72%. There was no significant hematologic toxicity, ATEs or DVTs. Three patients had wound complications that required surgical intervention, five minor complications were resolved without operative intervention (two delayed perineal wound healing, two anastomotic leaks, and a superficial wound infection) [120].

The studies by Willet *et al.* provided the rational basis for a phase I study to evaluate the combination of bevacizumab, capecitabine, oxaliplatin, and radiation therapy in patients with pathologically confirmed adenocarcinoma of the rectum. Patients received 50.4 Gy of external beam radiation therapy to the tumor in 28 fractions. Capecitabine, oxaliplatin, and bevacizumab were administered concurrently with radiation therapy. The recommended Phase II dose was bevacizumab

(15 mg kg^{-1} day 1 + 10 mg kg^{-1} days 8 and 22), oxaliplatin (50 mg m^{-2} weekly), and capecitabine (625 mg m^{-2} bid during radiation days). Six of 11 patients had clinical responses: two of them had a pathologic complete response, and three had microscopic disease only. One patient experienced a postoperative abscess, one a syncopal episode during adjuvant chemotherapy, and one a subclinical myocardial infarction during adjuvant chemotherapy [121]. The good safety profile and the encouraging response rates provided the basis for an ongoing phase II trial.

Another phase II trial of bevacizumab combined with capecitabine and RT as neoadjuvant treatment of patients with LARC is ongoing in The Netherlands. Both EORTC and NCI/ECOG are exploring the combination of bevacizumab + capecitabine, oxaliplatin + RT in patients affected by LARC [122].

In a phase I study, bevacizumab in combination with capecitabine and RT was evaluated as primary treatment in patients with locally advanced pancreatic adenocarcinoma. Bevacizumab was administered two weeks before RT (50.4 Gy), during RT (12 patients each at 2.5, 5.0, 7.5, and 10 mg kg^{-1}), and after RT until disease progression. Capecitabine was administered on days 14 through 52 (650 mg mq^{-1} orally twice daily for the first six patients; 825 mg mq^{-1} for the remaining patients). Nine (20%) of 46 assessable patients had confirmed PRs with median PFS of 6.2 months. Four patients have undergone radical pancreaticoduodenectomy without perioperative complication. The median survival was 11.6 months. Concurrent bevacizumab did not significantly increase the acute toxicity of the CRT regimen. However, ulceration and bleeding in the radiation field possibly related to bevacizumab occurred when the tumor involved the duodenal mucosa [123].

Bevacizumab (10 mg kg^{-1}) in combination with induction fixed-dose rate gemcitabine followed by bevacizumab (10 mg kg^{-1}) with accelerated RT as primary treatment of potentially resectable pancreatic adenocarcinoma has obtained an interesting resectability rate (8/10 patients had negative surgical margins) and 30% of patients were found to have only microscopic foci of residual tumor in surgical specimens. Four patients had grade 3 wound complications (29%) with one grade 2 pancreatic leak (10%) [124].

The combination of bevacizumab with RT and temozolomide as primary treatment of grade III–IV glioma was found feasible [125]. In another ongoing phase II trial bevacizumab (10 mg kg^{-1} every two weeks) was combined with temozolomide and RT after surgical resection of glioblastoma multiforme [126]. After completion of RT both drugs were continued. Interim results reported encouraging results in terms of PFS (over 8.8 months) associated with impaired wound healing and venous thromboembolisms in some patients.

A number of phase I–II clinical trials are exploring the activity of other AIAs, such as TKIs, vandetanib, sorafenib, sunitinib, and others, in combination with RT or CRT are at an early stage (*www.clinicaltrial.gov*).

Among drugs with antiangiogenic activity, thalidomide and celecoxib was employed in combination with RT in the treatment of brain metastases and locally advancer cervix cancer respectively. In both cases, the addition of the drugs did not result in substantial improvement of clinical outcome or clinical benefit. More recently, patients with poor prognosis lung cancer received high doses of celecoxib in combination with thoracic RT. The actuarial local PFS was 66% at one year and about 42% at two years [127].

11.6 Challenges of Study Design for Combined Antiangiogenic Therapy

The study design of most randomized clinical trials of first generation angiogenesis inhibitors consisted of the comparison of a standard chemotherapeutic regimen with or without the test compound.

Conventional phase I studies for chemotherapeutic agents are designed to find the MTD and DLT, based on the assumption that there are direct correlations among dose, activity, and toxicity. With AIAs there is no linear relationship between dose and activity, but toxicity may increase with the dose. In addition, MTD may not be defined if the drug has a wide therapeutic ratio. The study phase I design could be improved by the identification of the effects on the target, the measurement of "surrogates" for biological activity, and the quantification of pharmacokinetic parameters. As proposed by Korn *et al.* the identification of the target effect through pharmacodynamic studies is a key tool for the selection of the minimum target inhibiting dose (MTID) and the optimal schedule of administration [128]. However, this methodology presents several challenges, such as the difficulty in identifying the pivotal biological target and the lack of sensitive, specific, and validated pharmacodynamic assays.

The standard endpoint of phase II studies is to demonstrate the disease-oriented activity of a new drug, evaluating the objective response rate by standard criteria (e.g., Response Evaluation Criteria In Solid Tumors – RECISTs), but in most cases, this approach is not adequate for AIAs being cytostatic with a low evidence of tumor shrinkage [129, 130]. Therefore, considering response rate as the only endpoint for a phase II study with AIAs may increase the potential risk of underestimating or rejecting potentially active drugs. TTP, progression-free-survival, OS, early progression rate, and growth modulation index might be more appropriate end-points. A possible alternative methodological approach is the randomized discontinuation design that can distinguish the stabilization of disease due to natural history from that which is treatment-related, but this requires a large number of cases [131].

Phase III trials with AIAs have similar designs and the same end-points in terms of efficacy (OS, quality of life, etc.) to those with cytotoxic drugs. However, the key issue remains the selection of the patients. Molecular characterization of tumors may play a pivotal role in patient selection and might become an important stratification variable.

Many of the key issues hampering the development of targeted agents alone or in combination might be resolved by the exploratory or "target-development" clinical trial designs, also defined as phase 0 trials [132]. In this new approach, the evaluation of intended target(s) modulation is the primary endpoint. This type of "proof-of-concept" trial involves early integration of real-time biomarker assays to demonstrate the mechanism of action *in vivo* so as to be informative for subsequent drug development. The objectives include the development of pharmacodynamic assays in human tumors or surrogate tissues, the estimation of pharmacokinetic parameters, evaluation of the drug effect on its target(s), and imaging of the target of interest. Phase 0 trials of different agents with the same molecular target might facilitate prioritization with respect to further development in phase I and II trials or even help to select the lead agent among similar molecules. The pharmacokinetic and pharmacodynamic data for different schedules of administration will help to select the best one for optimal efficacy and toxicity, involving limited number of patients and doses over a relatively short period of time.

In conclusion, there are several potential ways to improve study design: (i) By the determination of validated surrogate biomarkers, such as the expression and/or the activity of the therapeutic target: prospective multiparametric studies should be planned in order to identify the method of choice which needs to be feasible, reproducible, and standardized prior to wide clinical use. (ii) By evaluating the direct effects of antiangiogenic therapy (i.e., on mature circulating endothelial cells and their progenitors) and testing the value of new vascular imaging techniques as potential methods for quantifying clinical and vascular responses to these agents. (iii) Pharmacodynamic indicators should confirm a therapeutic concentration of the drug at the molecular target.

References

1. Bergers, G. and Benjamin, L.E. (2003) Tumorigenesis and the angiogenic switch. *Nat. Rev. Cancer*, **3**, 401–410.
2. Folkman, J. and Hochberg, M. (1973) Self-regulation of growth in three dimensions. *J. Exp. Med.*, **138**, 745–753.
3. Longo, R., Sarmiento, R., Fanelli, M., Capaccetti, B., Gattuso, D., and Gasparini, G. (2002) Anti-angiogenic therapy: rationale, challenges and clinical studies. *Angiogenesis*, **5**, 237–256.
4. St Croix, B., Rago, C., Velculescu, V., Traverso, G., Romans, K.E., Montgomery, E., Lal, A., Riggins, G.J., Lengauer, C., Vogelstein, B., and Kinzler, K.W. (2000) Genes expressed in human tumor and endothelium. *Science*, **289**, 1197–1202.
5. Pepper, M.S. (2001) Role of the matrix metalloproteinases and plasminogen activator-plasmin systems in angiogenesis. *Arterioscler. Thrombovasc. Biol.*, **21**, 1104–1117.
6. Stupack, D.G. and Cheresh, D.A. (2004) Integrins and angiogenesis. *Curr. Top. Dev. Biol.*, **64**, 207–238.
7. Rini, B.I. and Small, E.J. (2005) Biology and clinical development of vascular endothelial growth factor-targeted therapy in renal cell carcinoma. *J. Clin. Oncol.*, **23**, 1028–1043.
8. Hicklin, D.J. and Ellis, L.M. (2005) Role of the vascular endothelial growth factor pathway in tumor growth and angiogenesis. *J. Clin. Oncol.*, **23**, 1011–1027.
9. Jain, R.K. *et al.* (2005) Normalization of tumor vasculature: an emerging concept in antiangiogenic therapy. *Science*, **307**, 58–62.
10. Escudier, B., Pluzanska, A., Koralewski, P., Ravaud, A., Bracarda, S., and Szczylik, C. (2007) Bevacizumab plus interferon α-2a for treatment of metastatic renal cell carcinoma: a randomised, double-blind phase III trial. *Lancet*, **379**, 2103–2111.
11. Rini, B.I., Halabi, S., Rosenberg, J.E., Stadler, W.M., Vaena, D.A.,

Ou, S.S., Archer, L., Atkins, J.N., Picus, J., Czaykowski, P., Dutcher, J., and Small, E.J. (2008) Bevacizumab plus interferon alfa compared with interferon alfa monotherapy in patients with metastatic renal cell carcinoma: CALGB 90206. *J. Clin. Oncol.*, **26** (33), 5422–5428.

12. Escudier, B., Lassau, N., Angevin, E., Soria, J.C., Chami, L., Lamuraglia, M., Zafarana, E., Landreau, V., Schwartz, B., Brendel, E., Armand, J.P., and Robert, C. (2007) Phase I trial of sorafenib in combination with IFN-2a in patients with unresectable and/or metastatic renal cell carcinoma or malignant melanoma. *Clin. Cancer Res.*, **13** (6), 1801–1809.
13. Gollob, J.A., Rathmell, W.K., Richmond, T.M., Marino, C.B., Miller, E.K., Grigson, G., Watkins, C., Gu, L., Peterson, B.L., and Wright, J.J. (2007) Phase II trial of sorafenib plus interferon alfa-2b as first- or second-line therapy in patients with metastatic renal cell cancer. *J. Clin. Oncol.*, **25** (22), 3288–3295.
14. Ryan, C.W., Goldman, B.H., Lara, P.N., Mack, P.C. Jr., Beer, T.M., Tangen, C.M., Lemmon, D., Pan, C.X., Drabkin, H.A., and Crawford, E.D. (2007) Sorafenib with interferon alfa-2b as first-line treatment of advanced renal carcinoma: a phase II study of the southwest oncology group. *J. Clin. Oncol.*, **25** (22), 3296–3301.
15. Yao, J.C., Phan, A., Hoff, P.M., Chen, H.X., Charnsangavej, C., Yeung, S.C.J., Hess, K., Chaan, N.G., Abruzzese, J.A., and Ajani, J.A. (2008) Targeting vascular endothelial growth factor in advanced carcinoid tumor: a random assignment phase II study of depot octreotide with bevacizumab and pegylated interferon alfa-2b. *J. Clin. Oncol.*, **26** (8), 1316–1323.
16. Tortora, G., Ciardiello, F., and Gasparini, G. (2008) Combined targeting of EGFR-dependent and VEGF-dependent pathways: rationale, preclinical studies and clinical applications. *Nat. Clin. Pract. Oncol.*, **5** (9), 521–530.
17. Viloria-Petit, A., Crombet, T., Jothy, S., Hicklin, D., Bohlen, P., Schlaeppi, J.M., Rak, J., and Kerbel, R.S. (2001) Acquired resistance to the antitumor effect of epidermal growth factor receptor-blocking antibodies in vivo: a role for altered tumor angiogenesis. *Cancer Res.*, **61**, 5090–5101.
18. Hainsworth, J.D., Sosman, J.A., Spigel, D.R., Edwards, D.L., and Baughman, C. (2005) Treatment of metastatic renal cell carcinoma with a combination of bevacizumab and erlotinib. *J. Clin. Oncol.*, **23** (31), 7889–7896.
19. Bukowski, R.M., Kabbinavar, F., Figlin, R.A., Flaherty, K., Srinivas, S., Vaishampayan, U., Drabkin, H.A., Dutcher, J., Ryba, S., Xia, Q., Scappaticci, F.A., and McDermott, D. (2007) Randomized phase II study of erlotinib combined with bevacizumab compared with bevacizumab alone in metastatic renal cell cancer. *J. Clin. Oncol.*, **25** (29), 4536–4541.
20. Dickler, M., Rugo, H., Caravelli, J., Brogi, E., Sachs, D., Panageas, K., Flores, S., Moasser, M., Norton, L., and Hudis, C. (2004) Phase II trial of erlotinib (OSI-774), an epidermal growth factor receptor (EGFR)-tyrosine kinase inhibitor, and bevacizumab, a recombinant humanized monoclonal antibody to vascular endothelial growth factor (VEGF), in patients (pts) with metastatic breast cancer (MBC). *J. Clin. Oncol.*, **22** (14S, July 15 Supplement), 2001.
21. Ordonez, J., Gomez Martin, C., and Cortes-Funes, H. (2006) Trastuzumab in combination with bevacizumab in advanced breast cancer patient resistant to chemotherapy. *J. Clin. Oncol.*, **24** (18S, June 20 Supplement), 10762.
22. Saltz, L.B., Lenz, H.J., Kindler, H.L., Hochster, H.S., Wadler, S., Hoff, P.M., Kemeny, N.E., Hollywood, E.M., Gonen, M., Quinones, M., Morse, M., and Chen, H.X. (2007) Randomized phase II trial of cetuximab, bevacizumab, and irinotecan compared with cetuximab and bevacizumab alone in irinotecan-refractory colorectal cancer:

the BOND-2 study. *J. Clin. Oncol.*, **25** (29), 4557–4561.

23. Tol, J., Koopman, M., Rodenburg, C.J., Cats, A., Creemers, G.J., Schrama, J.G., Erdkamp, F.L., Vos, A.H., Mol, L., Antonini, N.F., and Punt, C.J. (2008) A randomised phase III study on capecitabine, oxaliplatin and bevacizumab with or without cetuximab in first-line advanced colorectal cancer, the CAIRO2 study of the Dutch Colorectal Cancer Group (DCCG). An interim analysis of toxicity. *Ann. Oncol.*, **19** (4), 734–738.
24. Chu, E. (2008) Dual biologic therapy in the first-line mCRC setting: implications of the CAIRO2 study. *Clin. Colorectal Cancer*, **7** (4), 226.
25. Hecht, J.R., Mitchell, E., Chidiac, T., Scroggin, C., Hagenstad, C., Spigel, D., Marshall, J., Cohn, A., McCollum, D., Stella, P., Deeter, R., Shahin, S., and Amado, R.G. (2008) A randomized phase IIIB trial of chemotherapy, bevacizumab, and panitumumab compared with chemotherapy and bevacizumab alone for metastatic colorectal cancer. *J. Clin. Oncol.*, **27** (5), 672–680.
26. Herbst, R.S., Johnson, D.H., Mininberg, E., Carbone, D.P., Henderson, T., Kim, E.S., Blumenschein, G., Lee, J.G. Jr.,Liu, D.D., Truong, M.T., Hong, W.K., Tran, H., Tsao, A., Xie, D., Ramies, D.A., Mass, R., Seshagiri, S., Eberhard, D.A., Kelley, S.K., and Sandler, A. (2005) Phase I/II trial evaluating the anti-vascular endothelial growth factor monoclonal antibody bevacizumab in combination with the HER-1/epidermal growth factor receptor tyrosine kinase inhibitor erlotinib for patients with recurrent non–small-cell lung cancer. *J. Clin. Oncol.*, **10**, 2544–2555.
27. Herbst, R.S., O'Neill, V.J., Fehrenbacher, L., Belani, C.P., Bonomi, P.D., Hart, L., Melnyk, O., Ramies, D., Lin, M., and Sandler, A. (2007) Phase II study of efficacy and safety of bevacizumab in combination with chemotherapy or erlotinib compared with chemotherapy alone for treatment of recurrent or refractory non–small-cell lung cancer. *J. Clin. Oncol.*, **25** (30), 4743–4750.
28. Patel, P.H., Kondagunta, G.V., Redman, B.G., Hudes, G.R., Kim, S.T., Chen, I., and Motzer, R.J. (2007) Phase I/II study of sunitinib malate in combination with gefitinib in patients (pts) with metastatic renal cell carcinoma (mRCC). *J. Clin. Oncol.*, **25** (18S, June 20 Supplement), 5097.
29. Wedge, S.R., Ogilvie, D.J., Dukes, M., Kendrew, J., Chester, R., Jackson, J.A., Boffey, S.J., Valentine, P.J., Curwen, J.O., Musgrove, H.L., Graham, G.A., Hughes, G.D., Thomas, A.P., Stokes, E.S.E., Curry, B., Richmond, G.H.P., Wadsworth, P.F., Bigley, A.L., and Hennequin, L.F. (2002) ZD6474 inhibits vascular endothelial growth factor signaling, angiogenesis, and tumor growth following oral administration. *Cancer Res.*, **62**, 4645–4655.
30. Holden, S.N., Eckhardt, S.G., Basser, R., de Boer, R., Rischin, D., Green, M., Rosenthal, M.A., Wheeler, C., Barge, A., and Hurwitz, H.I. (2005) Clinical evaluation of ZD6474, an orally active inhibitor of VEGF and EGF receptor signaling, in patients with solid, malignant tumors. *Ann. Oncol.*, **16**, 1391–1397.
31. Tamura, T., Minami, H., Yamada, Y., Yamamoto, N., Shimoyama, T., Murakami, H., Horiike, A., Fujisaka, Y., Shinkai, T., Tahara, M., Kawada, K., Ebi, H., Sasaki, Y., Jiang, H., and Saijo, N. (2006) A phase I dose-escalation study of ZD6474 in Japanese patients with solid, malignant tumors. *J. Thorac. Oncol.*, **1**, 1002–1009.
32. Heymach, J.V., Johnson, B.E., Prager, D., Csada, E., Roubec, J., Pešek, M., Špásová, I., Belani, C.P., Bodrogi, I., Gadgeel, S., Kennedy, S.J., Hou, J., and Herbst, R.S. (2007) Randomized, placebo-controlled phase II study of vandetanib plus docetaxel in previously treated non small-cell lung cancer. *J. Clin. Oncol.*, **25**, 4270–4277.
33. Heymach, J.V., Paz-Ares, L., De Braud, F., Sebastian, M., Stewart, D.J., Eberhardt, W.E.E., Ranade, A.A., Cohen, G., Trigo, J.M., Sandler, A.B., Bonomi, P.D., Herbst, R.S.,

Krebs, A.D., Vasselli, J., and Johnson, B.E. (2008) Randomized phase II study of vandetanib alone or with paclitaxel and carboplatin as first-line treatment for advanced non–small-cell lung cancer. *J. Clin. Oncol.*, **26** (33), 5407–5415.

34. Azad, N.S., Posadas, E.M., Kwitkowski, V.E., Steinberg, S.M., Jain, L., Annunziata, C.M., Minasian, L., Sarosy, G., Kotz, H.L., Premkumar, A., Cao, L., McNally, D., Chow, C., Chen, H.X., Wright, J.J., Figg, W.J., and Kohn, E.C. (2008) Combination targeted therapy with sorafenib and bevacizumab results in enhanced toxicity and antitumor activity. *J. Clin. Oncol.*, **26** (22), 3709–3714.

35. Feldman, D.R., Ginsberg, M.S., Baum, M., Flombaum, C., Hassoun, H., Velasco, S., Fischer, P., Ishill, N.M., Ronnen, E.A., and Motzer, R.J. (2008) Phase I trial of bevacizumab plus sunitinib in patients with metastatic renal cell carcinoma. *J. Clin. Oncol.*, **26** (15S, May 20 Supplement), 5100.

36. Whorf, R.C., Hainsworth, J.D., Spigel, D.R., Yardley, D.A., Burris, H.A.III, Waterhouse, D.M., Vazquez, E.R., and Greco, F.A. (2008) Phase II study of bevacizumab and everolimus (RAD001) in the treatment of advanced renal cell carcinoma (RCC). *J. Clin. Oncol.*, **26**, 15S, Abstract 5010.

37. Rini, B.I., Michaelson, M.D., Rosenberg, J.E., Bukowski, R.M., Sosman, J.A., Stadler, W.M., Hutson, T.E., Margolin, K., Harmon, C.S., DePrimo, S.E., Kim, S.T., Chen, I., and George, D.J. (2008) Antitumor activity and biomarker analysis of sunitinib in patients with bevacizumab-refractory metastatic renal cell carcinoma. *J. Clin. Oncol.*, **26** (22), 3743–3748.

38. Tamaskar, I., Garcia, J.A., Elson, P., Wood, L., Mekhail, T., Dreicer, R., Rini, B.I., and Bukowski, R.M. (2008) Antitumor effects of sunitinib or sorafenib in patients with metastatic renal cell carcinoma who received prior antiangiogenic therapy. *J. Urol.*, **179** (1), 81–86.

39. Sablin, M.P., Bouaita, L., Balleyguier, C., Gautier, J., Celier, C., Balcaceres, J.L., Oudard, S., Ravaud, A., Négrier, S., and Escudier, B. (2007) Sequential use of sorafenib and sunitinib in renal cancer: retrospective analysis in 90 patients. *J. Clin. Oncol.*, **25** (18S, June 20 Supplement), 5038.

40. Dham, A. and Dudek, A.Z. (2007) Sequential therapy with sorafenib and sunitinib in renal cell carcinoma. *J. Clin. Oncol.*, **25** (18S, June 20 Supplement), 5106.

41. Rini, B.I., Wilding, G.T., Hudes, G., Stadler, W.M., Kim, S., Tarazi, J.C., Bycott, P.W., Liau, K.F., and Dutcher, J.P. (2007) Axitinib (AG-013736; AG) in patients (pts) with metastatic renal cell cancer (RCC) refractory to sorafenib. *J. Clin. Oncol.*, **25** (18S, June 20 Supplement), 5032.

42. Motzer, R.J., Escudier, B., Oudard, S., Hutson, T.E., Porta, C., Bracarda, S., Grünwald, V., Thompson, J.A., Figlin, R.A., Hollaender, N., Urbanowitz, G., Berg, W.J., Kay, A., Lebwohl, D., and Ravaud, A. (2008) Efficacy of everolimus in advanced renal cell carcinoma: a double-blind, randomised, placebo-controlled phase III trial. *Lancet*, **372** (9637), 449–456.

43. Owen, J.B., Coia, L.R., and Hanks, G.E. (1992) Recent patterns of growth in radiation therapy facilities in the United States: a patterns of care study report. *Int. J. Radiat. Oncol. Biol. Phys.*, **24**, 983–986.

44. McGinn, C.J., Shewach, D.S., and Lawrence, T.S. (1996) Radiosensitizing nucleosides. *J. Natl. Cancer Inst.*, **88**, 1193–1203.

45. Stratford, I.J. (1992) Concepts and developments in radiosensitization of mammalian cells. *Int. J. Radiat. Oncol. Biol. Phys.*, **22**, 529–532.

46. Nieder, C., Nieder, C., Wiedenmann, N., Andratschke, N.H., Astner, S.T., and Molls, M. (2007) Radiation therapy plus angiogenesis inhibition with bevacizumab: rationale and initial experience. *Rev. Recent Clin. Trials*, **2**, 163–168.

47. Wachsberger, P., Burd, R., and Dicker, A.P. (2004) Improving tumor response

to radiotherapy by targeting angiogenesis signaling pathways. *Hematol. Oncol. Clin. North Am.*, **18**, 1039–1057.

48. Gray, L.H., Conger, A.D., Ebert, M. *et al.* (1953) The concentration of oxygen dissolved in tissues at the time of irradiation as a factor in radiotherapy. *Br. J. Radiol.*, **26**, 638–648.

49. Horsman, M.R. and Overgaard, J. (2002) in *Basic Clinical Radiobiology for Radiation Oncologists*, 3rd edn (ed. G.G.Steel), Edward Arnold, London, pp. 158–168.

50. Hansen-Algenstaedt, N., Stoll, B.R., Padera, T.P. *et al.* (2000) Tumor oxygenation in hormone-dependent tumors during vascular endothelial growth factor receptor-2-blockade, hormone ablation, and chemotherapy. *Cancer Res.*, **60**, 4556–4560.

51. Fenton, B.M., Paoni, S.F., Grimwood, B.G., and Ding, I. (2003) Disparate effects of endostatin on tumor vascular perfusion and hypoxia in two murine mammary carcinomas. *Int. J. Radiat. Oncol. Biol. Phys.*, **57**, 1038–1046.

52. Bernsen, H.J.J.A., Rijken, P.F.J.W., Peters, J.P.W. *et al.* (1999) Suramin treatment of human glioma xenografts; effects on tumor vasculature and oxygenation status. *J. Neurooncol.*, **44**, 129–136.

53. Teicher, B.A., Holden, S.A., Ara, G. *et al.* (1995) Influence of an anti-angiogenic treatment on 9L gliosarcoma: oxygenation and response to cytotoxic therapy. *Int. J. Cancer*, **61**, 732–737.

54. Rofstad, E.K., Henriksen, K., Galappathi, K., and Mathiesen, B. (2003) Antiangiogenic treatment with thrombospondin-1 enhances primary tumor radiation response and prevents growth of dormant pulmonary micrometastases after curative radiation therapy in human melanoma xenografts. *Cancer Res.*, **63**, 4055–4061.

55. Winkler, F., Kozin, S.V., Tong, R. *et al.* (2004) Kinetics of vascular normalization by VEGFR2 blockade governs brain tumor response to radiation: role of oxygenation, angiopoietin-1, and matrix metalloproteinases. *Cancer Cell.*, **6**, 553–563.

56. Ansiaux, R., Baudelet, C., Jordan, B.F. *et al.* (2005) Thalidomide radiosensitizes tumors through early changes in the tumor microenvironment. *Clin. Cancer Res.*, **11**, 743–750.

57. Segers, J., Di Fazio, V., Ansiaux, R. *et al.* (2006) Potentiation of cyclophosphamide chemotherapy using the antiangiogenic drug thalidomide: importance of optimal scheduling to exploit the "normalization" window of the tumor vasculature. *Cancer Lett.*, **244**, 129–135.

58. Horsman, M.R. and Siemann, D.W. (2006) Pathophysiologic effects of vascular-targeting agents and theimplications for combination with conventional therapies. *Cancer Res.*, **66**, 11520–11539.

59. Höckel, M., Schlenger, K., Doctrow, S., Kissel, T., and Vaupel, P. (1993) Therapeutic angiogenesis. *Arch. Surg.*, **128**, 423–429.

60. Okunieff, P., Hoeckel, M., Dunphy, E.P., Schlenger, K., Knoop, C., and Vaupel, P. (1993) Oxygen tension distributions are sufficient to explain the local response of human breast tumors treated with radiation alone. *Int. J. Radiat. Oncol. Biol. Phys.*, **26**, 631–636.

61. Vaupel, P., Thews, O., and Hoeckel, M. (2001) Treatment resistance of solid tumors: role of hypoxia and anemia. *Med. Oncol.*, **8**, 243–259.

62. Shannon, A.M., Bouchier-Hayes, D.J., Condron, C.M., and Toomey, D. (2003) Tumour hypoxia, chemotherapeutic resistance and hypoxia-related therapies. *Cancer Treat. Rev.*, **29**, 297–307.

63. O'Donnell, J.L., Joyce, M.R., Shannon, A.M., Harmey, J., Geraghty, J., and Bouchier-Hayes, D. (2006) Oncological implications of hypoxia inducible factor-1alpha (HIF-1alpha) expression. *Cancer Treat. Rev.*, **32**, 407–416.

64. Rak, J., Yu, J.L., Kerbel, R.S., and Coomber, B.L. (2002) What do oncogenic mutations have to do with angiogenesis/vascular dependence of tumors? *Cancer Res.*, **62**, 1931–1934.

65. Semenza, G.L. (2000) HIF-1: mediator of physiological and pathophysiological responses to hypoxia. *J. Appl. Physiol.*, **88**, 1474–1480.
66. Forsythe, J.A., Jiang, B.H., Iyer, N.V., Agani, F., Leung, S.W., Koos, R.D., and Semenza, G.L. (1996) Activation of vascular endothelial growth factor gene transcription by hypoxia-inducible factor 1. *Mol. Cell. Biol.*, **16**, 4604–4613.
67. Sutherland, R.M. (1998) Tumor hypoxia and gene expression–implications for malignant progression and therapy. *Acta Oncol.*, **37**, 567–574.
68. Williams, K.J., Telfer, B.A., Xenaki, D., Sheridan, M.R., Desbaillets, I., Peters, H.J., Honess, D., Harris, A.L., Dachs, G.U., van der Kogel, A., and Stratford, I.J. (2005) Enhanced response to radiotherapy in tumours deficient in the function of hypoxia-inducible factor-1. *Radiother. Oncol.*, **75**, 89–98.
69. Koukourakis, M.I., Giatromanolaki, A., Sivridis, E., Simopoulos, K., Pissakas, G., Gatter, K.C., and Harris, A.L. (2001) Squamous cell head and neck cancer: evidence of angiogenic regeneration during radiotherapy. *Anticancer Res.*, **21**, 4301–4309.
70. Gorski, D.H., Beckett, M.A., Jaskowiak, N.T., Calvin, D.P., Mauceri, H.J., Salloum, R.M., Seetharam, S., Koons, A., Hari, D.M., Kufe, D.W., and Weichselbaum, R.R. (1999) Blockage of the vascular endothelial growth factor stress response increases the antitumor effects of ionizing radiation. *Cancer Res.*, **59**, 3374–3378.
71. McBride, W.H., Chiang, C.S., Olson, J.L., Wang, C.C., Hong, J.H., Pajonk, F., Dougherty, G.J., Iwamoto, K.S., Pervan, M., and Liao, Y.P. (2004) A sense of danger from radiation. *Radiat. Res.*, **162**, 1–19.
72. Sonveaux, P., Brouet, A., Havaux, X., Grégoire, V., Dessy, C., Balligand, J.L., and Feron, O. (2003) Irradiation-induced angiogenesis through the up-regulation of the nitric oxide pathway: implications for tumor radiotherapy. *Cancer Res.*, **63**, 1012–1019.
73. Garcia-Barros, M., Paris, F., Cordon-Cardo, C., Lyden, D., Rafii, S., Haimovitz-Friedman, A., Fuks, Z., and Kolesnick, R. (2003) Tumor response to radiotherapy regulated by endothelial cell apoptosis. *Science*, **300**, 1155–1159.
74. Hendry, J.H. (1992) Treatment acceleration in radiotherapy: the relative time factors and dose-response slopes for tumours and normal tissues. *Radiother. Oncol.*, **25**, 308–312.
75. Gong, H., Pöttgen, C., Stüben, G., Havers, W., Stuschke, M., and Schweigerer, L. (2003) Arginine deiminase and other antiangiogenic agents inhibit unfavorable neuroblastoma growth: potentiation by irradiation. *Int. J. Cancer*, **106**, 723–728.
76. Dings, R.P.M., Williams, B.W., Song, C.W., Griffioen, A.W., Mayo, K.H., and Griffin, R.J. (2005) Anginex synergizes with radiation therapy to inhibit tumor growth by radiosensitizing endothelial cells. *Int. J. Cancer*, **115**, 312–319.
77. Leith, J.T., Papa, G., Quaranto, L., and Michelson, S. (1992) Modifications of the volumetric growth response and steady-state hypoxic fractions of xenografted DLD-2 human colon carcinomas by administration of basic fibroblast growth factor or suramin. *Br. J. Cancer*, **66**, 345–348.
78. Murata, R., Nishimura, Y., and Hiraoka, M. (1997) An antiangiogenic agent (TNP-470) inhibited reoxygenation during fractionated radiotherapy of murine mammary carcinoma. *Int. J. Radiat. Oncol. Biol. Phys.*, **37**, 1107–1113.
79. Griffin, R.J., Williams, B.W., Wild, R., Cherrington, J.M., Park, H., and Song, C.W. (2002) Simultaneous inhibition of the receptor kinase activity of vascular endothelial, fibroblast, and platelet-derived growth factors suppresses tumor growth and enhances tumor radiation response. *Cancer Res.*, **62**, 1702–1706.
80. Jain, R.K., Duda, D.G., Clark, J.W. *et al.* (2006) Lessons from phase III clinical trials on anti-VEGF therapy

for cancer. *Nat. Clin. Pract. Oncol.*, **3**, 24–40.

81. Jain, R.K. (2005) Normalization of tumor vasculature: an emerging concept in antiangiogenic therapy. *Science*, **307**, 58–62.

82. Winkler, F., Kozin, S.V., Tong, R.T., Chae, S.S., Booth, M.F., Garkavtsev, I., Xu, L., Hicklin, D.J., Fukumura, D., di Tomaso, E., Munn, L.L., and Jain, R.K. (2004) Kinetics of vascular normalization by VEGFR2 blockade governs brain tumor response to radiation: role of oxygenation, angiopoietin-1, and matrix metalloproteinases. *Cancer Cell*, **6**, 553–563.

83. Ansiaux, R., Baudelet, C., Jordan, B.F., Beghein, N., Sonveaux, P., De Wever, J., Martinive, P., Grégoire, V., Feron, O., and Gallez, B. (2005) Thalidomide radiosensitizes tumors through early changes in the tumor microenvironment. *Clin. Cancer Res.*, **11**, 743–750.

84. Segers, J., Di Fazio, V., Ansiaux, R., Martinive, P., Feron, O., Wallemacq, P., and Gallez, B. (2006) Potentiation of cyclophosphamide chemotherapy using the anti-angiogenic drug thalidomide: importance of optimal scheduling to exploit the 'normalization' window of the tumor vasculature. *Cancer Lett.*, **244**, 129–135.

85. O'Reilly, M.S. (2008) The interaction of radiation therapy and antiangiogenic therapy. *Cancer J.*, **14**, 207–213.

86. Hess, C., Vuong, V., Hegyi, I. *et al.* (2001) Effect of VEGF receptor inhibitor PTK787/ZK222548 combined with ionizing radiation on endothelial cells and tumour growth. *Br. J. Cancer*, **85**, 2010–2016.

87. Griscelli, F., Li, H., Cheong, C. *et al.* (2000) Combined effects of radiotherapy and angiostatin gene therapy in glioma tumor model. *Proc. Natl. Acad. Sci. U.S.A.*, **97**, 6698–6703.

88. Hanna, N.N., Seetharam, S., Mauceri, H.J. *et al.* (2000) Antitumor interaction of short-course endostatin and ionizing radiation. *Cancer J.*, **6**, 287–293.

89. Guptra, V.K., Jaskowiak, N.T., Beckett, M.A. *et al.* (2002) Vascular endothelial growth factor enhances endothelial cell survival and tumor radioresistance. *Cancer J.*, **8**, 47–54.

90. Lu, B., Geng, L., Musiek, A. *et al.* (2004) Broad spectrum receptor tyrosine kinase inhibitor, SU6668, sensitizes radiation via targeting survival pathway of vascular endothelium. *Int. J. Radiat. Oncol. Biol. Phys.*, **58**, 844–850.

91. Schueneman, A.A., Himmelfarb, E., Geng, L. *et al.* (2003) SU11248 maintenance therapy prevents tumor regrowth after fractionated irradiation of murine tumor models. *Cancer Res.*, **63**, 4009–4016.

92. Shi, W., Teschendorf, C., Muzyczka, N., and Siemann, D.W. (2003) Gene therapy delivery of endostatin enhances the treatment efficacy of radiation. *Radiother. Oncol.*, **66**, 1–9.

93. Lund, E.L., Bastholm, L., and Kristjansen, P.E.G. (2000) Therapeutic synergy of TNP-470 and ionizing radiation: effects on tumor growth, vessel morphology, and angiogenesis in human glioblastoma multiforme xenografts. *Clin. Cancer Res.*, **6**, 971–978.

94. Huber, P.E., Bischof, M., Jenne, J. *et al.* (2005) Trimodal cancer treatment: beneficial effects of combined antiangiogenesis, radiation, and chemotherapy. *Cancer Res.*, **65**, 3643–3655.

95. Zips, D., Hessel, F., Krause, M. *et al.* (2005) Impact of adjuvant inhibition of vascular endothelial growth factor receptor tyrosine kinases on tumor growth delay and local tumor control after fractionated irradiation in human squamous cell carcinomas in nude mice. *Int. J. Radiat. Oncol. Biol. Phys.*, **61**, 908–914.

96. Schuuring, J., Bussink, J., Bernsen, H.J.J.A., Peeters, W., and van der Kogel, A.J. (2005) Irradiation combined with SU5416: microvascular changes and growth delay in a human xenograft glioblastoma tumor line. *Int. J. Radiat. Oncol. Biol. Phys.*, **61**, 529–534.

97. Kozin, S.V., Boucher, Y., Hicklin, D.J., Bohlen, P., Jain, R.K., and Suit, H.D. (2001) Vascular endothelial growth factor receptor-2-blocking antibody potentiates radiation-induced long-term control of human tumor xenografts. *Cancer Res.*, **61**, 39–44.

98. Rofstad, E.K., Henriksen, K., Galappathi, K., and Mathiesen, B. (2003) Antiangiogenic treatment with thrombospondin-1 enhances primary tumor radiation response and prevents growth of dormant pulmonary micrometastases after curative radiation therapy in human melanoma xenografts. *Cancer Res.*, **63**, 4055–4061.
99. Mauceri, H.J., Hanna, N.N., Beckett, M.A. *et al.* (1998) Combined effects of angiostatin and ionizing radiation in antitumor therapy. *Nature*, **394**, 287–291.
100. Geng, L., Donnelly, E., McMahon, G. *et al.* (2001) Inhibition of vascular endothelial growth factor receptor signaling leads to reversal of tumor resistance to radiotherapy. *Cancer Res.*, **61**, 2413–2419.
101. Ning, S., Laird, D., Cherrington, J.M., and Knox, S.J. (2002) The antiangiogenic agents SU5416 and SU6668 increase antitumor effects of fractionated irradiation. *Radiat. Res.*, **157**, 45–51.
102. Zips, D., Krause, M., Hessel, F. *et al.* (2003) Experimental study on different combination schedules of VEGF receptor inhibitor PTK787/ZK222584 and fractionated irradiation. *Anticancer Res.*, **23**, 3869–3876.
103. Brazelle, W.D., Shi, W., and Siemann, D.W. (2006) VEGF associated tyrosine kinase inhibition increases the tumor response to single and fractionated dose radiotherapy. *Int. J. Radiat. Oncol. Biol. Phys.*, **65**, 836–841.
104. Damiano, V., Melisi, D., Bianco, C. *et al.* (2005) Cooperative antitumor effect of multitargeted kinase inhibitor ZD6474 and ionizing radiation in glioblastoma. *Clin. Cancer Res.*, **11**, 5639–5644.
105. Kaliski, A., Maggiorella, L., Cengel, K.A. *et al.* (2005) Angiogenesis and tumor growth inhibition by a matrix metalloproteinase inhibitor targeting radiation-induced invasion. *Mol. Cancer Ther.*, **4**, 1717–1728.
106. Williams, K.J., Telfer, B.A., Shannon, A.M., Babur, M., Stratford, I.J., and Wedge, S.R. (2007) Combining radiotherapy with AZD2171, a potent inhibitor of vascular endothelial growth factor signaling: pathophysiologic effects and therapeutic benefit. *Mol. Cancer Ther.*, **6**, 599–606.
107. Cao, C., Albert, J.M., Geng, L., Ivy, P.S., Sandler, A., Johnson, D.H., and Lu, B. (2006) Vascular endothelial growth factor tyrosine kinase inhibitor AZD2171 and fractionated radiotherapy in mouse models of lung cancer. *Cancer Res.*, **66**, 11409–11415.
108. Solesvik, O.V., Rofstad, E.K., and Brustad, T. (1984) Vascular changes in a human malignant melanoma xenograft following single-dose irradiation. *Radiat. Res.*, **98**, 115–128.
109. Zywietz, F., Hahn, L.S., and Lierse, W. (1994) Ultrastructural studies on tumor capillaries of a rat rhabdomyosarcoma during fractionated radiotherapy. *Acta Anat. (Basel)*, **150**, 80–85.
110. Shannon, A.M. and Williams, K.J. (2008) Antiangiogenics and radiotherapy. *J. Pharm. Pharmacol.*, **60**, 1029–1036.
111. Murata, R., Nishimura, Y., and Hiraoka, M. (1997) An antiangiogenic agent (TNP-470) inhibited reoxygenation during fractionated radiotherapy of murine mammary carcinoma. *Int. J. Radiat. Oncol. Biol. Phys.*, **37**, 1107–1113.
112. Williams, K.J., Telfer, B.A., Brave, S., Kendrew, J., Whittaker, L., Stratford, I.J., and Wedge, S.R. (2004) ZD6474, a potent inhibitor of vascular endothelial growth factor signaling, combined with radiotherapy: schedule-dependent enhancement of antitumor activity. *Clin. Cancer Res.*, **10**, 8587–8593.
113. Kozin, S.V., Winkler, F., Garkavtsev, I., Hicklin, D.J., Jain, R.K., and Boucher, Y. (2007) Human tumor xenografts recurring after radiotherapy are more sensitive to anti-vascular endothelial growth factor receptor-2 treatment than treatment-naive tumors. *Cancer Res.*, **67**, 5076–5082.
114. Kumar, P., Benedict, R., Urzua, F. *et al.* (2005) Combination treatment

significantly enhances the efficacy of antitumor therapy by preferentially targeting angiogenesis. *Lab. Invest.*, **85**, 756–767.

115. Huber, P.E., Bischof, M., Jenne, J., Heiland, S., Peschke, P., Saffrich, R., Gröne, H.J., Debus, J., Lipson, K.E., and Abdollahi, A. (2005) Trimodal cancer treatment: beneficial effects of combined antiangiogenesis, radiation, and chemotherapy. *Cancer Res.*, **65** (9), 3643–3655.
116. Seiwert, T.Y., Haraf, D.J., Cohen, E.E., Stenson, K., Witt, M.E., Dekker, A., Kocherginsky, M., Weichselbaum, R.R., Chen, H.X., and Vokes, E.E. (2008) Phase I study of bevacizumab added to fluorouracil- and hydroxyurea-based concomitant chemoradiotherapy for poor-prognosis head and neck cancer. *J. Clin. Oncol.*, **26**, 1732–1741.
117. Willett, C.G., Boucher, Y., di Tomaso, E., Duda, D.G. *et al.* (2004) Direct evidence that the VEGF-specific antibody bevacizumab has antivascular effects in human rectal cancer. *Nat. Med.*, **10**, 145–147.
118. Willett, C.G., Boucher, Y., Duda, D.G., di Tomaso, E. *et al.* (2005) Surrogate markers for antiangiogenic therapy and dose-limiting toxicities for bevacizumab with radiation and chemotherapy: continued experience of a phase I trial in rectal cancer patients. *J. Clin. Oncol.*, **23**, 8136–8139.
119. Torino, F., Cascinu, S., Ciardiello, F., Ballestrero, A., Lencioni, M., Filippelli, G., Martignetti, A., Granetto, C., Sarmiento, R., and Gasparini, G. (2008) Neoadjuvant bevacizumab combined with capecitabine and standard radiation therapy in locally advanced rectal cancer: a phase II study. *Ann. Oncol.*, **19** (Suppl 6), vi29–vi105, Abstract 111.
120. Crane, C.H., Eng, C., Feig, B.W., Das, P., Skibber, J.M., Chang, J.G., Wolff, R.A. *et al.* (2009) Phase II trial of neoadjuvant bevacizumab (BEV), capecitabine (CAP), and radiotherapy (XRT) for locally advanced rectal cancer. *Int. J. Radiat. Oncol. Biol. Phys.*, (May21), Abstract 4091.
121. Czito, B.G., Bendell, J.C., Willett, C.G., and Morse, M.A. *et al.* (2007) Bevacizumab, oxaliplatin, and capecitabine with radiation therapy in rectal cancer: phase I trial results. *Int. J. Radiat. Oncol. Biol. Phys.*, **68**, 472–478.
122. Willett, C.G., Duda, D.G., Czito, B.G., Bendell, J.C., Clark, J.W., and Jain, R.K. (2007) Targeted therapy in rectal cancer. *Oncology (Williston Park)*, **21**, 1055–1065.
123. Crane, C.H., Ellis, L.M., Abbruzzese, J.L., Amos, C., Xiong, H.Q., Ho, L., Evans, D.B., Tamm, E.P., Ng, C., Pisters, P.W., Charnsangavej, C., Delclos, M.E., O'Reilly, M., Lee, J.E., and Wolff, R.A. (2006) Phase I trial evaluating the safety of bevacizumab with concurrent radiotherapy and capecitabine in locally advanced pancreatic cancer. *J. Clin. Oncol.*, **24**, 1145–1151.
124. Moser, A.J., Zeh, H.J., Ramanathan, R.K., Krasinksas, A.M., Tublin, M.E., Smith, R.P., Stover, F.S., Lee, K.K., Hughes, S.J., and Bartlett, D.L. (2008) Preoperative treatment of potentially-resectable pancreatic adenocarcinoma with fixed-dose rate gemcitabine (GEM), bevacizumab (BEV), and 30 Gy radiotherapy (RT). *J. Clin. Oncol.*, (Suppl 26), Abstract 4631.
125. Narayana, A., Golfinos, J.G., Fischer, I., Raza, S., Kelly, P., Parker, E., Knopp, E.A., Medabalmi, P., Zagzag, D., Eagan, P., and Gruber, M.L. (2008) Feasibility of using bevacizumab with radiation therapy and temozolomide in newly diagnosed high-grade glioma. *Int. J. Radiat. Oncol. Biol. Phys.*, **72**, 383–389.
126. Lai, A., Filka, E., McGibbon, B., Nghiemphu, P.L., Graham, C., Yong, W.H., Mischel, P., Liau, L.M., Bergsneider, M., Pope, W., Selch, M., and Cloughesy, T. (2008) Phase II pilot study of bevacizumab in combination with temozolomide and regional radiation therapy for up-front treatment of patients with newly diagnosed glioblastoma multiforme: interim analysis of safety and tolerability. *Int. J. Radiat. Oncol. Biol. Phys.*, **71**, 1372–1380.

127. Liao, Z., Komaki, R., Milas, L., Yuan, C., Kies, M., Chang, J.Y., Jeter, M., Guerrero, T., Blumenschien, G., Smith, C.M., Fossella, F., Brown, B., and Cox, J.D. (2005) A phase I clinical trial of thoracic radiotherapy and concurrent celecoxib for patients with unfavorable performance status inoperable/unresectable non-small cell lung cancer. *Clin. Cancer Res.*, **11**, 3342–3348.

128. Korn, E.L., Arbuck, S.G., Pluda, J.M. *et al.* (2001) Clinical trial designs for cytostatic agents: are new approaches needed? *J. Clin. Oncol.*, **19**, 265–272.

129. Therasse, P., Eisenhauer, E.A., and Buyse, M. (2006) Update in methodology and conduct of clinical cancer. *Eur. J. Clin. Oncol.*, **42**, 1322–1330.

130. Michaelis, L.C. and Ratain, M. (2006) Measuring response in a post-RECIST world : from black and white to shades of grey. *Nature*, **6**, 409–414.

131. Rosner, G.L., Stadler, W., and Ratain, M.J. (2C02) Randomized discontinuation design. Application to cytostatic antineoplastic agents. *J. Clin. Oncol.*, **20**, 4478–4484.

132. Kummar, S., Kinders, R., Rubinstein, L., Parchment, R.E., Murgo, A.J., Collins, J., Pickeral, O., Low, J., Steinberg, S.M., Gutierrez, M., Yang, S., Helman, L., Wiltrout, R., Tomaszewski, J.E., and Doroshow, J.H. (2007) Compressing drug development timelines in oncology using phase '0' trials. *Nat. Rev. Cancer*, **7**, 131–139.

12
Tumor Specificity of Antiangiogenic Agents

Olivier Rixe, Ronan J. Kelly, and Giuseppe Giaccone

12.1
Introduction

VEGF inhibition has the theoretical potential to cause regression of immature tumor blood vessels and to inhibit the development of new tumor vasculature without compromising healthy vasculature. Additional factors make VEGF an attractive therapeutic target: (i) VEGF circulates in the blood and acts directly on vascular endothelial cells, (ii) drugs that target VEGF do not need to penetrate tumor tissue to inhibit tumor angiogenesis, and (iii) endothelial cells do not mutate to a treatment-resistant phenotype unlike genetically unstable tumor cells [1].

High intratumoral and/or circulating VEGF levels are associated with a poor prognosis in patients with several solid tumors including breast, lung, and renal carcinomas [2]. In addition to promoting tumor growth and survival by localized angiogenic effects, evidence suggests that VEGF plays a role in the systemic dissemination of a large number of solid tumors [3]. In addition to inhibiting tumor angiogenesis, studies in solid tumor models suggest that antagonism of VEGF signaling promotes tumor cell apoptosis [4] and enhances the antitumor activity of chemotherapy [5]. Collectively, these findings suggest that targeting VEGF is a rational therapeutic approach in solid tumors.

Angiogenesis is one of the most important pathways involved in cancer development. The angiogenic phenomenon is extremely complex and has been described as a "true cytokine storm" with many molecular events and biological steps involved in new blood vessel formation. Angiogenesis is not merely restricted to the production of VEGF, or to changes in endothelial or tumor cells. Several additional cells are involved including stromal cells, cancer stem cells, and pericytes. This high level of complexity has several consequences for drug development. In recent years, different strategies in drug design have emerged. Drugs have been developed that target a single kinase pathway or drugs may interact with several molecular targets within the tumor. Targeted treatments must either specifically inhibit a critical biological process or have a wider spectrum of activity interfering with several nonspecific effectors of the pathway. The latter strategy leads to the absence of selectivity and may expose subjects to putative toxicities.

Tumor Angiogenesis – From Molecular Mechanisms to Targeted Therapy. Edited by Francis S. Markland, Stephen Swenson, and Radu Minea
Copyright © 2010 WILEY-VCH Verlag GmbH & Co. KGaA, Weinheim
ISBN: 978-3-527-32091-2

Effectors of tumor angiogenesis are expressed on normal tissues and help explain drug limiting toxicities observed with anti-VEGF therapies. Renal expression of VEGF and other proteins such as Delta-like ligand 4 (Dll4), hypoxic-inducible factors (HIF-1α and HIF-2α), fibroblast growth factor receptors (FGF-Rs), platelet-derived growth factor receptor-beta (PDGFR-β) are involved in the vascular development and maintenance of homeostasis within endothelial cells in the glomerulus. This wide spectrum of proteins is one of the determining factors explaining specific class side effects seen with antiangiogenic compounds.

A drug combining drug selectivity versus target specificity represents the current and future quest for an ideal antiangiogenic molecule. Elucidating the role of VEGF in normal and tumor tissue is a critical step in developing strategies that effectively target the tumor, while minimizing side effects and morbidity.

12.2
Tumor Specificity

An ideal cancer target is present solely in tumor cells [6]. Frequently, targets are present in tumor cells and the targeted therapy, given at a therapeutic level, can hit the tumor cells with minimal effect on normal tissues. Another problem to be considered is that of varying genetic forms of VEGF, including alternatively spliced forms or posttranscriptionally modified forms of membrane-bound targets [7]. The pathological role of VEGF is well documented, but its function in normal tissues is poorly defined [8]. An improved understanding of the function of VEGF may be needed for a more efficacious antiangiogenic therapy.

Adverse events observed with anti-VEGF treatments have begun to reveal potential functions of VEGF in quiescent vasculature. VEGF-A and its receptors are not specific to endothelial cells but can be vital for the function and maintenance of nonendothelial cells and tissues.

12.2.1
VEGF Expression in the Adult

VEGF plays a prominent role in new blood vessel formation during embryogenesis [9]. In adults, VEGF is a key regulator of many physiologic processes including the reproductive cycle and it is involved in the monthly menstrual changes in the uterus, ovary, and breast of premenopausal women. It is also involved in wound healing, bone repair, and in skeletal muscle response to exercise [10].

Maharaj *et al.* have proposed a classification system describing three levels of expression of VEGF in normal tissues. In the "sparse cellular" expression group, including brain, retina, and testes, VEGF is expressed primarily in pericytes and vascular stromal cells [11]. The "intermediate group," with moderate expression of VEGF consists of cardiac and skeletal muscles. In the "highest density" group, VEGF is expressed in the epithelia of fenestrated vasculature tissues such as the glomerulus, choroid plexus, liver, and pancreas (see Figure 12.1). Interestingly,

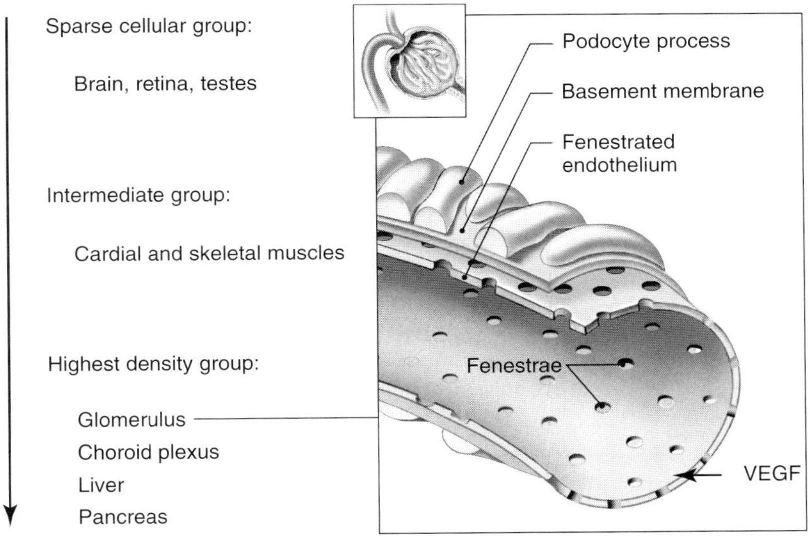

Figure 12.1 Expression of VEGF in normal tissues. There are varying levels of VEGF in normal tissues. In the "sparse cellular" expression group, including brain, retina, and testes, VEGF is expressed primarily in pericytes and vascular stromal cells. The "intermediate group," with moderate expression of VEGF consists of cardiac and skeletal muscles. In the "highest density" group, VEGF is expressed in the epithelia of fenestrated vasculature tissues such as the glomerulus, choroid plexus, liver, and pancreas.

endothelial cells do not generally express VEGF [12]. Aortic endothelial cells are an exception and they do express VEGF whereas those of the vena cava do not [13]. The effectiveness of this paracrine signaling system is highlighted by the fact that the ligand VEGF and its functional receptors VEGFR2 and VEGFR3 are expressed in close proximity to one another in many tissues throughout the body [14, 15].

12.2.2
Function of VEGF in the Adult

The function of VEGF appears to be pleiotropic. The low expression of VEGF in the vasculature of tissues that have barrier functions (retina, brain, testes) may account for the impermeability of these vessels [16]. Inhibition of VEGF leads to an alteration of the microvasculature as well as to vessel regression in a number of normal tissues, emphasizing the tropic role of this growth factor.

The role of VEGF on fenestrated endothelial cells is crucial. Fenestrae are discontinuities that facilitate movement of particles in and out of the circulation (see Figure 12.1). Fenestrated endothelial cells are found in endocrine tissues as well as in the kidney, choroid plexus, and gastrointestinal tract. The permeability of endothelial cells is mediated via both VEGFR1 and VEGFR2. VEGF both induces and maintains fenestration.

VEGF also has an effect on nonvascular cells, based on the presence of either VEGFR1 or VEGFR2 on those cells. VEGF-A was shown to be neuroprotective, helping both neuronal growth and survival [17]. In addition, several reports have demonstrated the action of VEGF on dendritic cells, monocytes, macrophages, eosinophils, megakaryocytes, and type II lung alveolar epithelial cells. VEGF may also support the rich vasculature of the muscle, particularly after training induced hypertrophy [18].

12.2.3 VEGF and Tumor Cells

VEGF-A binds and activates two tyrosine kinase receptors, VEGFR1 and VEGFR2. VEGFR2 plays a major role in tumor angiogenesis. VEGFR2 is mostly expressed on vascular endothelial cells and its distribution grossly follows that of accompanying VEGF-A levels. VEGFR2 expression is observed at a low level in neuronal cells, osteoblasts, pancreatic duct cells, and megakaryocytes. Expression levels are notably three- to fivefold higher in the tumor vasculature than in the normal vasculature.

The affinity of VEGFR1 for VEGF-A is higher than VEGFR2. However, VEGF-A does not stimulate the proliferation of NIH3T3 cells overexpressing VEGFR1. In addition, the rate of growth of lung carcinomas in VEGFR1 (−/−) mice is similar to that in the wild-type mice, suggesting a minor contribution of VEGFR1 in tumor progression and angiogenesis. Immunohistochemical staining and RT-PCR analysis conducted in dogs showed that VEGFR1 is expressed in many normal tissues [19].

Within the tumor tissue, the tumor cells and the surrounding stroma express high levels of VEGF-A. In contrast, most tumors (except Kaposi sarcomas) do not express VEGFR2.

These data underline the absence of specific distribution of VEGF-A (and its receptors) in the tumor compared to normal tissues, and provides insight into the occurrence of toxicities attributed to the anti-VEGF agents.

12.2.4 Metastasis and VEGF Selectivity

Recent studies have investigated the cell homing effect of tumor cells toward metastatic sites [20]. Kaplan *et al.* have demonstrated the critical role that bone marrow–derived cellular infiltrates play prior to the arrival of tumor cells. This initial preliminary step is required to enable tumor cell spread and the establishment of metastasis. Interestingly, bone marrow-derived hematopoietic progenitors express VEGFR1 and are activated by VEGF produced by the tumor. The development of this premetastatic niche appears to be selectively driven by VEGF/VEGFR1. This observation supports the introduction of a specific anti-VEGF therapy to prevent and treat metastatic spread.

12.3 Drug Selectivity

Many selective drugs have been brought into the clinic. Strategies have been developed in two opposite directions: the use of drugs with a broad spectrum of activity and the development of very specific compounds that target only one key protein involved in the angiogenic process [21]. Broad versus selective inhibition is not a drug development issue that is restricted to VEGF-VEGFR inhibition, but can be applied to other targeted agents including anti-EGFR molecules [22]. In clinical trials, highly specific or selective inhibition of only one kinase involved in a signaling pathway has been associated with modest response rates. Selective versus multiple inhibitions have potential significant implications in terms of drug activity and toxicity.

12.3.1 VEGF-Targeted Approaches

VEGF-targeted agents include neutralizing monoclonal antibodies or soluble VEGF receptors or receptor hybrids, and small molecules that target the VEGF receptors and other angiogenic proteins.

12.3.1.1 Specificity of Anti-VEGF Monoclonal Antibodies: Are Toxicities Related to the Absence of Selectivity?

Bevacizumab is an anti-$VEGF_{165}$ humanized monoclonal antibody approved by the US Food and Drug Administration for the treatment of metastatic kidney, colon, breast, and non–small cell lung carcinomas in combination with chemotherapy. Bevacizumab is specific for VEGF (also referred as VEGF-A) but does not neutralize other members of the VEGF family (VEGF-B to E) [23]. Monoclonal antibodies, including bevacizumab, have emerged as a class of specific and selective anticancer molecules [24]. The mechanism of action of bevacizumab has been extensively studied (see Figure 12.2). The current view of its angiogenesis inhibition involves a complex series of events such as (i) inhibition of the classic sprouting angiogenesis (including inhibition of new blood vessel growth and vascular regression), (ii) modification of vessel permeability and vascular constriction, (iii) direct effect on tumor cell function, (iv) offsetting the effects of chemotherapy induction of VEGF levels, and (v) and inhibition of bone marrow–derived endothelial progenitor cells. Induction of endothelial cell apoptosis is one of the major effects of bevacizumab. VEGF also plays a significant role in numerous prosurvival pathways in endothelial cell homeostasis [25] including activation of BCl-2, Akt, and survivin.

The safety profile of bevacizumab has been studied extensively in many randomized controlled clinical trials. Clinically relevant side effects include hypertension, bleeding, proteinuria, arterial thromboembolic events, wound healing complications, and gastrointestinal perforation (see Table 12.1). One of the most common side effects, reported in 21–64% of patients, is proteinuria. Nephrotic-range proteinuria denotes structural damage to the glomerular filtration

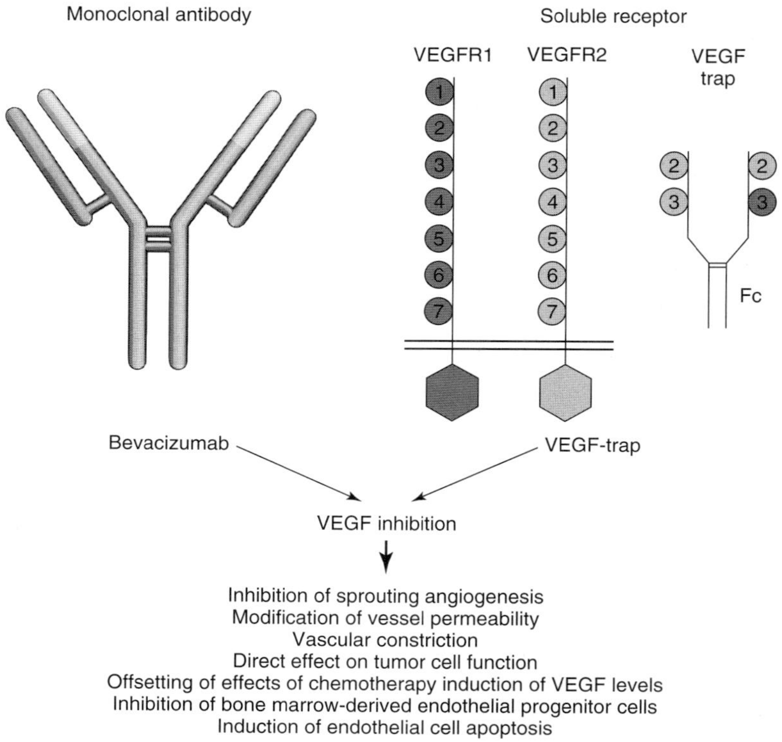

Figure 12.2 VEGF inhibitors: mechanism of action. Bevacizumab induced angiogenesis inhibition involves a complex series of events as listed above. Induction of endothelial cell apoptosis is one of the major effects of bevacizumab.

barrier and occurs in 1–2% of bevacizumab-treated patients [26]. The occurrence of renal toxicities demonstrates the absence of total selectivity for tumor cell targeting [27]. Putative causes of this renal toxicity have been suggested to include an indirect effect of the therapy (called *off-target toxicity* related to immunologic disorders induced by the monoclonal antibody) or *on-target* effects due to the overexpression of VEGF in the normal kidney.

In a very elegant publication, Eremina *et al.* reported six cases of patients with proteinuria associated with thrombotic microangiopathy (TMA) after bevacizumab administration [28]. Using knockout (KO) mice, the authors demonstrated that local genetic ablation of VEGF production in the kidney led to the development of lesions similar to the TMA observed in humans. In this study, decreased VEGF production by podocytes resulted in the mice developing proteinuria after a period of four weeks, which was subsequently followed by the occurrence of hypertension. Kidneys of all the KO mice were pale and shrunken, while electron microscopy of the glomeruli displayed typical features of TMA. Upon disease progression endothelial cells became swollen. Alterations of both podocytes and

Table 12.1 Class toxicities of anti-VEGF agents.

Symptoms	Mechanism
Hypertension	Glomerulonephritis Capillary rarefaction
Proteinuria	Glomerulonephritis
Thrombosis	Unknown
Wound healing complication	Unknown
Gastro-intestinal perforation	Tumor necrosis Others

Clinically relevant side effects of bevacizumab include hypertension, bleeding, proteinuria, arterial/venous thromboembolic events, wound healing complications, and gastrointestinal perforation. The proposed mechanisms of these class toxicities are listed above.

endothelial cell fenestrations were observed. Interestingly, schistocytes were found in blood smears from more than 50% of the animals. These findings are similar to those reported in humans. The authors suggested that VEGF plays a critical role within the glomerulus. VEGF has no apparent involvement in the peripheral microvasculature, as extrarenal circulating VEGF was not affected in this KO model. The authors concluded that the renal damage induced by bevacizumab is a direct reduction of glomerular VEGF production, previously described as the "*on-target*" effect of the drug. Bevacizumab induced damage to the glomerular endothelium suggests that local VEGF-A production is required for the appropriate maintenance of this specialized vascular bed. Bevacizumab also interferes with the critical cross talk between podocytes and glomerular endothelial cells. In parallel, a similar observation has been reported with preeclampsia, a disorder that complicates approximately 5% of all pregnancies, and it is characterized by new-onset hypertension and proteinuria, in association with a characteristic glomerular lesion similar to that described with bevacizumab. Recent evidence suggests that this unusual glomerular lesion is mediated by a soluble VEGF receptor that deprives glomerular endothelial cells of the VEGF that they require, leading to cellular injury and disruption of the filtration apparatus with subsequent proteinuria [29].

Severe hypertension is likely related to the renal toxicity induced by bevacizumab and can be a clinical symptom associated with TMA (see Table 12.1). In addition, VEGF starvation reduces endothelial nitric oxide production, which further exacerbates high blood pressure by causing vasoconstriction.

An increased incidence of arterial thromboembolic events in patients receiving bevacizumab was found (3.8%) compared to the placebo arms (1.7%). A reduction in the protective effect of nitric oxide on endothelial cells may be part of the mechanism for the development of this prothrombotic state.

After bevacizumab treatment, vessel regression was also noted in the pancreas, trachea, thyroid, and small intestine [30]. These observations strongly support the absence of absolute tumor selectivity of anti-VEGF agents, emphasizing the critical role of VEGF in normal tissue homeostasis.

12.3.1.2 **Association with Cytotoxics: Loss of Specificity or Increase of Drug Targeting?**

The addition of bevacizumab to chemotherapeutic agents or to interferon-α improves survival rates among patients with cancers of the colon, breast, lung, and the kidney. While bevacizumab is the first antiangiogenic agent to demonstrate a survival benefit in patients with many types of cancer, there is no evidence that bevacizumab as a single agent has a significant activity in patients with metastatic carcinomas (with the exception of renal cell carcinomas). Several clinical trials have established the efficacy and tolerability of bevacizumab in metastatic non–small cell lung cancer (NSCLC) [31]. A large phase III trial (Eastern Cooperative Oncology Group (ECOG) study 4599) evaluated bevacizumab plus chemotherapy in 878 treatment-naive patients with advanced nonsquamous NSCLC [32]. Patients were randomized to receive carboplatin (AUC of 6)/paclitaxel (200 mg m^{-2}) $\pm$ bevacizumab (15 mg kg^{-1}) every three weeks for six cycles; bevacizumab monotherapy (15 mg kg^{-1} every three weeks) was then continued until progressive disease or intolerable toxicity up to one year in the bevacizumab arm. Patients receiving bevacizumab had a significantly improved median overall survival (12.5 versus 10.2 months; $P = 0.007$), progression-free survival (6.4 versus 4.5 months; $P < 0.0001$), and response rates (27.2% versus 10.0%; $P < 0.0001$) compared with chemotherapy alone.

Similar observations were reported in metastatic colorectal cancers (CRCs). In the 3200 ECOG trial, bevacizumab had no activity in CRC as a single agent [33]. By contrast, the antitumor activity of bevacizumab when added to first-line chemotherapy with metastatic CRC has been demonstrated in phase II and phase III trials. In a pivotal phase III study of first-line bevacizumab in combination with irinotecan and bolus 5-FU/LV (IFL) conducted in 813 patients, the addition of bevacizumab significantly improved median overall survival by 30% (20 vs. 15.6 months, $p < 0.0001$).

These trials conducted not only in lung and colon cancers but also in breast carcinomas highlight an important point: a specific targeted compound, bevacizumab, must be associated with nonspecific tools and cytotoxics, to demonstrate a significant antitumor efficacy [34]. Many preclinical studies have provided some insight into the mechanisms by which anti-VEGF therapy increases the effectiveness of chemotherapy. Jain *et al.*, have developed the concept of vessel normalization [35], whereby antibodies to VEGF lead to a redistribution of tumor blood flow, resulting in increasing delivery of chemotherapy to the tumor. This model supports the hypothesis that tumor vasculature is disorganized, with significant functional alterations (vascular hyperpermeability and blood flow heterogeneity). This structural and functional tumor vessel impairment forms a physiological barrier that hinders the delivery of anticancer agents to tumors. This hypothesis awaits confirmation in humans and is currently being investigated in prospective trials [36]. Recent animal models, however, did not find any increase in tumor drug uptake in mice bearing human cancer xenografts after anti-VEGF treatment [37], although an increase in vascular (endothelial) damage was demonstrated with the combination. It has been proposed that VEGF, via its tyrosine kinase receptors, triggers multiple cytoplasmic kinases, for example, stress-activated protein kinase-2/p38 (SAPK2/p38) or phosphorylation of focal adhesion kinase (FAK) [38], and survival pathways in endothelial cells. VEGF inhibition sensitizes the tumor endothelial cells but not the tumor cells, to the cell-damage induced by cytotoxics [39]. At present, anti-VEGF therapy can be considered as a chemotherapy sensitizer to the tumor vasculature. These findings lead to the conclusion that bevacizumab, beyond its known specific mechanism of action, can also induce an increase in drug delivery of nonspecific agents.

12.3.1.3 **Paradoxical Specificity against Antiproliferative Targets?**

The first VEGF-A isoform described by Ferrara and Plouet, called VEGF-A_{165} has been extensively investigated for its function and roles in cancer. Since 1989, other isoforms such as VEGF-A_{121}, VEGF-A_{145}, VEGF-A_{148} have been identified and are the results of alternative splicing of exons 6 and 7. In addition, Bates *et al.* identified, from the exon 8 distal splice site selection, an isoform called VEGF-A_{165}b. VEGF-A_{165}b is widely expressed in many normal tissues including plasma, urine, renal cortex, lung, and pancreatic islets [40]. Experimental evidence suggests that VEGF-A_{165}b plays a negative role in the regulation of angiogenesis. In cancer cells (melanoma, colon, and bladder carcinoma), VEGF-A_{165}b isoforms represent the majority of VEGF-A. The key structural specificity of this isoform has profound implications for receptor interaction and function. VEGF-A_{165}b interacts with VEGFR2 but does not bind neurophilin 1. *In vitro*, VEGF-A_{165}b inhibits endothelial cell migration and proliferation [41]. *In vivo*, VEGF-A_{165}b does not increase vascular permeability, it significantly inhibits angiogenesis when VEGF-A is overexpressed, and fails to stimulate angiogenesis in several mice models [42]. Transfection of VEGF-A_{165}b delays the growth of several tumor models and recombinant VEGF-A_{165}b reduces the growth of disseminated metastatic melanoma tumors. The profound and opposite effects in the functional role of VEGF-A and VEGF-A_{165}b could be related to competitive binding to VEGFR2. This data supports the dual role of VEGF-A and its isoform, and provides evidence for a regulatory function between pro- and antiangiogenic factors at the molecular level by alternative splicing [43]. VEGF-A_{165}b contains binding domains for the vast majority of anti-VEGF-A antibodies, including bevacizumab. A phase I study of intravenous Vascular Endothelial Growth Factor Trap, Aflibercept, has been completed in patients with solid tumors. Dose-limiting toxicities were rectal ulceration and proteinuria. Other mechanism specific toxicities included hypertension [44]. The absence of specificity of bevacizumab for VEGF-A and its binding to VEGF-A_{165}b has therapeutic implications. Interestingly, bevacizumab has no effect on VEGF-A_{165}b-expressing tumors in experimental colorectal carcinoma

[45]. Accordingly, the treatment of patients whose tumors express high levels of VEGF-A_{165}b with bevacizumab may not be effective. This hypothesis remains to be demonstrated in patients. The impact of the inhibition of VEGF-A_{165}b on normal tissues also needs to be further investigated.

12.3.1.4 **VEGF-Trap**

VEGF-trap (aflibercept) is a novel antiangiogenic protein [44] consisting of human VEGF receptor extracellular domains fused to the Fc portion of human IgG1. This molecule interferes with the biological actions of VEGF by complexing with circulating VEGF and preventing it from interacting with its receptors on endothelial cells. Aflibercept is anticipated to be more active than other anti-VEGF agents because of its particularly high binding affinity to VEGF and its ability to bind other related proangiogenic factors such as VEGF-B and the placental growth factors, PlGF1 and PlGF2 [46, 47]. In preclinical studies, aflibercept inhibited the growth of a variety of tumor types with a broad pharmacological index. Histological analysis indicated that treatment with aflibercept resulted in the formation of largely avascular and necrotic tumors, demonstrating that tumor-induced angiogenesis was blocked. Aflibercept has been investigated in association with approved cytotoxic agents (CPT-11, docetaxel, oxaliplatin) in several ongoing phase I, II, and III trials. Hypertension and proteinuria have been associated with this specific VEGF blocker both in preclinical and clinical studies. The mechanism of renal toxicity is similar to that of bevacizumab, with the occurrence of TMA and podocyte alterations [48]. We can conclude, based on these observations, that renal VEGF starvation related to specific anti-VEGF agents induces class-specific collateral damage on normal tissues.

12.3.2
Kinases Signaling Pathways as a Target for Antiangiogenic Activity

As described previously, two main drug design approaches to VEGFR kinase inhibition have been developed: selective-targeted versus multiple-targeted agents. LoRusso defined a new therapeutic classification [21] by introducing the terms (i) single-targeted kinase inhibitors, (ii) multiple-targeted single-spectrum kinase inhibitors (inhibition of mechanistically and structurally related class of kinase targets), and (iii) "extended-spectrum selective kinase inhibitors" (ESSKIs, inhibition of several different target pathways). In a recent publication [49], Karaman *et al.* have provided the most comprehensive description of selectivity (or nonselectivity) for small molecule kinase inhibitors including anti-VEGFRs. They presented interaction maps for 38 kinase inhibitors across a panel of 317 representative kinases. The authors introduced a selectivity score correlated with the observed interaction patterns. This analysis revealed a diversity of interaction patterns within the human kinome. Interestingly, most of the compounds tested did not bind many "off-targets" with affinities similar to that of the primary targets. Most of the compounds bind only a relatively small number of kinases with a high affinity. However, the selectivity score of the seven small molecules approved for the treatment of solid

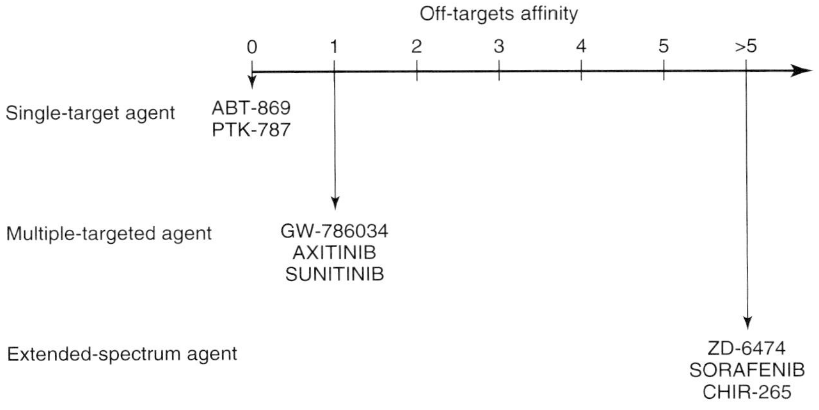

Figure 12.3 Selectivity of VEGFR inhibitors. Two main drug design approaches to VEGFR kinase inhibition have been developed: selective-targeted versus multiple-targeted agents. The selectivity score of small molecules can vary depending on whether an agent is (i) a single-targeted kinase inhibitor, (ii) a multip e-targeted single-spectrum kinase inhibitor, or (iii) an extended-spectrum selective kinase inhibitors (inhibition of several different target pathways).

tumors vary more than 50-fold, between lapatinib (the most selective) to sunitinib (the least selective). Of the VEGFR inhibitors, PTK-787 (valatinib) appears to have a high selective affinity for VEGFR2, while the highest affinity off-targets are closely related receptor tyrosine kinases. In contrast, sunitinib, originally designed to be a selective VEGFR, PDGFR, and c-kit inhibitor, binds with comparable affinity to multiple groups of kinases with no strong preference between either tyrosine or serine-threonine kinases (see Figure 12.3).

The biological sequelae of multikinase inhibition remains undefined. In terms of toxicity, differences have been observed between VEGFR2 selective agents (e.g., axitinib, pazopanib, valatinib) compared to sunitinib. For instance, sunitinib has demonstrated significant hematological toxicity [50], including dose-limiting thrombocytopenia, while the more selective anti-VEGFR2 agents appear to have less hematotoxic effects. The spectrum of target inhibition does not seem to affect the antitumor efficacy. In two phase II studies conducted in metastatic renal cell carcinoma, axitinib and sunitinib demonstrated similar antitumor activity and had comparable objective response rates (42 and 40%, respectively) [51].

Interestingly, several ESSKIs have recently been investigated in clinical trials. These agents have been designed to have several kinase targets, including kinases affecting angiogenic signaling. The choice of the pathways appears to be empiric but may offer a better chance of inhibition of cancer cell growth. One such agent, ZD6474 (vandetanib) targets the VEGFR2 and EGFR [52]. Combined targeting of VEGF and EGFR signaling pathways represents an attractive approach to treating malignancies such as NSCLC. Preclinical studies have shown that combined dual pathway inhibitors have at least additive antitumor activity [53, 54]. In addition, EGFR inhibitors may act to sensitize cells to antiangiogenic therapy by lowering the tumor survival threshold as one of the signaling pathways leading to cell growth and survival is shut down. Moreover, it has been postulated that "overactivation" of non-EGFR pathways driving VEGF expression results in an increased resistance to EGFR inhibitors. This could potentially be overcome by combining EGFR inhibition with an anti-VEGF agent.

ZD6474 inhibits the growth of a variety of human tumor xenografts [55], including A549 lung adenocarcinoma and Calu-6 NSCLC. Studies in PC-9 human lung adenocarcinoma xenografts suggest that the antitumor activity of ZD6474 may largely reflect inhibition of the EGFR as evidenced by its cross-resistance with gefitinib. A phase I trial in solid tumors established the maximum tolerated dose (MTD) of 300 mg d^{-1}; the most common dose-limiting toxicities were diarrhea, hypertension, and rash. In a recent phase III study evaluating docetaxel $\pm$ ZD6474 in patients with previously treated locally advanced or metastatic NSCLC, ZD6474 demonstrated significant evidence of clinical benefit (improvement in progression-free survival with a positive trend for overall survival) [56]. ZD6474 (300 mg per day) versus gefitinib(250 mg per day) as second- or third-line treatment in patients with advanced NSCLC showed that estimated median time to progression was significantly longer in patients receiving ZD6474 compared with gefitinib (11.9 vs. 8.1 weeks; HR, 0.632; $p = 0.011$). This drug is currently being investigated in several phase III studies in advanced NSCLC, and the impact of its dual inhibition versus single EGFR or VEGFR inhibitors remains to be determined.

12.3.3 Non-VEGF/VEGFR Inhibition

Tumor angiogenesis is not restricted to VEGF/VEGFRs activation. There are many alternative VEGF independent pathways including the Notch–Dll4 activation system. Other mechanisms stimulating new vessel growth include the release of angiogenic cytokines (e.g., angiopoietins, PDGF, SDF-1, and TGF-β). Broader signaling cascades also play a predominant role (e.g., HIF activation). An individual discussion on each of these kinases is beyond the scope of this chapter but we have focused on the HIF-1α and Notch pathways due to the close relationship/cross talk existing between these pathways and VEGF.

12.3.3.1 HIF-1α Inhibition

HIFs are heterodimeric basic helix-loop-helix/PAS proteins that induce the transcription of a diverse array of genes to affect the hypoxic response. HIFs are composed of an α subunit (HIF-1α, HIF-2α, and HIF-3α) and a β subunit (HIF-1β or ARNT) [57]. There are multiple splice variants of HIF-1α and HIF-3α, with dominant negative proteins being produced in some tissues. HIFs are key elements that mediate cellular adaptation to low O_2 by activating the transcription of target genes involved in angiogenesis, metabolism, and extracellular matrix remodeling [58]. HIF-2α expression increases with sporadic *VHL*-deficient clear cell renal carcinoma, via c-Myc activation [57]. Overexpression of HIFs promotes tumorigenesis, but paradoxically, there is genetic evidence for a tumor suppressor role of HIF-2α. The critical role of HIFs in renal cell carcinoma is the basis for the current development of anticancer targeted therapies in this setting.

In human tissues, HIF-1α messenger RNA expression is generally ubiquitous with high expression in the kidneys (predominantly in the cortical and medullary ducts, and glomerular cells) [59]. The expression of HIF-2α is detectable in podocytes, as well as in cortical and medullary endothelial and interstitial cells, but is absent in the fully developed kidney. HIF-2α may play a role in the development of the tubular system, but the real role of HIFs in nephrogenesis remains to be determined. Chronic hypoxia is the final common pathway to end-stage renal failure, and HIFs are the master switch of the hypoxic adaptation response. HIFs have, however, contrasting effects on the kidneys, from a cytoprotective effect in acute ischemia-reperfusion injury to a role in inflammation with a profibrotic response in chronic renal hypoxia.

Recently, the first phase I study in humans with an anti-HIF1-α RNA antagonist has been reported. EZN-2968 is a third-generation oligonucleotide (locked nucleic acid). *In vivo*, this compound has a broad distribution in major organs including the kidneys [60]. No renal toxicities have been reported in mice. In the phase I study conducted in 18 patients, 1 patient experienced grade 3 hypertension. Interestingly, no proteinuria or renal dysfunction has been reported to this point [61]. This preliminary report suggests, despite the expression of the target on normal tissue, a favorable toxicity profile.

12.3.3.2 **Notch Inhibition**

Notch receptors comprise a family of transmembrane receptors, their ligands and their transcription factors. They are expressed by various cell types. They are involved in cell differentiation, cell fate, and proliferation. The Notch signaling pathway functions as another angiogenesis pathway that appears to run in parallel to and is independent of VEGF/VEGFR activation [62]. The four membrane receptors (Notch 1, 2, 3, and 4) interact with transmembrane ligands (i.e., Jagged 1 and Dll4). Dll4 is expressed exclusively by endothelial cells and represents a major stimulus (via its notch receptor) of angiogenesis [63]. Knockout of only Dll4 in mice is lethal to the embryo and represents an essential signal for vascular development. The Notch–Dll4 complex is upregulated in tumor vasculature and drugs targeting this pathway are in an early phase of preclinical development. These drugs include neutralizing antibodies against Dll4, and inhibitors of the Notch receptors. Drugs that target Dll4 can paradoxically increase tumor angiogenesis, but most of these neovessels are functionally abnormal and compromise blood delivery [63].

As previously described, podocytes represent a type of epithelial glomerular cell involved in the ultrafiltration process. Alterations of several podocyte proteins have been described in glomerulosclerosis [64]. The Notch pathway is crucial during nephrogenesis, as Notch1 activity is very high in the kidney. However, little activity of Notch1 can be detected in mature kidneys under normal conditions [65]. Activation of the Notch pathway in mature podocytes is associated with glomerular disease, and pharmacological inhibition of Notch cleavage can prevent podocyte apoptosis and albuminuria [66]. Similarly, the Notch signaling pathway is overexpressed in the proximal tubules in a model of acute injury such as tubular necrosis. It is postulated that Notch pathway stimulation in healthy adult kidneys is likely to lead to complications. Finally, further studies are warranted to determine whether Notch activation is a response to or a primary cause of renal disease. The specificity of Notch receptor inhibition also represents an important issue, as Notch2, but not Notch1, is required for fate acquisition in the mammalian nephron [67]. Inhibition of the Notch pathway as an anticancer strategy is in its early stages (phase I studies are ongoing investigating MK-0752, a specific Notch inhibitor [68]). A precise preclinical and clinical evaluation of its putative renal toxicity/protection needs to be addressed and closely monitored.

12.4
Conclusion

In his original observation published in the New England Journal of Medicine in 1971, Folkman stated that tumor cells and endothelial cells of new blood vessels "constitute a highly integrated ecosystem" [69]. This observation suggests a certain level of specificity of tumor angiogenesis, and by consequence a potent specific targeting of antiangiogenic molecules. Since the report on the initial phase I study with the "specific" VEGF inhibitor (bevacizumab) [70], thousands of publications have demonstrated the absence of complete specificity of this strategy. The lack of specificity is related to (i) the absence of a selective spectrum of antiangiogenics, illustrated by the tyrosine kinase inhibitors and (ii) the absence of specific expression of the target to tumor tissue only.

Even in a situation of absolute specificity, antiangiogenic agents would have limited efficacy. This effect was predicted by Folkman when he realized that antiangiogenesis would merely arrest tumor growth but would not generally eliminate all the tumor cells. The limited percentage of complete remissions achieved by antiangiogenics in the clinic when used as single agent therapies, have confirmed this observation [71]. On the basis of these reports, the combination of antiangiogenics given with other therapeutic modalities has been explored. The association with cytotoxics leads to the obvious conclusion of a loss of specificity. The challenge in the coming years will be to combine antiangiogenics with other specific targeted therapies to produce a relevant clinical benefit in specific tumor types or in an individualized patient.

The utilization of the molecular differences displayed between the vasculature of normal tissue and tumor tissue may herald the beginning of a new frontier. The "vascular zip code" has been used to describe the unique expression of cell-surface molecules found in each vascular bed. The characterization of tumor blood vessels includes selective overexpression of a heterogenous group of proteins (proteases, integrins, growth factor receptors, and proteoglycans). The use of phage display libraries expressing random peptides or protein fragments have been useful in analyzing vascular heterogeneity amongst normal tissues and tumors [72]. These phage-screening studies have revealed a previously unsuspected degree of vascular specialization. Peptides that home to a specific site in the vasculature are attractive as carriers of both diagnostic and therapeutic agents. The future of antiangiogenics may include homing peptide directed drug delivery, thereby concentrating the drug at the targeted site leading to improved efficacy and less toxicity.

References

1. Ellis, L.M. and Hicklin, D.J. (2008) Pathways mediating resistance to vascular endothelial growth factor-targeted therapy. *Clin. Cancer Res.*, **14** (20), 6371–6375.
2. Dvorak, H.F. (2002) Vascular permeability factor/vascular endothelial growth factor: a critical cytokine in tumor angiogenesis and a potential target for diagnosis and therapy. *J. Clin. Oncol.*, **20** (21), 4368–4380.
3. Takayama, K. *et al.* (2000) Suppression of tumor angiogenesis and growth by gene transfer of a soluble form of vascular endothelial growth factor receptor into a remote organ. *Cancer Res.*, **60** (8), 2169–2177.
4. Ge, Y.L., Zhang, X., Zhang, J.Y, Hou, L., *et al.* (2009) The mechanisms on apoptosis by inhibiting VEGF expression in human breast cancer cells. *Int. Immunopharmacol.*, **9** (4), 389–395.
5. Hurwitz, H., Fehrenbacher, L., Novotny, W., Cartwright, T., Hainsworth, J., Heim, W., Berlin, J., Baron, A., Griffing, S., Holmgren, E., Ferrara, N., Fyfe, G., Rogers, B., Ross, R., and Kabbinavar, F. (2004) Bevacizumab plus irinotecan, fluorouracil, and leucovorin for metastatic colorectal cancer. *N. Engl. J. Med.*, **350** (23), 2335–2342.
6. Ross, J.S., Schenkein, D.P., Pietrusko, R., Rolfe, M., Linette, G.P., Stec, J., Stagliano, N.E., Ginsburg, G.S., Symmans, W.F., Pusztai, L., and Hortobagyi, G.N. (2004) Targeted therapies for cancer 2004. *Am. J. Clin. Pathol.*, **122** (4), 598–609.
7. Woolard, J., Bevan, H.S., Harper, S.J., and Bates, D.O. (2009) Molecular diversity of VEGF-a as a regulator of its biological activity. *Microcirculation,* June 10, 1–21.
8. Ferrara, N. (2004) Vascular endothelial growth factor: basic science and clinical progress. *Endocr. Rev.*, **25** (4), 581–611.
9. Hiratsuka, S., Minowa, O., Kuno, J., Noda, T., and Shibuya, M. (1998) Flt-1 lacking the tyrosine kinase domain is sufficient for normal development and angiogenesis in mice. *Proc. Natl. Acad. Sci. U.S.A.*, **95** (16), 9349–9354.
10. Milkiewicz, M. and Haas, T.L. (2005) Effect of mechanical stretch on HIF-1α and MMP-2 expression in capillaries isolated from overloaded skeletal muscles: laser capture microdissection study. *Am. J. Physiol. Heart Circ. Physiol.*, **289** (3), H1315–H1320.
11. Maharaj, A.S. and D'Amore, P.A. (2007) Roles for VEGF in the adult. *Microvasc. Res.*, **74** (2-3), 100–113.
12. Coultas, L., Chawengsaksophak, K., and Rossant, J. (2005) Endothelial cells and VEGF in vascular development. *Nature,* **438** (7070), 937–945.
13. Maharaj, A.S., Saint-Geniez, M., Maldonado, A.E., and D'Amore, P.A. (2006) Vascular endothelial growth factor localization in the adult. *Am. J. Pathol.*, **168** (2), 639–648.
14. Monacci, W.T., Merrill, M.J., and Oldfield, E.H. (1993) Expression of vascular permeability factor/vascular endothelial growth factor in normal rat tissues. *Am. J. Physiol.*, **264** (4 Pt 1), C995–1002.
15. Fong, G.H., Rossant, J., Gertsenstein, M., and Breitman, M.L. (1995) Role of the Flt-1 receptor tyrosine kinase in regulating the assembly of vascular endothelium. *Nature,* **376** (6535), 66–70.
16. Senger, D.R., Galli, S.J., Dvorak, A.M., Perruzzi, C.A., Harvey, V.S., and Dvorak, H.F. (1983) Tumor cells secrete a vascular permeability factor that

promotes accumulation of ascites fluid. *Science*, **219** (4587), 983–985.
17. Kilic, U., Kilic, E., Järve, A., Guo, Z., Spudich, A., Bieber, K., Barzena, U., Bassetti, C.L., Marti, H.H., and Hermann, D.M. (2006) Human vascular endothelial growth factor protects axotomized retinal ganglion cells in vivo by activating ERK-1/2 and Akt pathways. *J. Neurosci.*, **26** (48), 12439–12446.
18. Li, J., Hampton, T., Morgan, J.P., and Simons, M. (1997) Stretch-induced VEGF expression in the heart. *J. Clin. Invest.*, **100** (1), 18–24.
19. Uchida, N., Nagai, K., Sakurada, Y., and Shirota, K. (2008) Distribution of VEGF and flt-1 in the normal dog tissues. *J. Vet. Med. Sci.*, **70** (11), 1273–1276.
20. Kaplan, R.N., Rafii, S., and Lyden, D. (2006) Preparing the "soil": the premetastatic niche. *Cancer Res.*, **66** (23), 11089–11093.
21. Lorusso, P.M. and Eder, J.P. (2008) Therapeutic potential of novel selective-spectrum kinase inhibitors in oncology. *Expert Opin. Invest. Drugs*, **17** (7), 1013–1028.
22. Cascone, T., Gridelli, C., and Ciardiello, F. (2007) Combined targeted therapies in non-small cell lung cancer: a winner strategy? *Curr. Opin. Oncol.*, **19** (2), 98–102.
23. Krämer, I. and Lipp, H.P. (2007) Bevacizumab, a humanized anti-angiogenic monoclonal antibody for the treatment of colorectal cancer. *J. Clin. Pharm. Ther.*, **32** (1), 1–14.
24. Presta, L.G. (2002) Engineering antibodies for therapy. *Curr. Pharm. Biotechnol.*, **3** (3), 237–256.
25. Ellis, L.M. and Hicklin, D.J. (2008) VEGF-targeted therapy: mechanisms of anti-tumour activity. *Nat. Rev. Cancer*, **8** (8), 579–591.
26. Ostendorf, T., De Vriese, A.S., and Floege, J. (2007) Renal side effects of anti-VEGF therapy in man: a new test system. *Nephrol. Dial. Transplant.*, **22** (10), 2778–2780.
27. Eremina, V., Baelde, H.J., and Quaggin, S.E. (2007) Role of the VEGF–a signaling pathway in the glomerulus: evidence for crosstalk between components of the glomerular filtration barrier. *Nephron Physiol.*, **106** (2), p32–p37.
28. Eremina, V., Jefferson, J.A., Kowalewska, J., Hochster, H., Haas, M., Weisstuch, J., Richardson, C., Kopp, J.B., Kabir, M.G., Backx, P.H., Gerber, H.P., Ferrara, N., Barisoni, L., Alpers, C.E., and Quaggin, S.E. (2008) VEGF inhibition and renal thrombotic microangiopathy. *N. Engl. J. Med.*, **358** (11), 1129–1136.
29. Sela, S., Itin, A., Natanson-Yaron, S., Greenfield, C., Goldman-Wohl, D., Yagel, S., and Keshet, E. (2008) A novel human-specific soluble vascular endothelial growth factor receptor 1: cell-type-specific splicing and implications to vascular endothelial growth factor homeostasis and preeclampsia. *Circ. Res.*, **102** (12), 1566–1574.
30. Zachary, I. (2001) Signaling mechanisms mediating vascular protective actions of vascular endothelial growth factor. *Am. J. Physiol. Cell Physiol.*, **280** (6), C1375–C1386.
31. Johnson, D.H. *et al.* (2004) Randomized phase II trial comparing bevacizumab plus carboplatin and paclitaxel with carboplatin and paclitaxel alone in previously untreated locally advanced or metastatic non-small-cell lung cancer. *J. Clin. Oncol.*, **22** (11), 2184–2191.
32. Sandler, A., Gray, R., Perry, M.C., Brahmer, J., Schiller, J.H., Dowlati, A., Lilenbaum, R., and Johnson, D.H. (2006) Paclitaxel-carboplatin alone or with bevacizumab for non-small-cell lung cancer. *N. Engl. J. Med.*, **355** (24), 2542–2550.
33. Giantonio, B.J., Catalano, P.J., Meropol, N.J., O'Dwyer, P.J., Mitchell, E.P., Alberts, S.R., Schwartz, M.A., and Benson, A.B.III, Eastern Cooperative Oncology Group Study E3200 (2007) Bevacizumab in combination with oxaliplatin, fluorouracil, and leucovorin (FOLFOX4) for reviously treated metastatic colorectal cancer: results from the Eastern Cooperative Oncology Group Study E3200. *J. Clin. Oncol.*, **25** (12), 1539–1544.
34. Browder, T., Butterfield, C.E., Kräling, B.M., Shi, B., Marshall, B., O'Reilly,

M.S., and Folkman, J. (2000) Antiangiogenic scheduling of chemotherapy improves efficacy against experimental drug-resistant cancer. *Cancer Res.*, **60** (7), 1878–1886.

35. Jain, R.K. (2005) Normalization of tumor vasculature: an emerging concept in antiangiogenic therapy. *Science*, **307** (5706), 58–62.
36. A phase II trial of Bevacizumab and Ixabepilone in treating patients with advanced kidney cancer, *www.clinicaltrials.gov*. Trial identifier: NCT00820209.
37. Kasman, I., Bagri, A., Mak, J. *et al.* (2008) Mechanistic evaluation of the combination effect of anti-VEGF and chemotherapy. Proceedings of the 100th Annual Meeting of the American Association for Cancer Research, April 12–16, San Diego, Abstract A2494.
38. Rousseau, S., Houle, F., Kotanides, H., Witte, L., Waltenberger, J., Landry, J., and Huot, J. (2000) Vascular endothelial growth factor (VEGF)-driven actin-based motility is mediated by VEGFR2 and requires concerted activation of stress-activated protein kinase 2 (SAPK2/p38) and geldanamycin-sensitive phosphorylation of focal adhesion kinase. *J. Biol. Chem.*, **275** (14), 10661–10672.
39. Pietras, K. and Hanahan, D. (2005) A multitargeted, metronomic, and maximum-tolerated dose "chemo-switch" regimen is antiangiogenic, producing objective responses and survival benefit in a mouse model of cancer. *J. Clin. Oncol.*, **23** (5), 939–952.
40. Bates, D.O., Cui, T.G., Doughty, J.M., Winkler, M., Sugiono, M., Shields, J.D., Peat, D., Gillatt, D., and Harper, S.J. (2002) VEGF165b, an inhibitory splice variant of vascular endothelial growth factor, is down-regulated in renal cell carcinoma. *Cancer Res.*, **62** (14), 4123–4131.
41. Woolard, J., Wang, W.Y., Bevan, H.S., Qiu, Y., Morbidelli, L., Pritchard-Jones, R.O., Cui, T.G., Sugiono, M., Waine, E., Perrin, R., Foster, R., Digby-Bell, J., Shields, J.D., Whittles, C.E., Mushens, R.E., Gillatt, D.A., Ziche, M., Harper, S.J., and Bates, D.O. (2004) VEGF165b, an inhibitory vascular endothelial growth factor splice variant: mechanism of action, in vivo effect on angiogenesis and endogenous protein expression. *Cancer Res.*, **64** (21), 7822–7835.
42. Bates, D.O. and Harper, S.J. (2005) Therapeutic potential of inhibitory VEGF splice variants. *Future Oncol.*, **1** (4), 467–473.
43. Nowak, D.G., Woolard, J., Amin, E.M., Konopatskaya, O., Saleem, M.A., Churchill, A.J., Ladomery, M.R., Harper, S.J., and Bates, D.O. (2008) Expression of pro- and anti-angiogenic isoforms of VEGF is differentially regulated by splicing and growth factors. *J. Cell Sci.*, **121** (Pt 20), 3487–3495.
44. Lockhart, A., Rothenberg, M., Dupoent, J., *et al.* (2010) A phase I study of intravenous Vascular Endothelial Growth Factor Trap, Aflibercept, in patients with advanced solid tumors. *J. Clin. Oncol.*, **28** (2), 207–214.
45. Varey, A.H., Rennel, E.S., Qiu, Y., Bevan, H.S., Perrin, R.M., Raffy, S., Dixon, A.R., Paraskeva, C., Zaccheo, O., Hassan, A.B., Harper, S.J., and Bates, D.O. (2008) VEGF 165 b, an antiangiogenic VEGF-A isoform, binds and inhibits bevacizumab treatment in experimental colorectal carcinoma: balance of pro- and antiangiogenic VEGF-A isoforms has implications for therapy. *Br. J. Cancer*, **98** (8), 1366–1379.
46. Holash, J., Davis, S., Papadopoulos, N., Croll, S.D., Ho, L., Russell, M., Boland, P., Leidich, R., Hylton, D., Burova, E., Ioffe, E., Huang, T., Radziejewski, C., Bailey, K., Fandl, J.P., Daly, T., Wiegand, S.J., Yancopoulos, G.D., and Rudge, J.S. (2002) VEGF-trap: a VEGF blocker with potent antitumor effects. *Proc. Natl. Acad. Sci. U.S.A.*, **99** (17), 11393–11398.
47. Lau, S.C., Rosa, D.D., and Jayson, G. (2005) Technology evaluation: VEGF trap (cancer), Regeneron/sanofi-aventis. *Curr. Opin. Mol. Ther.*, **7** (5), 493–501.
48. Izzedine, H., Brocheriou, I., Deray, G., and Rixe, O. (2007) Thrombotic microangiopathy and anti-VEGF agents. *Nephrol. Dial. Transplant.*, **22** (5), 1481–1482.

49. Karaman, M.W., Herrgard, S., Treiber, D.K., Gallant, P., Atteridge, C.E., Campbell, B.T., Chan, K.W., Ciceri, P., Davis, M.I., Edeen, P.T., Faraoni, R., Floyd, M., Hunt, J.P., Lockhart, D.J., Milanov, Z.V., Morrison, M.J., Pallares, G., Patel, H.K., Pritchard, S., Wodicka, L.M., and Zarrinkar, P.P. (2008) A quantitative analysis of kinase inhibitor selectivity. *Nat. Biotechnol.*, **26** (1), 127–132.
50. Billemont, B., Izzedine, H., and Rixe, O. (2007) Macrocytosis due to treatment with sunitinib. *N. Engl. J. Med.*, **357** (13), 1351–1352.
51. Rixe, O., Bukowski, R.M., Michaelson, M.D., Wilding, G., Hudes, G.R., Bolte, O., Motzer, R.J., Bycott, P., Liau, K.F., Freddo, J., Trask, P.C., Kim, S., and Rini, B.I. (2007) Axitinib treatment in patients with cytokine-refractory metastatic renal-cell cancer: a phase II study. *Lancet Oncol.*, **8** (11), 975–984.
52. Drevs, J., Konerding, M.A., Wolloscheck, T., Wedge, S.R., Ryan, A.J., Ogilvie, D.J., and Esser, N. (2004) The VEGF receptor tyrosine kinase inhibitor, ZD6474, inhibits angiogenesis and effects microvascular architecture within an orthotopically implanted renal cell carcinoma. *Angiogenesis*, **7** (4), 347–354.
53. Ciardiello, F., Caputo, R., Damiano, V. *et al.* (2003) Antitumor effects of ZD6474, a small molecule vascular endothelial growth factor receptor tyrosine kinase inhibitor, with additional activity against epidermal growth factor receptor tyrosine kinase. *Clin. Cancer Res.*, 1546–1556.
54. Wedge, S.R., Ogilvie, D.J., Dukes, M. *et al.* (2002) ZD6474 inhibits vascular endothelial growth factor signaling, angiogenesis, and tumor growth following oral administration. *Cancer Res.*, **62** 4645–4655.
55. Taguchi, F., Koh, Y., Koizumi, F. *et al.* (2004) Anticancer effects of ZD6474, a VEGF receptor tyrosine kinase inhibitor, in gefitinib ("Iressa")-sensitive and resistant xenograft models. *Cancer Sci.*, **95** (12), 984–989.
56. Herbst, R.S., Sun, Y., Korfee, S., Germonpré, P., Saijo, N., Zhou, C., Wang, J., Langmuir, P., Kennedy, S.J., and Johnson, B.E. (2009) Vandetanib plus docetaxel versus docetaxel as second-line treatment for patients with advanced non-small cell lung cancer (NSCLC): a randomized, double-blind phase III trial (ZODIAC). *J. Clin. Oncol.*, **27**, 18s (Supplement; Abstract CRA8003).
57. Haase, V.H. (2006) Hypoxia-inducible factors in the kidney. *Am. J. Physiol. Renal Physiol.*, **291** (2), F271–F281.
58. Schofield, C.J. and Ratcliffe, P.J. (2004) Oxygen sensing by HIF hydroxylases. *Nat. Rev. Mol. Cell Biol.*, **5** (5), 343–354.
59. Bernhardt, W.M., Schmitt, R., Rosenberger, C., Münchenhagen, P.M., Gröne, H.J., Frei, U., Warnecke, C., Bachmann, S., Wiesener, M.S., Willam, C., and Eckardt, K.U. (2006) Expression of hypoxia-inducible transcription factors in developing human and rat kidneys. *Kidney Int.*, **69** (1), 114–122.
60. Greenberger, L.M., Horak, I.D., Filpula, D., Sapra, P., Westergaard, M., Frydenlund, H.F., Albaek, C., Schroder, H., and Orum, H. (2008) A RNA antagonist of hypoxia-inducible factor-1 alpha, EZN-2968, inhibits tumor cell growth. *Mol. Cancer Ther.*, **7** (11), 3598–3608.
61. Lewis, N., Cohen, R.B., Nishida, Y. *et al.* (2008) Phase I, pharmacokinetic (PK), dose-escalation study of EZN 2968, a novel hypoxia-inducible factor-1 alpha RNA antagonist, administered weekly in patients with solid tumors. 20th EORTC-NCI-AACR Symposium on Molecular Targets and Cancer Therapeutics, Abstract #398, Geneva, 2008.
62. Sainson, R.C. and Harris, A.L. (2008) Regulation of angiogenesis by homotypic and heterotypic notch signalling in endothelial cells and pericytes: from basic research to potential therapies. *Angiogenesis*, **11** (1), 41–51.
63. Noguera-Troise, I., Daly, C., Papadopoulos, N.J., Coetzee, S., Boland, P., Gale, N.W., Lin, H.C., Yancopoulos, G.D., and Thurston, G. (2006) Blockade of Dll4 inhibits tumour growth by promoting non-productive angiogenesis. *Nature*, **444** (7122), 1032–1037.
64. Pollak, M.R. (2008) Focal segmental glomerulosclerosis: recent advances.

Curr. Opin. Nephrol. Hypertens., **17** (2), 138–142.

65. Kretzler, M. and Allred, L. (2008) Notch inhibition reverses kidney failure. *Nat. Med.*, **14** (3), 246–247.

66. Niranjan, T., Bielesz, B., Gruenwald, A., Ponda, M.P., Kopp, J.B., Thomas, D.B., and Susztak, K. (2008) The Notch pathway in podocytes plays a role in the development of glomerular disease. *Nat. Med.*, **14** (3), 290–298.

67. Cheng, H.T., Kim, M., Valerius, M.T., Surendran, K., Schuster-Gossler, K., Gossler, A., McMahon, A.P., and Kopan, R. (2007) Notch2, but not Notch1, is required for proximal fate acquisition in the mammalian nephron. *Development*, **134** (4), 801–811.

68. A Phase I/II study of MK-0752 followed by docetaxel in advanced or metastatic breast cancer, *www.clinicaltrials.gov*. Trial identifier: NCT0064533.

69. Folkman, J. (1971) Tumor angiogenesis: therapeutic implications. *N. Engl. J. Med.*, **285** (21), 1182–1186.

70. Gordon, M.S., Margolin, K., Talpaz, M., Sledge, G.W. Jr., Holmgren, E., Benjamin, R., Stalter, S., Shak, S., and Adelman, D. (2001) Phase I safety and pharmacokinetic study of recombinant human anti-vascular endothelial growth factor in patients with advanced cancer. *J. Clin. Oncol.*, **19** (3), 843–850.

71. Presta, L.G. (2008) Molecular engineering and design of therapeutic antibodies. *Curr. Opin. Immunol.*, **20** (4), 460–470.

72. Ruoslahti, E. *et al.* (2004) Vascular zip codes in angiogenesis and metastasis. *Biochem. Soc. Trans.*, **32** (Pt3), 397–402.

Part V
Imaging and Biomarkers in Angiogenesis

Tumor Angiogenesis – From Molecular Mechanisms to Targeted Therapy. Edited by Francis S. Markland, Stephen Swenson, and Radu Minea
Copyright © 2010 WILEY-VCH Verlag GmbH & Co. KGaA, Weinheim
ISBN: 978-3-527-32091-2

13
In vivo Imaging of Tumor Angiogenesis

Baris Turkbey, Gregory Ravizzini, and Peter L. Choyke

13.1
Introduction

Angiogenesis is the process of new vessel formation from preexisting vasculature and it occurs physiologically during embryogenesis, wound healing, and menstruation. Angiogenesis also plays a role in pathological conditions such as diabetic retinopathy, rheumatoid arthritis, psoriasis, cardiovascular insufficiency, infections, and cancer. With regard to cancer, angiogenesis is an essential step for growth; tumors can only grow to a diameter of 2–3 mm based on the diffusion limits of oxygen in tissue as first proposed by Folkman [1, 2].

Normal microcirculation within tissues is organized in a hierarchical system of arterioles, capillaries, and venules, although individual organ variations occur. In the case of tumor microcirculation, however, this hierarchy is lost; the microcirculation is disorganized, with irregular, tortuous, and fragile vessels demonstrating increased permeability. Permeability is increased because of structural abnormalities in the endothelium including large gaps between endothelial cells, discontinuous basement membranes, and loosely adherent pericytes, resulting in large fenestrations capable of leaking large proteins. Moreover, such vessels also express a variety of surface markers indicative of their activated angiogenic state.

13.2
Mechanisms of Tumor Angiogenesis

Physiological angiogenesis is a highly regulated process, whereas in tumor angiogenesis, the same regulatory mechanisms are defective resulting in haphazard vessel growth, which is typically less efficient than the normal microcirculation in delivering nutrients and removing toxins. For instance, in the transition of adenomas to carcinomas, K-ras mutations lead to upregulation of vascular endothelial growth factor (VEGF), which is a highly potent mitogen [3, 4]. VEGF interacts with several high-affinity tyrosine kinase receptors, VEGFR1 and VEGFR2 [5]. VEGFR1 is critical for physiologic and developmental angiogenesis, such as that

Tumor Angiogenesis – From Molecular Mechanisms to Targeted Therapy. Edited by Francis S. Markland, Stephen Swenson, and Radu Minea
Copyright © 2010 WILEY-VCH Verlag GmbH & Co. KGaA, Weinheim
ISBN: 978-3-527-32091-2

occurring during embryogenesis or somatogenesis, and is highly expressed in specific cell types during these processes [6]. Tumoral angiogenesis is mainly mediated through VEGFR2, which is preferentially expressed in tumor cells and their associated endothelial cells [7]. VEGFR2 is responsible for the mitogenesis and increased vascular permeability of endothelium in tumor-associated vessels. The presence of VEGFR2 within cancer tissue is associated with a poor prognosis. Thus, therapeutic agents targeting either VEGF or VEGFR or both are being investigated for the treatment of some metastatic tumors [8–10].

Several cytokines, oncogenes, and tumor suppressor genes regulate the release of VEGF, which activates endothelial cell proliferation and leads to the recruitment of circulating endothelial precursor cells [11]. Hypoxia, which is frequently found within tumors, is another important stimulus for VEGF. While the VEGF pathway plays the dominant role in angiogenesis, several other mediators such as, integrins, matrix metalloproteinases (MMPs) other tyrosine kinase receptors and G protein–coupled receptors also contribute to complex regulation of angiogenesis [12].

Integrins are cell adhesion molecules which are comprised of two noncovalently bound transmembrane subunits (alpha and beta). Integrins are expressed on endothelial cells and tumor cells; in endothelial cells they facilitate endothelial proliferation, migration, and survival, whereas in tumor cells they modulate cell invasion and migration along with vasculature, thus promoting distant metastases [13, 14]. The $\alpha v\beta_3$ integrin, which is one of the most extensively studied integrins, is significantly overexpressed in the endothelium during tumor angiogenesis. The $\alpha v\beta_3$ integrin in the cell membrane binds to the interstitial matrix, which specifically contains arginine-glycine-aspartic acid (RGD) amino acid sequence [15]. The presence of $\alpha v\beta_3$ on tumors may correlate with grade of the tumor [16, 17]. Besides $\alpha v\beta_3$; other important integrins in angiogenesis include $\alpha 5\beta_1$ and $\alpha v\beta_5$ [18–20].

MMPs are enzymes that are involved in tissue remodeling based on the proteolytic breakdown of the extracellular matrix (ECM) and basement membranes, contributing to tumor invasion and metastatic disease. Additionally, MMPs directly result in cell proliferation and angiogenesis stimulation [21]. Several MMPs, specifically MMP-2, MMP-9, and membrane type 1 MMP have been linked to angiogenesis [22]. MMPs are overexpressed in a wide range of tumor types [23–27].

13.3 Imaging

13.3.1 *Ex vivo* Tumor Imaging

It is important to standardize *in vivo* imaging methods for angiogenesis. A standard method for quantification of angiogenesis within tissue specimens is microvessel density (MVD), which is determined by counting the maximal number of stained (typically with CD31 antibody) blood vessels per unit area on a histological section. Moreover, an index for minimal intercapillary distance within tumoral tissues can be established and this has been linked to prognosis in some cancer types [28–32]. However, MVD is not always a reliable predictor of prognosis [33–36]. Additionally, MVD cannot depict functional aspects of tumoral vessels such as their permeability [37]. Despite its limitations, MVD remains the most common standardized tissue measurement method for angiogenesis. Tissue VEGF and VEGFR levels have also been used to assess angiogenesis.

Next, we consider *in vivo* imaging techniques and their correlation with angiogenesis.

13.3.2
In vivo Imaging Techniques

Ultrasound (US) is a widely available imaging modality, which has long been used to evaluate the patency and flow dynamics within vessels. A major advantage of US is that it can be performed repeatedly since it is harmless, portable, and relatively inexpensive yet produces real-time images. Gray-scale US images provide anatomic information about tumor size and location. Functional flow information is provided by color and/or power Doppler US, which demonstrate flow kinetics without exogenous contrast media (Figure 13.1). In color Doppler US, quantitative analysis of spectral waveforms derived from flowing blood can differentiate benign and malignant lesions, in some cases. However, both low- and high-resistive index (RI) vessels can be found in tumors, and thus the specificity of color Doppler US mode is limited [38, 39]. Quantitative measurements of color Doppler US modes were reported to correlate with MVD measurements in several studies [40–42]. Contrast enhanced color Doppler US performed after the injection of microbubbles

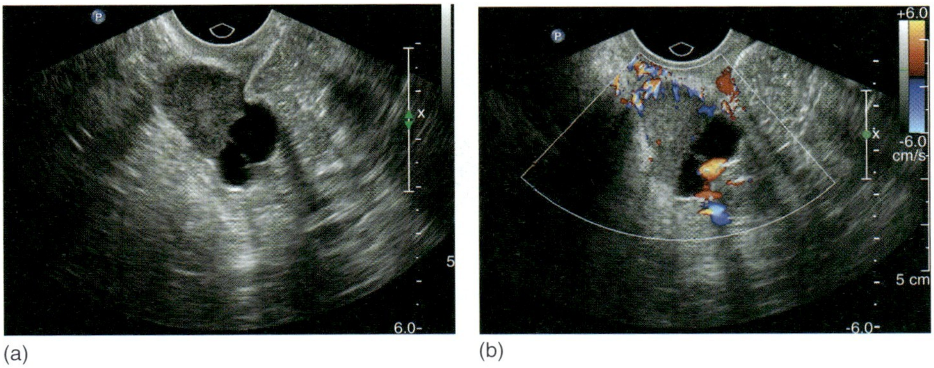

Figure 13.1 51-year-old female patient with ovarian cancer. Gray-scale pelvis sonography image demonstrates a left ovarian mass with solid and cystic components (a); color Doppler mode shows increased vascularity of the mass consistent with tumor-induced angiogenesis (b).

allows the depiction of vessels smaller than can be seen with color Doppler US. Microbubbles are acoustically active micron-sized emulsions that encapsulate gas bubbles in a lipid polymer or protein shell and are compressible with acoustic pressure. At low megahertz US frequency, acoustic energy results in oscillation of the bubbles causing a strong acoustic signal; therefore, they appear as echogenic (bright) particles on images [43, 44]. Contrast enhanced Doppler US was shown to be effective in the detection of tumor angiogenesis and in the assessment of tumor response to antiangiogenic treatment in several types of cancer [45–52]. US has several limitations. It is an operator-dependent imaging modality and results vary with skill and experience. Moreover, vessels smaller than 50 μm cannot be visualized due to inherent limitations in resolution and sensitivity [53].

Computed tomography (CT) has been widely used in cancer imaging. Contrast media enhanced CT with low molecular weight iodinated agents enables visualization of tumors and their relationship to surrounding anatomical structures. By obtaining images at different phases after injection (arterial, early venous, and late venous) it is possible to derive information about tumor vascularization. With the advent of multidetector technology, faster image acquisition is possible. On early arterial phase images, tumoral vascularity can be readily demonstrated with high detail; moreover isotropic imaging enables visualization of the tumor vascularity in any plane without loss of spatial information. By obtaining serial images through the same location over time, dynamic contrast enhanced computed tomography (DCE-CT) can be obtained. DCE-CT was first used in the emergency work-up of acute ischemic stroke where it is known as *CT perfusion*; by using several analysis models, kinetic parameters such as blood volume, blood flow, and mean transit time can be calculated on CT perfusion studies. This is only possible because of the intact blood–brain barrier that constrains the contrast media to the vasculature, thus assuring accurate measurements of blood flow and volume. DCE-CT can be used to assess tumor angiogenesis by exploiting the differences in contrast kinetics between normal and tumor tissues [54, 55] (Figure 13.2). An advantage of CT perfusion is that the relationship between the concentration of the iodinated contrast material within the tissue and the X-ray attenuation of the tissue is almost always linear; this results in a simplified analysis of the images and fewer errors [56]. DCE-CT was shown to be useful in evaluating the response of tumors to chemotherapeutic agents [57–63]. Among the major disadvantages of CT perfusion are its potentially high radiation exposure due to repetitive scanning, and the need for iodinated contrast material, which can be nephrotoxic and allergenic. Motion artifacts are also difficult to correct. Currently only low molecular weight iodinated contrast agents are available for studies thus limiting the repertoire of DCE-CT.

Magnetic resonance imaging (MRI) is also used as an anatomic imaging method for tumor detection, staging, and treatment follow-up. Conventional enhanced MRI is directly analogous to contrast enhanced CT in that it displays a snapshot of enhancement within a lesion and its surrounding normal vessels; although the anatomical information from conventional enhanced MRI is valuable, it lacks functional information. Dynamic contrast enhanced magnetic resonance imaging (DCE-MRI), which relies on fast MRI sequences obtained after the rapid intravenous

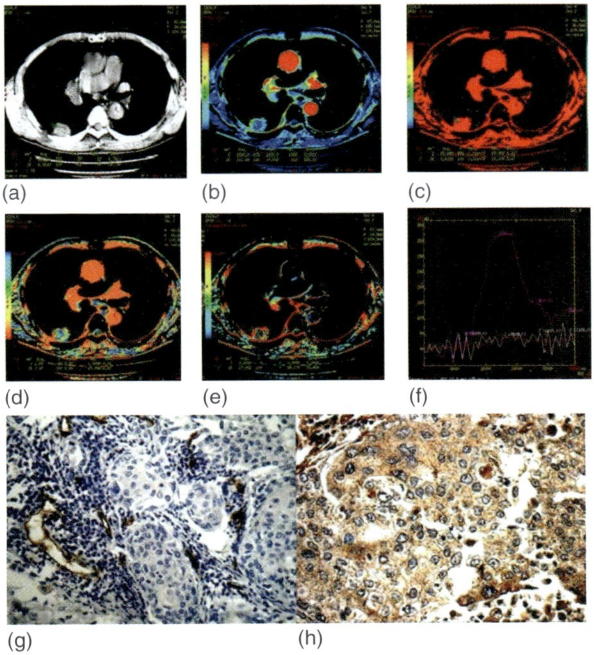

Figure 13.2 61-year-old male patient with well-differentiated squamous carcinoma of lung in the posterior basal segment of right lung. Axial enhanced computed tomography image shows the right lung mass (a), blood flow (b), blood volume (c), mean transit time (d), permeability surface (e) maps and time density curve (f) of CT perfusion confirms the diagnosis of lung cancer. CD34 staining shows presence of immature tumor microvessels (g) and VEGF expression is also strongly positive (h). (Reprinted from Ma *et al.* Peripheral pulmonary nodules: relationship between multislice spiral CT perfusion imaging and tumor angiogenesis and VEGF expression. *BMC Cancer* 2008 with permission.)

administration of a gadolinium (Gd)-based contrast agent is an emerging imaging modality to assess tumor angiogenesis [64]. Unlike conventional enhanced MRI, DCE-MRI enables the depiction of physiologic alterations as well as morphologic changes. Intravenous contrast agents used for both conventional enhanced and DCE-MRI are mainly low molecular weight contrast media (LMCM) (e.g. Gd-DTPA, molecular weight = 567 Da). During DCE-MRI tumors demonstrate fast and high amplitude enhancement followed by a relatively rapid wash out compared to normal tissue. These imaging properties can be readily detected in larger lesions, but it is sometimes impossible to differentiate tumoral enhancement from normal tissue enhancement in small lesions or in all parts of large lesions. The dynamic imaging data obtained with DCE-MRI can generate wash in and wash out curves, which are mathematically fit to two-compartment pharmacokinetic models. For functional analysis of tissue microcirculation, Tofts *et al.* proposed the kinetic parameters K^{trans}, k_{ep}, fpV, and v_e to describe tumor and tissue permeability [65]. K^{trans} describes the transendothelial transport of the contrast medium from the vascular compartment to the tumor interstitium (wash in) and is dominated by plasma flow in low flow conditions and by permeability in high flow conditions. The second parameter, k_{ep} is the reverse transport parameter of contrast material back into the vascular space (wash out). v_e is the extravascular, extracellular volume fraction of the tumor (also known as the fraction of tumor volume occupied by the extracellular extravascular space) and fpV represents the plasma volume fraction compared to the whole tissue volume [64]. In the two-compartment model, an arterial input function (AIF) is defined in order to insure that the parameters are not influenced by contrast injection rates or the cardiovascular condition of the patient. The AIF optimizes the kinetic analysis by measuring the signal from an artery near the tumor and, if performed correctly, can normalize the scans so that they are directly comparable. Additionally, tissue T1 and the native T1 relaxivity of the tissue prior to contrast administration should be determined in order to convert the MRI signal intensity into Gd concentration, since there is no fixed relationship between the concentration of Gd-based contrast material and tissue signal intensity as there is with CT scans. After defining an AIF and establishing the precontrast T1 map, two-compartment DCE-MRI analyzes can be performed using dedicated analysis tools. Another semiquantitative parameter helpful for evaluating DCE-MRI data is the area under the gadolinium concentration curve (AUGC). The AUGC is not based on a model and therefore may be more accurate in some cases.

The parameters (K^{trans}, k_{ep}, v_e, fpV) can be obtained numerically and as color-coded tissue maps. All of these parameters appear to be higher in tumors than in surrounding tissues and they decrease dramatically after antiangiogenic treatment (Figure 13.3). DCE-MRI is not limited by ionizing radiation; however, Gd-based contrast agents are not without risk as nephrogenic systemic fibrosis has recently been reported in patients with poor renal function. DCE-MRI is difficult to compare on different MR units since different units vary in field strength and the availability of pulse sequences that influence the Gd-concentration-to-signal intensity relationship. Furthermore, there is a lack of consensus on the optimal analysis technique. This has resulted in a plethora of overlapping and methods of analysis. Despite these problems, DCE-MRI is the most widely used method for monitoring response to antiangiogenic cancer treatment because of its widespread availability and repeatability in the context of human clinical trials.

DCE-MRI can also be performed with macromolecular contrast media (MMCM), which have molecular weights ranging from 5 to 90 kDa [66]. Ultrasmall paramagnetic iron oxides, Gd albumin, Gd liposomes, Gd dendrimers, and Gd viral particles are examples of MMCM. Their large molecular diameters prolong their intravascular retention and slow their leakage from angiogenic vessels. In contrast to LMCM, MMCM can not pass through normal endothelium, but passes readily, albeit more slowly, through tumor vessels. MMCM have different pharmacokinetic properties than LMCM, therefore K^{trans} values obtained with MMCM are often lower but are less influenced by flow effects. These properties make MMCM not only potentially selective contrast agents for imaging tumor angiogenesis but also plausible drug delivery systems for more selective treatment (Figure 13.4). The

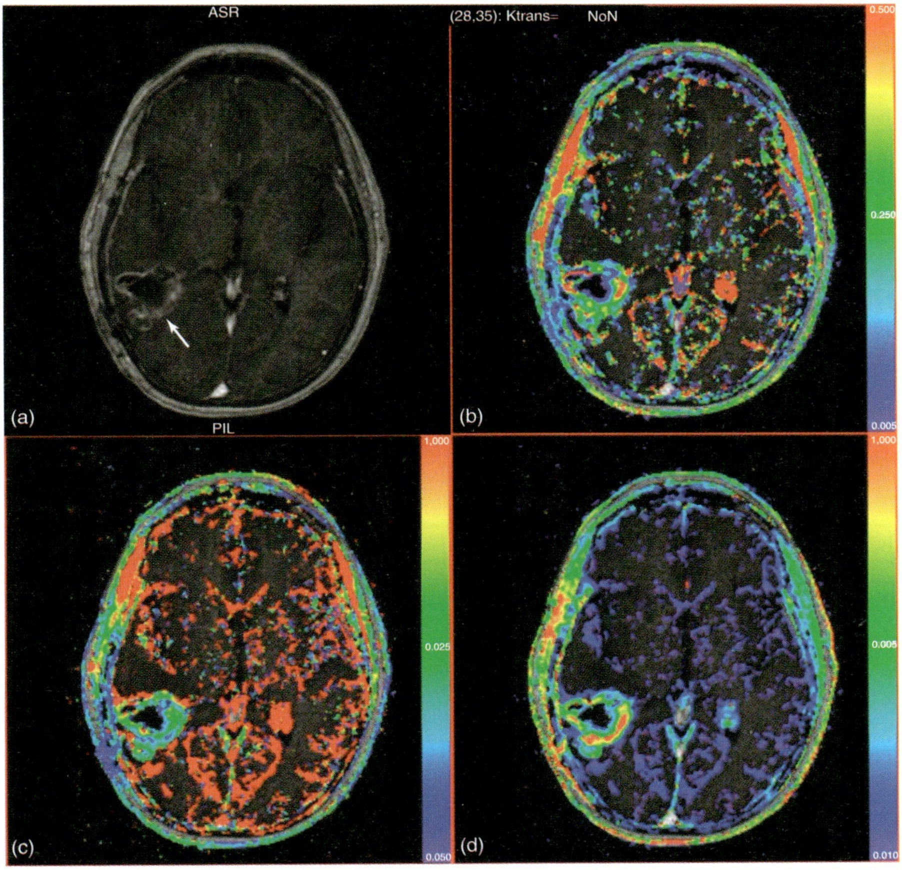

Figure 13.3 55-year-old female patient with right temporal lobe glioma. Axial Gd-enhanced MR image demonstrates the mass with peripheral enhancement (arrow) (a), K^{trans} (b), K_{ep} (c), and v_e (d) color-coded maps obtained from dynamic contrast enhanced MR imaging clearly depict the vascular changes within the tumor compared with the normal tissues.

major problem with MMCM is that they have a delayed clearance from the body, which may result in prolonged retention and more toxicity [67]. To date, only a few MMCMs have been tested in clinical trials in humans.

Positron emission tomography (PET) relies on an unstable radionuclide that emits a positron. The positron after traveling a short distance will "annihilate" with an electron from the sorrounding environment, resulting in the emission of two γ rays emerging in opposite direction which are used for generating images. By administering a small amount of a radioactive compound (radiotracer) to a subject and measuring the radioacitivity emanating from the body, three-dimensional spatial reconstructions of the radiotracer location at the time of imaging can be obtained. The intensity of the imaging signal is proportional to the amount of

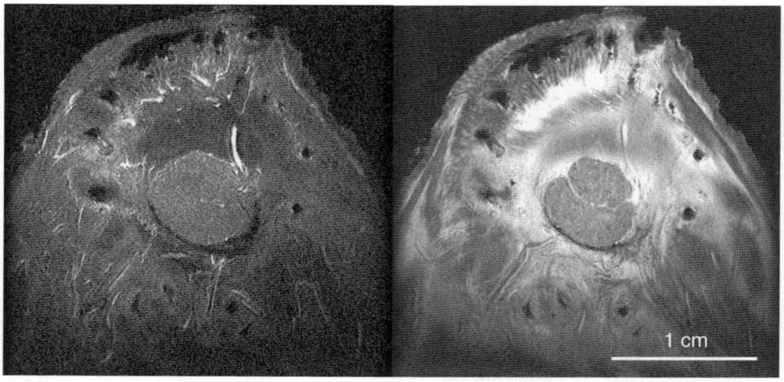

Figure 13.4 Comparison of tumor delineation and vessel enhancement between an MR image obtained with a macromolecular contrast agent, albumin-(Gd-DTPA)$_{45}$ (a), and an MR image obtained with a combination of albumin-(Gd-DTPA)$_{45}$ and a small molecular agent, gadopentetate dimeglumine (Magnevist; Schering) (b). In (b), massive extravasation of gadopentetate dimeglumine is visible, and the depiction of the internally fragmented tumor is improved. (Reprinted from Van Vliet *et al.* MR angiography of tumor-related vasculature: from the clinic to the microenvironment. *Radiographics* 2005 with permission of RSNA.)

tracer accumulating in specific lesions or tissues and semiquantitative in nature by means of calculation of a standardized uptake value [68]. For blood volume imaging using PET, red blood cells (RBCs) are labeled with inhaled ^{11}C-carbon monoxide, which binds tightly to RBCs [69]. Additionally, injection of ^{15}O-labeled water and inhalation of ^{15}O-labeled carbon dioxide have been used to study perfusion of normal and tumor tissues [70, 71]. ^{15}O-labeled water has also been used in clinical trials involving antiangiogenic therapy [72–74]. However, radioisotopes such as ^{11}C and ^{15}O have short half-life (20 and 2 min, respectively), requiring a nearby cyclotron for their production. In contrast, ^{18}Fluorine has a much longer half-life (110 min) allowing for off-site production and regional distribution. ^{18}F fluorodeoxyglucose (FDG) is the most widely used radiotracer in PET imaging, and reflects glucose metabolism of tumor cells that is increased in a variety of cancer types due to a higher rate of aerobic and anaerobic glycolysis and accelerated glucose membrane transport. The evidence supporting a link between FDG uptake and MVD is contradictory and more studies are necessary [75–80] (Figure 13.5).

13.3.3 Targeted Imaging Techniques

To this point, both anatomic and functional imaging tests have been discussed. The third approach is molecular imaging in which targeted molecules bind to highly expressed markers on the endothelium of tumor vasculature. These cell surface markers have been identified by extensive immunohistochemical analysis of cancers; however, many of them are not available as *in vivo* imaging targets

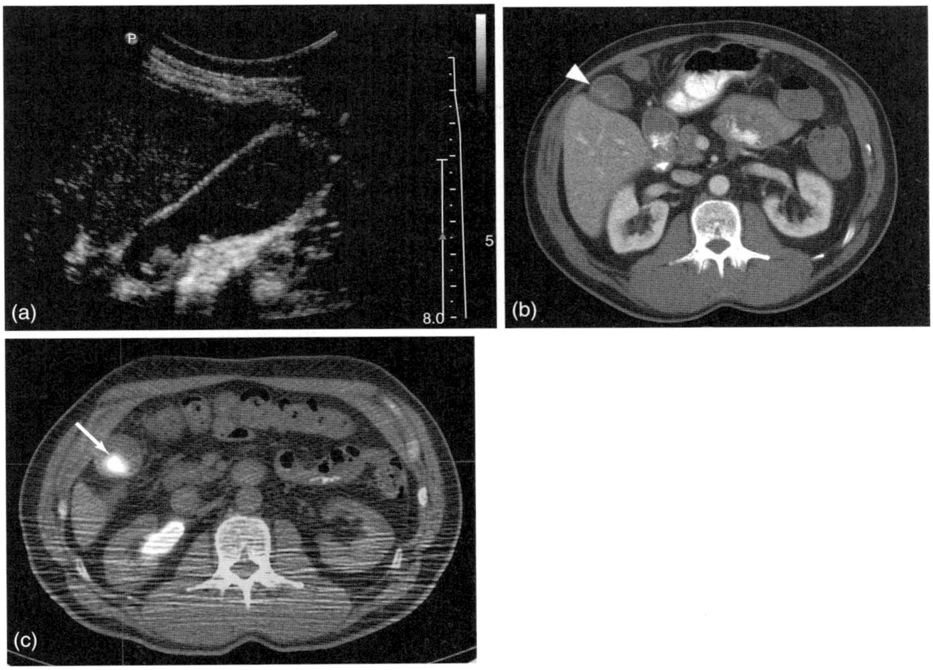

Figure 13.5 57-year-old male with melanoma. Sonography shows an echogenic mass within the gallbladder (a), axial contrast enhanced computed tomography image demonstrates mild asymmetric enhancement of posterior wall of the gall bladder suspicious for melanoma involvement (arrowhead) (b), axial PET/CT image demonstrates intense FDG uptake of the gallbladder consistent with melanoma involvement (arrow) (c).

because of limited access. Among the molecules that are potential targets for imaging are VEGF and its receptor (VEGFR-2), and several molecules from the integrin, MMP, and ECM families.

13.3.3.1 Targeted Imaging of VEGF and VEGFR-2

As previously noted, VEGF has a central role during embryonic vascular development and is an important regulator of angiogenesis. The VEGF/VEGFR interaction is one of the most extensively studied angiogenesis-related signaling pathways. VEGFR1 is important for physiologic and developmental angiogenesis, whereas VEGFR2 is vital for the mitogenic, angiogenic, and permeability-enhancing effects of VEGF. Overexpression of VEGF is a poor prognostic indicator for several cancer types.

Targeted contrast enhanced ultrasonography was reported to be effective for both *in vivo* molecular imaging and quantification of VEGFR-2 in human breast cancer, mice angiosarcoma, rat malignant glioma, and human melanoma tumor models [81–84] (Figure 13.6). Backer *et al.* tested a targeted agent in which VEGF

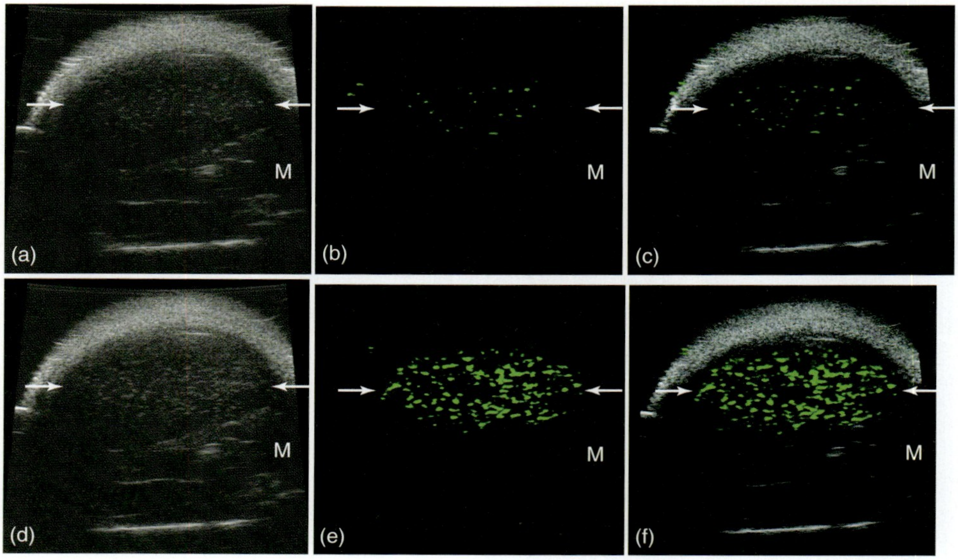

Figure 13.6 Microbubble sonography of a 67NR tumor (nonmetastatic breast cancer). Control ultrasound contrast agent (UCA) enhanced sonogram (a), color-coded background-subtracted image (b), and background-subtracted image imposed over a predestruction image (c). Vascular endothelial growth factor receptor 2–targeted UCA enhanced sonogram (d), color-coded background-subtracted image (e), and background-subtracted image imposed over a predestruction image (f). These images show a significant difference between retention of the control and VEGFR2-targeted UCAs in a small subcutaneous tumor (arrows). A difference in UCA retention between the tumor and surrounding muscle (M) was also noted. In addition, the significantly higher retention of the VEGFR2-targeted UCA in the 67NR tumor compared with the 4T1 tumor was observed. (Reprinted from Lyshchik A, *et al.* Molecular imaging of vascular endothelial growth factor receptor 2 expression using targeted contrast enhanced high-frequency ultrasonography. *J Ultrasound Med.* 2007 with permission.)

was a targeting ligand conjugated to a fifth-generation polyamidoamine dendrimer labeled with a near infrared (NIR) fluorophore. This agent demonstrated sites of overexpression of VEGFR-2 within breast tumor xenografts in mice [85].

Additionally, ^{124}I radiolabeled anti-VEGF antibodies showed limited cross-immunoreactivity in mice [86]; however, Cai *et al.* labeled a variant of VEGF, VEGF121, with ^{64}Cu in order to image tumor vascularity in mice with human glioblastoma xenografts and demonstrated that *in vivo* uptake correlated with VEGFR expression [87]. Nagengast *et al.* developed a chelated form of bevacizumab that permitted both ^{111}In (for gamma imaging) and ^{89}Zr labeling (for PET imaging) and found that there was high uptake in the human SKOV-3 ovarian tumor xenograft microenvironment as compared to the tumor itself [88]. Recently, Wang *et al.* developed a VEGFR-2-specific PET tracer, ^{64}Cu-DOTA-VEGF (DEE) in a murine breast tumor model and demonstrated its tumor-specific uptake [89].

13.3.3.2 Targeted Imaging of Integrins

The $\alpha v\beta_3$ integrin is significantly upregulated in activated endothelial cells during angiogenesis, whereas it is normally expressed in quiescent endothelial cells (AD). It binds to the arginine-glycine-aspartic acid (RGD) sequence, thereby mediating its activities. Inhibition of this interaction with monoclonal antibodies or RGD antagonists induces cellular death and inhibits angiogenesis [90]. Moreover, the interaction between the $\alpha v\beta_3$ integrin and the RGD peptide emerges as a potential candidate for targeted imaging and therapy since RGD ligands can be linked to contrast agents of imaging modalities.

Ultrasound Contrast enhanced US imaging with microbubbles labeled with peptides targeting the $\alpha v\beta_3$ integrin provide noninvasive detection of angiogenic vessels in malignant glioma, and hind limb ischemia models in mice and rats [91–93]. Such agents could be readily translated into clinical use.

Magnetic Resonance Imaging MRI can also demonstrate $\alpha v\beta_3$ integrin expression within tumors. For instance, using an $\alpha v\beta_3$-specific antibody, LM609, conjugated to a Gd containing paramagnetic liposome, Sipkins *et al.* demonstrated integrin expression in squamous cell carcinoma in a rabbit model [94]. Using a different approach, Winter *et al.* used a Gd-labeled perfluorocarbon emulsion to target the $\alpha v\beta_3$ integrin in a VX-2 rabbit tumor model on a 1.5T MR scanner. They showed increased contrast uptake within the tumor and they confirmed histopathologically that the particles were preferentially localized to the tumor [95]. Schmedider *et al.* used $\alpha v\beta_3$ integrin-targeted paramagnetic nanoparticles in mice bearing human melanoma xenografts and also demonstrated differential tumor uptake [96]. Mulder *et al.* used fluorescently labeled liposomes targeted with RGD and showed increased enhancement within areas of angiogenesis using *ex vivo* fluorescence microscopy [97]. Zhang *et al.* used ultrasmall superparamagnetic particles of iron oxides (USPIOs) coated with 3-aminopropyltrimethoxysilane (APTMS) and conjugated to RGD imaged with a 1.5T MR scanner and demonstrated tumor-specific uptake, which correlated with $\alpha v\beta_3$ integrin expression [98]. Recently, Lee *et al.* developed an iron-oxide-based nanoprobe for simultaneous PET and MRI using a polyaspartic acid (PASP)-coated iron oxide (IO) (PASP-IO) and demonstrated integrin-dependent uptake [99]. Despite these recent developments in targeted MRI of the $\alpha v\beta_3$ integrin, the relatively lower sensitivity of MRI remains a limitation; moreover, since only a small percentage of tumors are made up of vessels, it can be challenging to image integrins on the tumor neovasculature using MRI. Macromolecular MR contrast agents, higher field strengths, and newly designed coils may contribute to better targeted angiogenesis imaging on MRI.

PET/Radionuclide PET imaging, owing to its superior sensitivity, has been the leading modality for $\alpha v\beta_3$ integrin imaging. Extensive preclinical trials were performed using ^{18}F galacto RGD. Initially, Haubner *et al.* labeled RGD peptide with ^{125}I and demonstrated a high binding affinity and selectivity for $\alpha v\beta_3$ integrin both *in vitro* and *in vivo* [100]. The same group then described ^{18}F-labeling of a RGD-containing glycopeptide with 4-nitrophenyl 2-[^{18}F]fluoropropionate [101–104]. Chen *et al.* used PEGylated and polyvalenced versions of this tracer thus, improving its tumor targeting efficacy and pharmacokinetic properties [105–108] (Figure 13.7). Beer *et al.* concluded that [^{18}F]galacto RGD PET provided specific imaging of $\alpha v\beta_3$ integrin expression in neck squamous cell carcinoma and this could be used to assess angiogenesis response evaluation of integrin-targeted therapies [109]. In a study comparing tumor uptake of ^{18}F galacto RGD PET, a fluorinated integrin agonist to $\alpha v\beta_3$, and ^{18}F FDG PET, no correlation was found, but ^{18}FDG PET was more sensitive for tumor staging whereas ^{18}F galacto RGD PET was more predictive [110]. Kenny *et al.* have recently demonstrated success of a radiolabeled cyclic RGD, ^{18}F-AH111585 in detecting breast cancers in a phase 1 trial including seven patients with metastatic breast cancer [111] (Figure 13.8). ^{64}Cu-labeled RGD peptides have also been used for integrin imaging. Chen *et al.* demonstrated that the radiolabeled dimeric RGD peptides ^{64}Cu-DOTA-E[c(RGDyK)]2 and ^{64}Cu-DOTA-E[c(RGDfK)]2 had high specific tumor uptake in a human breast cancer tumor xenograft [112]. PEGylation of this ^{64}Cu-labeled RGD peptide resulted in better visualization and quantification of $\alpha v\beta_3$ integrin expression [113]. Wu *et al.* developed a tetrameric RGD peptide tracer, ^{64}Cu-DOTA-E{E[c(RGDfK)](2)}(2) with higher integrin avidity and favorable pharmacokinetics in mouse models of glioma [114]. Li *et al.* proposed that polyvalency has a profound effect on the receptor-binding affinity and *in vivo* kinetics of radiolabeled RGD multimers, which may lead to improved diagnostic and therapeutic agents [115]. The tetrameric structure of these ^{64}Cu-labeled RGD peptides not only results in higher receptor-binding affinity than does the monomer, but also results in more rapid clearance, higher metabolic stability, and significant receptor mediated tumor uptake, all of which contribute to better targeting of $\alpha v\beta_3$ integrin expression.

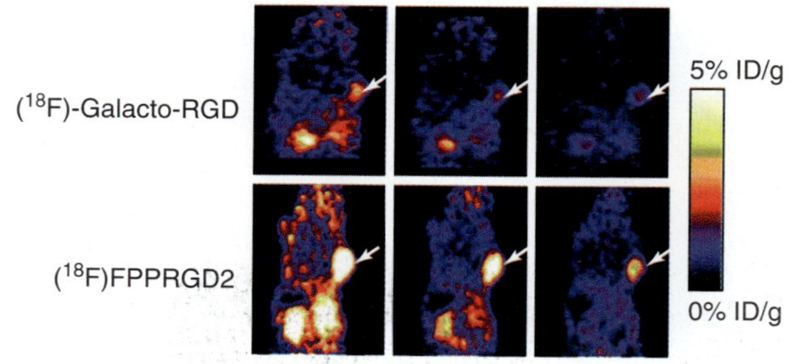

Figure 13.7 Comparison between 18F-FPPRGD2 and 18F-galacto-RGD. Decay-corrected whole body coronal images of female nude mice bearing U87MG tumors at 20 min, 1 h, and 2 h after tracer injection. The dimeric RGD peptide tracer 18F-FPPRGD2 showed significantly higher tumor activity accumulation and retention than the monomeric analog 18F-galacto-RGD (arrows). Both of these are now being tested in humans. (Courtesy of Dr. Shawn Chen from National Institute of Biomedical Imaging and Bioengineering, NIH [Liu, S., Liu, Z., Chen, K., Yan, Y., Watzlowik, P., Wester, H.-J., Chin, F.T., Chen, X. 18F-labeled galacto- and PEGylated RGD dimers for PET imaging of $\alpha v\beta 3$ integrin expression. *Mol Imaging Biol*, in press.])

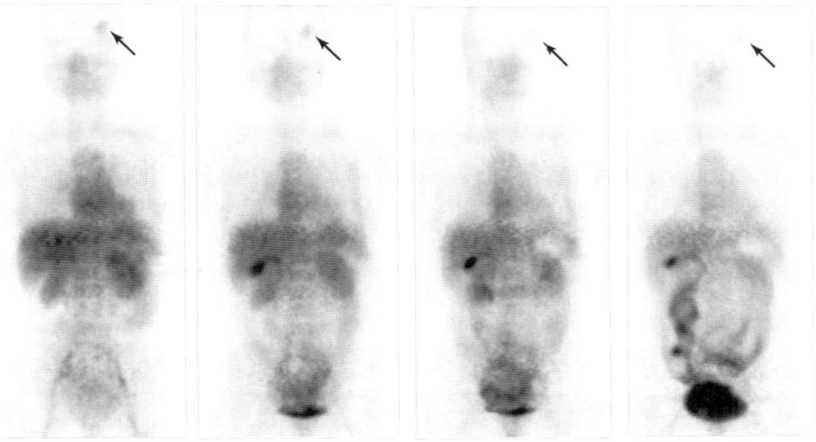

Figure 13.8 Coronal sequential ^{18}F-AH111858 PET scans in a patient shows increased uptake of the agent within the left cerebrum consistent with a primary tumor (arrows) (pathologic diagnosis: meningioma). (Courtesy of Dr. Ai Min Hui, GE Healthcare.)

^{99m}Tc-labeled RGD analogs/derivatives have also been used for imaging $\alpha v\beta_3$ integrin–positive tumors *in vivo* [116–121], but due to the chemistry required to link to ^{99m}Tc their pharmacokinetics, elimination pathways, and metabolic stability are less satisfactory than the current PET agents. ^{68}Ga- and ^{111}In-labeled DOTA-RGD peptides were recently shown to be able to visualize $\alpha v\beta_3$ integrin expression and these tracers were proposed as alternatives to ^{18}F galacto RGD peptides since their radiosynthesis is easier, and the blood pool activity is much lower than ^{18}F galacto RGD peptides [122].

Optical Imaging Optical imaging is a rapidly advancing subset of medical imaging that does not require ionizing radiation exposure and utilizes relatively low cost and portable equipment. Optical imaging probes utilize fluorophores that first must be excited with a photon before they release a photon, typically of longer wavelength (lower energy) than the excitation light. NIR fluorophores have the best tissue penetration and therefore, are preferred for transcutaneous imaging. Light of lower wavelength (e.g., green, red) does not penetrate well and light of longer wavelength (e.g., infrared) can damage tissue by heat deposition. Notwithstanding the optimal tissue penetration of NIR probes, the major limitation of optical imaging as a whole body imaging technique is its limited depth penetration, typically limited to several centimeters below the surface and even then only after considerable blurring due to light scattering. Therefore optical imaging is reserved for superficial tissues or during endoscopy or surgery. Another limitation of optical imaging is autofluorescence, which is light emitted by normal endogenous and exogenous molecules in the body. Several angiogenesis-specific optical agents have been developed for optical imaging; among them are Cy5.5-RGD peptides and the multivalent Cy5.5-regioselectively addressable functionalized template-(RAFT)-cRGD peptides [123–127]. A special kind of optical fluorophore is known as a *quantum dot*. Quantum dots (Qdots) are semiconductor crystals ranging between 5 and 20 nm in diameter, and the larger the size the longer the wavelength. They have high quantum yield and thus, are very bright. Targeting ligands can be attached to Qdots for angiogenesis imaging. Cai *et al.* targeted the tumor vasculature using RGD-labeled Qdots and demonstrated integrin expression in human glioblastoma bearing mice models [128]. However, clinical translation of Qdots is doubtful because they are often composed of heavy metals such as cadmium and selenium. Newer Qdots have potentially better biocompatibility.

13.3.3.3 Targeted Imaging of MMPs

MMPs are zinc-dependent enzymes that can selectively break down ECM and nonmatrix proteins. Various types of metastatic and invasive tumor cells such as lung, colon, breast, and pancreas cancers typically secrete MMPs, whereas normally, its secretion is at an extremely low level [129]. The best known MMPs are collagenase, stromelysins, metalloelastase, gelatinases, and membrane type MMPs. Targeted PET imaging of MMPs can be achieved by labeling MMP inhibitors with several radiotracers [130]. However, results of *in vivo* imaging trials using the MMPs as targeting moieties have been disappointing. [131–134].

13.3.3.4 Targeted Imaging of the Extracellular Matrix (ECM)

The ECM, supports cellular migration and proliferation thus promoting survival of the tumor. ECM targets include adhesion proteins, which are recruited from plasma during tumor growth. Fibronectin and vitronectin are among these adhesion proteins, and they play an important role in embryogenesis and carcinogenesis, but they are undetectable within mature vessels and normal adult tissues except those undergoing remodeling. This selective expression can be used to target tumor angiogenesis. Borsi *et al.* performed biodistribution studies in tumor-bearing mice to compare the blood clearance rate, *in vivo* stability, and performance of constructs capable of reacting with the ED-B domain of fibronectin. They showed that these tracers might be used for targeted imaging of fibronectin expression within tumor lesions [135]. Santimara *et al.* reported the use of I^{123}-labeled dimeric L19 (scFv)(2) in the detection of ED-B domain of fibronectin [136]. Berndorff *et al.* successfully targeted ED-B fibronectin expression in murine teratocarcinoma by using ^{99m}Tc-AP39 radioscintigraphy [137]. Recently, Rossin *et al.* targeted the ED-B domain of fibronectin in teratoma-bearing mice using ^{76}Br-labeled L19-SIP (fibronectin binding human antibody derivative) and demonstrated specific accumulation of the tracer within tumors [138].

Vitronectin is another adhesion protein important in ECM remodeling during angiogenesis. Its biological functions depend upon its conformation; in its native form in the plasma, vitronectin is inactive, but after tissue injury or tumor growth it is recruited from the ECM and becomes activated leading to tumor support and growth [139]. Bloemendal *et al.* used ^{123}I-labeled huMabVN18 targeting activated vitronectin in Rous sarcoma virus-induced tumors in chickens, and postulated that activated vitronectin can be selectively targeted [140]. Edwards *et al.* used the high-affinity peptide, ^{99m}Tc-NC100692 for vitronectin receptor imaging and proposed this labeled tracer as a targeting agent for vitronectin expression [141].

13.4 Conclusion

Imaging plays an important role in the early detection and treatment monitoring of tumor angiogenesis. Recent developments in conventional imaging modalities and contrast agents have shifted the focus of imaging from traditional anatomic methods to functional and molecular techniques. Preclinical testing suggests that such imaging approaches, including targeted angiogenesis-specific molecules will soon enable noninvasive surveillance of angiogenesis and may aid in the development of therapeutic agents capable of targeted and thus, personalized, cancer therapy.

References

1. Folkman, J. (1971) Tumor angiogenesis: therapeutic implications. *N. Engl. J. Med.*, **285**, 1182–1186.
2. Folkman, J. (1996) New perspectives in clinical oncology from angiogenesis research. *Eur. J. Cancer*, **32A**, 2534–2539.
3. Rak, J., Mitsuhashi, Y., Bayko, L. *et al.* (1995) Mutant ras oncogenes upregulate VEGF/VPF expression: implications for induction and inhibition of tumor angiogenesis. *Cancer Res.*, **55**, 4575–4580.
4. Tischer, E., Mitchell, R., Hartman, T. *et al.* (1991) The human gene for vascular endothelial growth factor. Multiple protein forms are encoded through alternative exon splicing. *Biol. Chem.*, **266**, 11947–11954.
5. Ferrara, N., Gerber, H.P., and LeCouter, J. (2003) The biology of VEGF and its receptors. *Nat. Med.*, **9**, 669–676.
6. Hicklin, D.J. and Ellis, L.M. (2005) Role of the vascular endothelial growth factor pathway in tumor growth and angiogenesis. *J. Clin. Oncol.*, **23**, 1011–1027.
7. Brown, L.F., Berse, B., Jackman, R.W. *et al.* (1995) Expression of vascular permeability factor (vascular endothelial growth factor) and its receptors in breast cancer. *Hum. Pathol.*, **26**, 86–91.
8. Middleton, G. and Lapka, D.V. (2004) Bevacizumab (Avastin). *Clin. J. Oncol. Nurs.*, **8**, 666–669.
9. Aita, M., Fasola, G., Defferrari, C. *et al.* (2008) Targeting the VEGF pathway: antiangiogenic strategies in the treatment of non-small cell lung cancer. *Crit. Rev. Oncol. Hematol.*, **68**, 183–196.
10. Lee, S.Y., Kim, D.K., Cho, J.H., Koh, J.Y., and Yoon, Y.H. (2008) Inhibitory effect of bevacizumab on the angiogenesis and growth of retinoblastoma. *Arch. Ophthalmol.*, **126**, 953–958.
11. Ferrara, N. (1995) The role of vascular endothelial growth factor in pathological angiogenesis. *Breast Cancer Res. Treat.*, **36**, 127–137.
12. Ferrara, N. (2002) VEGF and the quest for tumour angiogenesis factors. *Nat. Rev. Cancer*, **2**, 795–803.

13. Stupack, D.G. and Cheresh, D.A. (2004) Integrins and angiogenesis. *Curr. Top. Dev. Biol.*, **64**, 207–238.
14. Hood, J.D. and Cheresh, D.A. (2002) Role of integrins in cell invasion and migration. *Nat. Rev. Cancer*, **2**, 91–100.
15. Xiong, J.P., Stehle, T., Zhang, R. *et al.* (2002) Crystal structure of the extracellular segment of integrin alpha Vbeta3 in complex with an Arg-Gly-Asp ligand. *Science*, **296**, 151–155.
16. Gladson, C.L. (1996) Expression of integrin alpha v beta 3 in small blood vessels of glioblastoma tumors. *J. Neuropathol. Exp. Neurol.*, **55**, 1143–1149.
17. Goldberg, I., Davidson, B., Reich, R. *et al.* (2001) Alphav integrin expression is a novel marker of poor prognosis in advanced-stage ovarian carcinoma. *Clin. Cancer Res.*, **7**, 4073–4079.
18. Davidson, B., Goldberg, I., Reich, R. *et al.* (2003) AlphaV- and beta1-integrin subunits are commonly expressed in malignant effusions from ovarian carcinoma patients. *Gynecol. Oncol.*, **90**, 248–257.
19. Erdreich-Epstein, A., Shimada, H., Groshen, S. *et al.* (2000) Integrins alpha(v)beta3 and alpha(v)beta5 are expressed by endothelium of high-risk neuroblastoma and their inhibition is associated with increased endogenous ceramide. *Cancer Res.*, **60**, 712–721.
20. Parsons-Wingerter, P., Kasman, I.M., Norberg, S. *et al.* (2005) Uniform overexpression and rapid accessibility of alpha5beta1 integrin on blood vessels in tumors. *Am. J. Pathol.*, **167**, 193–211.
21. Freije, J.M., Balbín, M., Pendás, A.M. *et al.* (2003) Matrix metalloproteinases and tumor progression. *Adv. Exp. Med. Biol.*, **532**, 91–107.
22. Lafleur, M.A., Handsley, M.M., and Edwards, D.R. (2003) Metalloproteinases and their inhibitors in angiogenesis. *Expert Rev. Mol. Med.*, **5**, 1–39.
23. Ornstein, D.L. and Cohn, K.H. (2002) Balance between activation and inhibition of matrix metalloproteinase-2 (MMP-2) is altered in colorectal tumors compared to normal colonic epithelium. *Dig. Dis. Sci.*, **47**, 1821–1830.
24. Ellenrieder, V., Alber, B., Lacher, U. *et al.* (2000) Role of MT-MMPs and MMP-2 in pancreatic cancer progression. *Int. J. Cancer*, **85**, 14–20.
25. Jinga, D.C., Blidaru, A., Condrea, I. *et al.* (2006) MMP-9 and MMP-2 gelatinases and TIMP-1 and TIMP-2 inhibitors in breast cancer: correlations with prognostic factors. *J. Cell Mol. Med.*, **10**, 499–510.
26. Murray, G.I., Duncan, M.E., O'Neil, P. *et al.* (1998) Matrix metalloproteinase-1 is associated with poor prognosis in oesophageal cancer. *J. Pathol.*, **185**, 256–261.
27. Murray, G.I., Duncan, M.E., Arbuckle, E., Melvin, W.T., and Fothergill, J.E. (1998) Matrix metalloproteinases and their inhibitors in gastric cancer. *Gut*, **43**, 791–797.
28. Weidner, N. (1995) Intratumor microvessel density as a prognostic factor in cancer. *Am. J. Pathol.*, **147**, 9–19.
29. Hlatky, L., Hahnfeldt, P., and Folkman, J. (2002) Clinical application of antiangiogenic therapy: microvessel density, what it does and doesn't tell us. *J. Natl. Cancer Inst.*, **94**, 883–893.
30. Tanigawa, N., Matsumura, M., Amaya, H. *et al.* (1997) Tumor vascularity correlates with the prognosis of patients with esophageal squamous cell carcinoma. *Cancer*, **79**, 220–225.
31. Cooper, R.A., Wilks, D.P., Logue, J.P. *et al.* (1998) High tumor angiogenesis is associated with poorer survival in carcinoma of the cervix treated with radiotherapy. *Clin. Cancer Res.*, **4**, 2795–2800.
32. Inoue, K., Kamada, M., Slaton, J.W. *et al.* (2002) The prognostic value of angiogenesis and metastasis-related genes for progression of transitional cell carcinoma of the renal pelvis and ureter. *Clin. Cancer Res.*, **8**, 1863–1870.
33. Rubin, M.A., Buyyounouski, M., Bagiella, E. *et al.* (1999) Microvessel density in prostate cancer: lack of correlation with tumor grade, pathologic stage, and clinical outcome. *Urology*, **53**, 542–547.

34. MacLennan, G.T. and Bostwick, D.G. (1995) Microvessel density in renal cell carcinoma: lack of prognostic significance. *Urology*, **46**, 27–30.
35. Akslen, L.A. and Livolsi, V.A. (2000) Increased angiogenesis in papillary thyroid carcinoma but lack of prognostic importance. *Hum. Pathol.*, **31**, 439–442.
36. Paradiso, A., Ranieri, G., Silvestris, N. *et al.* (2001) Failure of primary breast cancer neoangiogenesis to predict pattern of distant metastasis. *Clin. Exp. Med.*, **1**, 127–132.
37. Barrett, T., Kobayashi, H., Brechbiel, M., and Choyke, P.L. (2006) Macromolecular MRI contrast agents for imaging tumor angiogenesis. *Eur. J. Radiol.*, **60**, 353–366.
38. Cosgrove, D. (2003) Angiogenesis imaging--ultrasound. *Br. J. Radiol.*, **76** (Spec. No. 1), S43–S49.
39. Ferrara, K.W., Merritt, C.R., Burns, P.N., Foster, F.S., Mattrey, R.F., and Wickline, S.A. (2000) Evaluation of tumor angiogenesis with US: imaging, Doppler, and contrast agents. *Acad. Radiol.*, **7**, 824–839.
40. Yang, W.T., Tse, G.M., Lam, P.K., Metreweli, C., and Chang, J. (2002) Correlation between color power Doppler sonographic measurement of breast tumor vasculature and immunohistochemical analysis of microvessel density for the quantitation of angiogenesis. *J. Ultrasound. Med.*, **21**, 1227–1235.
41. Hata, K., Nagami, H., Iida, K., Miyazaki, K., and Collins, W.P. (1998) Expression of thymidine phosphorylase in malignant ovarian tumors: correlation with microvessel density and an ultrasound-derived index of angiogenesis. *Ultrasound. Obstet. Gynecol.*, **12**, 201–206.
42. Cheng, W.F., Lee, C.N., Chu, J.S. *et al.* (1999) Vascularity index as a novel parameter for the in vivo assessment of angiogenesis in patients with cervical carcinoma. *Cancer*, **85**, 651–657.
43. Dayton, P.A., Chomas, J.E., Lum, A.F. *et al.* (2001) Optical and acoustical dynamics of microbubble contrast agents inside neutrophils. *Biophys. J.*, **80**, 1547–1556.
44. Chomas, J.E., Dayton, P., May, D., and Ferrara, K. (2001) Threshold of fragmentation for ultrasonic contrast agents. *J. Biomed. Opt.*, **6**, 141–150.
45. Kedar, R.P., Cosgrove, D., McCready, V.R., Bamber, J.C., and Carter, E.R. (1996) Microbubble contrast agent for color Doppler US: effect on breast masses. Work in progress. *Radiology*, **198**, 679–686.
46. Lagalla, R., Caruso, G., Urso, R., Bizzini, G., Marasà, L., and Miceli, V. (2000) The correlations between color Doppler using a contrast medium and the neoangiogenesis of small prostatic carcinomas. *Radiol. Med. (Torino)*, **99**, 270–275.
47. D'Arcy, T.J., Jayaram, V., Lynch, M. *et al.* (2004) Ovarian cancer detected non-invasively by contrast-enhanced power Doppler ultrasound. *BJOG*, **111**, 619–622.
48. Hotta, N., Tagaya, T., Maeno, T. *et al.* (2005) Advanced dynamic flow imaging with contrast-enhanced ultrasonography for the evaluation of tumor vascularity in liver tumors. *Clin. Imaging*, **29**, 34–41.
49. Wang, Z., Tang, J., An, L., Wang, W., Luo, Y., Li, J., and Xu, J. (2007) Contrast-enhanced ultrasonography for assessment of tumor vascularity in hepatocellular carcinoma. *J. Ultrasound Med.*, **26**, 757–762.
50. Lamuraglia, M., Escudier, B., Chami, L. *et al.* (2006) To predict progression-free survival and overall survival in metastatic renal cancer treated with sorafenib: pilot study using dynamic contrast-enhanced Doppler ultrasound. *Eur. J. Cancer*, **42**, 2472–2479.
51. Korpanty, G., Carbon, J.G., Grayburn, P.A., Fleming, J.B., and Brekken, R.A. (2007) Monitoring response to anticancer therapy by targeting microbubbles to tumor vasculature. *Clin. Cancer Res.*, **13**, 323–330.
52. Bertolotto, M., Pozzato, G., Crocè, L.S. *et al.* (2006) Blood flow changes in hepatocellular carcinoma after the administration of thalidomide assessed by

reperfusion kinetics during microbubble infusion: preliminary results. *Invest. Radiol.*, **41**, 15–21.

53. Meyerowitz, C.B., Fleischer, A.C., Pickens, D.R. *et al.* (1996) Quantification of tumor vascularity and flow with amplitude color Doppler sonography in an experimental model: preliminary results. *J. Ultrasound Med.*, **15**, 827–833.
54. Goh, V., Padhani, A.R., and Rasheed, S. (2007) Functional imaging of colorectal cancer angiogenesis. *Lancet Oncol.*, **8**, 245–255.
55. Miles, K.A. (2003) Perfusion CT for the assessment of tumour vascularity: which protocol?. *Br. J. Radiol.*, **76** (Spec No. 1), S36–S42.
56. Perini, R., Choe, R., Yodh, A.G. *et al.* (2008) Non-invasive assessment of tumor neovasculature: techniques and clinical applications. *Cancer Metastasis Rev.*, **27**, 615–630.
57. Ma, S.H., Le, H.B., Jia, B.H. *et al.* (2008) Peripheral pulmonary nodules: relationship between multi-slice spiral CT perfusion imaging and tumor angiogenesis and VEGF expression. *BMC Cancer*, **8**, 186.
58. Ippolito, D., Sironi, S., Pozzi, M. *et al.* (2008) Hepatocellular carcinoma in cirrhotic liver disease: functional computed tomography with perfusion imaging in the assessment of tumor vascularization. *Acad. Radiol.*, **15**, 919–927.
59. Sabir, A., Schor-Bardach, R., Wilcox, C.J. *et al.* (2008) Perfusion MDCT enables early detection of therapeutic response to antiangiogenic therapy. *AJR Am. J. Roentgenol.*, **191**, 133–139.
60. Li, Z.P., Meng, Q.F., Sun, C.H. *et al.* (2005) Tumor angiogenesis and dynamic CT in colorectal carcinoma: radiologic-pathologic correlation. *World J. Gastroenterol.*, **11**, 1287–1291.
61. Haider, M.A., Milosevic, M., Fyles, A. *et al.* (2005) Assessment of the tumor microenvironment in cervix cancer using dynamic contrast enhanced CT, interstitial fluid pressure and oxygen measurements. *Int. J. Radiat. Oncol. Biol. Phys.*, **62**, 1100–1107.
62. Koukourakis, M.I., Mavanis, I., Kouklakis, G. *et al.* (2007) Early antivascular effects of bevacizumab anti-VEGF monoclonal antibody on colorectal carcinomas assessed with functional CT imaging. *Am. J. Clin. Oncol.*, **30**, 315–318.
63. Makari, Y., Yasuda, T., Doki, Y. *et al.* (2007) Correlation between tumor blood flow assessed by perfusion CT and effect of neoadjuvant therapy in advanced esophageal cancers. *J. Surg. Oncol.*, **96**, 220–229.
64. Choyke, P.L., Dwyer, A.J., and Knopp, M.V. (2003) Functional tumor imaging with dynamic contrast-enhanced magnetic resonance imaging. *J. Magn. Reson. Imaging*, **17**, 509–520.
65. Tofts, P.S., Brix, G., Buckley, D.L. *et al.* (1999) Estimating kinetic parameters from dynamic contrast-enhanced T(1)-weighted MRI of a diffusable tracer: standardized quantities and symbols. *J. Magn. Reson. Imaging*, **10**, 223–232.
66. Padhani, A.R. (2003) MRI for assessing antivascular cancer treatments. *Br. J. Radiol.*, **76** (Spec. No. 1), S60–S80.
67. Barrett, T., Brechbiel, M., Bernardo, M., and Choyke, P.L. (2007) MRI of tumor angiogenesis. *J. Magn. Reson. Imaging*, **26**, 235–249.
68. Huang, S.C. (2000) Anatomy of SUV. Standardized uptake value. *Nucl. Med. Biol.*, **27**, 643–646.
69. Mintun, M.A., Raichle, M.E., Martin, W.R., and Herscovitch, P. (1984) Brain oxygen utilization measured with O-15 radiotracers and positron emission tomography. *J. Nucl. Med.*, **25**, 177–187.
70. West, J.B. and Dollery, C.T. (1962) Uptake of oxygen-15-labeled CO2 compared with carbon-11-labeled CO2 in the lung. *J. Appl. Physiol.*, **17**, 9–13.
71. Taniguchi, H., Kunishima, S., and Koh, T. (2003) The reproducibility of independently measuring human regional hepatic arterial, portal and total hepatic blood flow using [15O]water and positron emission tomography. *Nucl. Med. Commun.*, **24**, 497–501.
72. Herbst, R.S., Mullani, N.A., Davis, D.W. *et al.* (2002) Development of

biologic markers of response and assessment of antiangiogenic activity in a clinical trial of human recombinant endostatin. *J. Clin. Oncol.*, **20**, 3804–3814.

73. Lara, P.N. Jr., Quinn, D.I., Margolin, K. *et al.* (2003) SU5416 plus interferon alpha in advanced renal cell carcinoma: a phase II California Cancer Consortium Study with biological and imaging correlates of angiogenesis inhibition. *Clin. Cancer Res.*, **9**, 4772–4781.

74. Anderson, H., Yap, J.T., Wells, P. *et al.* (2003) Measurement of renal tumour and normal tissue perfusion using positron emission tomography in a phase II clinical trial of razoxane. *Br. J. Cancer*, **89**, 262–267.

75. Cho, S.M., Park, Y.G., Lee, J.M. *et al.* (2007) 18F-fluorodeoxyglucose positron emission tomography in patients with recurrent ovarian cancer: in comparison with vascularity, Ki-67, p53, and histologic grade. *Eur. Radiol.*, **17**, 409–417.

76. Bos, R., van Der Hoeven, J.J., van Der Wall, E. *et al.* (2002) Biologic correlates of (18)fluorodeoxyglucose uptake in human breast cancer measured by positron emission tomography. *J. Clin. Oncol.*, **20**, 379–387.

77. Zasadny, K.R., Tatsumi, M., and Wahl, R.L. (2003) FDG metabolism and uptake versus blood flow in women with untreated primary breast cancers. *Eur. J. Nucl. Med. Mol. Imaging*, **30**, 274–280.

78. Guo, J., Higashi, K., Ueda, Y. *et al.* (2006) Microvessel density: correlation with 18F-FDG uptake and prognostic impact in lung adenocarcinomas. *J. Nucl. Med.*, **47**, 419–425.

79. Buck, A.K. and Reske, S.N. (2004) Cellular origin and molecular mechanisms of 18F-FDG uptake: is there a contribution of the endothelium? *J. Nucl. Med.*, **45**, 461–463.

80. Veronesi, G., Landoni, C., Pelosi, G. *et al.* (2002) Fluoro-deoxi-glucose uptake and angiogenesis are independent biological features in lung metastases. *Br. J. Cancer*, **86**, 1391–1395.

81. Lyshchik, A., Fleischer, A.C., Huamani, J., Hallahan, D.E., Brissova, M., and Gore, J.C. (2007) Molecular imaging of vascular endothelial growth factor receptor 2 expression using targeted contrast-enhanced high-frequency ultrasonography. *J. Ultrasound Med.*, **26**, 1575–1586.

82. Lee, D.J., Lyshchik, A., Huamani, J., Hallahan, D.E., and Fleischer, A.C. (2008) Relationship between retention of a vascular endothelial growth factor receptor 2 (VEGFR2)-targeted ultrasonographic contrast agent and the level of VEGFR2 expression in an in vivo breast cancer model. *J. Ultrasound Med.*, **27**, 855–866.

83. Willmann, J.K., Paulmurugan, R., Chen, K. *et al.* (2008) US imaging of tumor angiogenesis with microbubbles targeted to vascular endothelial growth factor receptor type 2 in mice. *Radiology*, **246**, 508–518.

84. Rychak, J.J., Graba, J., Cheung, A.M. *et al.* (2007) Microultrasound molecular imaging of vascular endothelial growth factor receptor 2 in a mouse model of tumor angiogenesis. *Mol. Imaging*, **6**, 289–296.

85. Backer, M.V., Gaynutdinov, T.I., Patel, V. *et al.* (2005) Vascular endothelial growth factor selectively targets boronated dendrimers to tumor vasculature. *Mol. Cancer Ther.*, **4**, 1423–1429.

86. Collingridge, D.R., Carroll, V.A., Glaser, M. *et al.* (2002) The development of [(124)I]iodinated-VG76e: a novel tracer for imaging vascular endothelial growth factor in vivo using positron emission tomography. *Cancer Res.*, **62**, 5912–5919.

87. Cai, W., Chen, K., Mohamedali, K.A. *et al.* (2006) PET of vascular endothelial growth factor receptor expression. *J. Nucl. Med.*, **47**, 2048–2056.

88. Nagengast, W.B., de Vries, E.G., Hospers, G.A. *et al.* (2007) In vivo VEGF imaging with radiolabeled bevacizumab in a human ovarian tumor xenograft. *J. Nucl. Med.*, **48**, 1313–1319.

89. Wang, H., Cai, W., Chen, K. *et al.* (2007) A new PET tracer specific for vascular endothelial growth factor

receptor 2. *Eur. J. Nucl. Med. Mol. Imaging*, **34**, 2001–2010.

90. Cai, W. and Chen, X. (2006) Anti-angiogenic cancer therapy based on integrin alphavbeta3 antagonism. *Anticancer Agents Med. Chem.*, **6**, 407–428.

91. Ellegala, D.B., Leong-Poi, H., Carpenter, J.E. *et al.* (2003) Imaging tumor angiogenesis with contrast ultrasound and microbubbles targeted to alpha(v)beta3. *Circulation*, **108**, 336–341.

92. Leong-Poi, H., Christiansen, J., Klibanov, A.L., Kaul, S., and Lindner, J.R. (2003) Noninvasive assessment of angiogenesis by ultrasound and microbubbles targeted to alpha(v)-integrins. *Circulation*, **107**, 455–460.

93. Leong-Poi, H., Christiansen, J., Heppner, P. *et al.* (2005) Assessment of endogenous and therapeutic arteriogenesis by contrast ultrasound molecular imaging of integrin expression. *Circulation*, **111**, 3248–3254.

94. Sipkins, D.A., Cheresh, D.A., Kazemi, M.R., Nevin, L.M., Bednarski, M.D., and Li, K.C. (1998) Detection of tumor angiogenesis in vivo by alphaVbeta3-targeted magnetic resonance imaging. *Nat. Med.*, **4**, 623–626.

95. Winter, P.M., Caruthers, S.D., Kassner, A. *et al.* (2003) Molecular imaging of angiogenesis in nascent Vx-2 rabbit tumors using a novel alpha(nu)beta3-targeted nanoparticle and 1.5 tesla magnetic resonance imaging. *Cancer Res.*, **63**, 5838–5843.

96. Schmieder, A.H., Winter, P.M., Caruthers, S.D. *et al.* (2005) Molecular MR imaging of melanoma angiogenesis with alphanubeta3-targeted paramagnetic nanoparticles. *Magn. Reson. Med.*, **53**, 621–627.

97. Mulder, W.J., Strijkers, G.J., Habets, J.W. *et al.* (2005) MR molecular imaging and fluorescence microscopy for identification of activated tumor endothelium using a bimodal lipidic nanoparticle. *FASEB J.*, **19**, 2008–2010.

98. Zhang, C., Jugold, M., Woenne, E.C. *et al.* (2007) Specific targeting of tumor angiogenesis by RGD-conjugated ultrasmall superparamagnetic iron oxide particles using a clinical 1.5-T magnetic resonance scanner. *Cancer Res.*, **67**, 1555–1562.

99. Lee, H.Y., Li, Z., Chen, K. *et al.* (2008) PET/MRI dual-modality tumor imaging using arginine-glycine-aspartic (RGD)-conjugated radiolabeled iron oxide nanoparticles. *J. Nucl. Med.*, **49**, 1371–1379.

100. Haubner, R., Wester, H.J., Reuning, U. *et al.* (1999) Radiolabeled alpha(v)beta3 integrin antagonists: a new class of tracers for tumor targeting. *J. Nucl. Med.*, **40**, 1061–1071.

101. Haubner, R., Wester, H.J., Weber, W.A. *et al.* (2001) Noninvasive imaging of alpha(v)beta3 integrin expression using 18F-labeled RGD-containing glycopeptide and positron emission tomography. *Cancer Res.*, **61**, 1781–1785.

102. Haubner, R., Weber, W.A., Beer, A.J. *et al.* (2005) Noninvasive visualization of the activated alphavbeta3 integrin in cancer patients by positron emission tomography and [18F]Galacto-RGD. *PLoS Med.*, **2**, e70.

103. Beer, A.J., Haubner, R., Goebel, M., Luderschmidt, S. *et al.* (2005) Biodistribution and pharmacokinetics of the alphavbeta3-selective tracer 18F-galacto-RGD in cancer patients. *J. Nucl. Med.*, **46**, 1333–1341.

104. Beer, A.J., Haubner, R., Sarbia, M. *et al.* (2006) Positron emission tomography using [18F]Galacto-RGD identifies the level of integrin alpha(v)beta3 expression in man. *Clin. Cancer Res.*, **12**, 3942–3949.

105. Chen, X., Park, R., Hou, Y. *et al.* (2004) MicroPET imaging of brain tumor angiogenesis with 18F-labeled PEGylated RGD peptide. *Eur. J. Nucl. Med. Mol. Imaging*, **31**, 1081–1089.

106. Chen, X., Park, R., Shahinian, A.H. *et al.* (2004) 18F-labeled RGD peptide: initial evaluation for imaging brain tumor angiogenesis. *Nucl. Med. Biol.*, **31**, 179–189.

107. Chen, X., Park, R., Tohme, M., Shahinian, A.H., Bading, J.R., and Conti, P.S. (2004) MicroPET and autoradiographic imaging of breast cancer

alpha v-integrin expression using 18F- and 64Cu-labeled RGD peptide. *Bioconjug. Chem.*, **15**, 41–49.

108. Chen, X., Tohme, M., Park, R., Hou, Y., Bading, J.R., and Conti, P.S. (2004) Micro-PET imaging of alphavbeta3-integrin expression with 18F-labeled dimeric RGD peptide. *Mol. Imaging*, **3**, 96–104.

109. Beer, A.J., Grosu, A.L., Carlsen, J. *et al.* (2007) [18F]galacto-RGD positron emission tomography for imaging of alphavbeta3 expression on the neovasculature in patients with squamous cell carcinoma of the head and neck. *Clin. Cancer Res.*, **13**, 6610–6616.

110. Beer, A.J., Lorenzen, S., Metz, S. *et al.* (2008) Comparison of integrin alphaVbeta3 expression and glucose metabolism in primary and metastatic lesions in cancer patients: a PET study using 18F-galacto-RGD and 18F-FDG. *J. Nucl. Med.*, **49**, 22–29.

111. Kenny, L.M., Coombes, R.C., Oulie, I. *et al.* (2008) Phase I trial of the positron-emitting Arg-Gly-Asp (RGD) peptide radioligand 18F-AH111585 in breast cancer patients. *J. Nucl. Med.*, **49**, 879–886.

112. Chen, X., Liu, S., Hou, Y. *et al.* (2004) MicroPET imaging of breast cancer alphav-integrin expression with 64Cu-labeled dimeric RGD peptides. *Mol. Imaging Biol.*, **6**, 350–359.

113. Chen, X., Hou, Y., Tohme, M. *et al.* (2004) Pegylated Arg-Gly-Asp peptide: 64Cu labeling and PET imaging of brain tumor alphavbeta3-integrin expression. *J. Nucl. Med.*, **45**, 1776–1783.

114. Wu, Y., Zhang, X., Xiong, Z. *et al.* (2005) MicroPET imaging of glioma integrin $\alpha v\beta 3$ expression using (64)Cu-labeled tetrameric RGD peptide. *J. Nucl. Med.*, **46**, 1707–1718.

115. Li, Z.B., Cai, W., Cao, Q. *et al.* (2007) (64)Cu-labeled tetrameric and octameric RGD peptides for small-animal PET of tumor alpha(v)beta(3) integrin expression. *J. Nucl. Med.*, **48**, 1162–1171.

116. Su, Z.F., He, J., Rusckowski, M., and Hnatowich, D.J. (2003) In vitro cell studies of technetium-99m labeled RGD-HYNIC peptide, a comparison of tricine and EDDA as co-ligands. *Nucl. Med. Biol.*, **30**, 141–149.

117. Haubner, R., Bruchertseifer, F., Bock, M., Kessler, H., Schwaiger, M., and Wester, H.J. (2004) Synthesis and biological evaluation of a (99m)Tc-labeled cyclic RGD peptide for imaging the alphavbeta3 expression. *Nuklearmedizin*, **43**, 26–32.

118. Psimadas, D., Fani, M., Zikos, C., Xanthopoulos, S., Archimandritis, S.C., and Varvarigou, A.D. (2006) Study of the labeling of two novel RGD-peptidic derivatives with the precursor [99mTc(H2O)3(CO)3]+ and evaluation for early angiogenesis detection in cancer. *Appl. Radiat. Isot.*, **64**, 151–159.

119. Fani, M., Psimadas, D., Zikos, C. *et al.* (2006) Comparative evaluation of linear and cyclic 99mTc-RGD peptides for targeting of integrins in tumor angiogenesis. *Anticancer Res.*, **26**, 431–434.

120. Jung, K.H., Lee, K.H., Paik, J.Y. *et al.* (2006) Favorable biokinetic and tumor-targeting properties of 99mTc-labeled glucosamino RGD and effect of paclitaxel therapy. *J. Nucl. Med.*, **47**, 2000–2007.

121. Sancey, L., Ardisson, V., Riou, L.M. *et al.* (2007) In vivo imaging of tumour angiogenesis in mice with the alpha(v)beta (3) integrin-targeted tracer 99mTc-RAFT-RGD. *Eur. J. Nucl. Med. Mol. Imaging*, **34**, 2037–2047.

122. Decristoforo, C., Hernandez Gonzalez, I. *et al.* (2008) (68)Ga- and (111)In-labeled DOTA-RGD peptides for imaging of alphavbeta3 integrin expression. *Eur. J. Nucl. Med. Mol. Imaging*, **35**, 1507–1515.

123. Chen, X., Conti, P.S., and Moats, R.A. (2004) In vivo near-infrared fluorescence imaging of integrin alphavbeta3 in brain tumor xenografts. *Cancer Res.*, **64**, 8009–8014.

124. Jin, Z.H., Josserand, V., Foillard, S. *et al.* (2007) In vivo optical imaging of integrin alphaV-beta3 in mice using multivalent or monovalent cRGD targeting vectors. *Mol. Cancer*, **6**, 41.

125. Hsu, A.R., Hou, L.C., Veeravagu, A. *et al.* (2006) In vivo near-infrared

fluorescence imaging of integrin alphavbeta3 in an orthotopic glioblastoma model. *Mol. Imaging Biol.*, **8**, 315–323.

126. Cheng, Z., Wu, Y., Xiong, Z., Gambhir, S.S., and Chen, X. (2005) Near-infrared fluorescent RGD peptides for optical imaging of integrin alphavbeta3 expression in living mice. *Bioconjug. Chem.*, **16**, 1433–1441.

127. Aina, O.H., Marik, J., Gandour-Edwards, R., and Lam, K.S. (2005) Near-infrared optical imaging of ovarian cancer xenografts with novel alpha 3-integrin binding peptide "OA02". *Mol. Imaging*, **4**, 439–447.

128. Cai, W., Shin, D.W., Chen, K. *et al.* (2006) Peptide-labeled near-infrared quantum dots for imaging tumor vasculature in living subjects. *Nano Lett.*, **6**, 669–676.

129. Li, W.P. and Anderson, C.J. (2003) Imaging matrix metalloproteinase expression in tumors. *Q. J. Nucl. Med.*, **47**, 201–208.

130. Cai, W., Rao, J., Gambhir, S.S., and Chen, X. (2006) How molecular imaging is speeding up antiangiogenic drug development. *Mol. Cancer Ther.*, **5**, 2624–2633.

131. Furumoto, S., Takashima, K., Kubota, K., Ido, T., Iwata, R., and Fukuda, H. (2003) Tumor detection using 18F-labeled matrix metalloproteinase-2 inhibitor. *Nucl. Med. Biol.*, **30**, 119–125.

132. Giersing, B.K., Rae, M.T., CarballidoBrea, M., Williamson, R.A., and Blower, P.J. (2001) Synthesis and characterization of 111In-DTPA-N-TIMP-2: a radiopharmaceutical for imaging matrix metalloproteinase expression. *Bioconjug. Chem.*, **12**, 964–971.

133. Medina, O.P., Kairemo, K., Valtanen, H. *et al.* (2005) Radionuclide imaging of tumor xenografts in mice using a gelatinase-targeting peptide. *Anticancer Res.*, **25**, 33–42.

134. Zheng, Q.H., Fei, X., Liu, X. *et al.* (2004) Comparative studies of potential cancer biomarkers carbon-11 labeled MMP inhibitors (S)-2-(4'-[11C]methoxybiphenyl-4-sulfonylamino)-3-methylbutyric acid and N-hydroxy-(R)-2-[[(4'-[11C]methoxyphenyl)sulfonyl]benzylamino]-3-methylbutanamide. *Nucl. Med. Biol.*, **31**, 77–85.

135. Borsi, L., Balza, E., Bestagno, M. *et al.* (2002) Selective targeting of tumoral vasculature: comparison of different formats of an antibody (L19) to the ED-B domain of fibronectin. *Int. J. Cancer*, **102**, 75–85.

136. Santimaria, M., Moscatelli, G., Viale, G.L. *et al.* (2003) Immunoscintigraphic detection of the ED-B domain of fibronectin, a marker of angiogenesis, in patients with cancer. *Clin. Cancer Res.*, **9**, 571–579.

137. Berndorff, D., Borkowski, S., Moosmayer, D. *et al.* (2006) Imaging of tumor angiogenesis using 99mTc-labeled human recombinant anti-ED-B fibronectin antibody fragments. *J. Nucl. Med.*, **47**, 1707–1716.

138. Rossin, R., Berndorff, D., Friebe, M., Dinkelborg, L.M., and Welch, M.J. (2007) Small-animal PET of tumor angiogenesis using a (76)Br-labeled human recombinant antibody fragment to the ED-B domain of fibronectin. *J. Nucl. Med.*, **48**, 1172–1179.

139. Preissner, K.T. and Seiffert, D. (1998) Role of vitronectin and its receptors in haemostasis and vascular remodeling. *Thromb. Res.*, **89**, 1–21.

140. Bloemendal, H.J., de Boer, H.C., Koop, E.A. *et al.* (2004) Activated vitronectin as a target for anticancer therapy with human antibodies. *Cancer Immunol. Immunother.*, **53**, 799–808.

141. Edwards, D., Jones, P., Haramis, H. *et al.* (2008) 99mTc-NC100692–a tracer for imaging vitronectin receptors associated with angiogenesis: a preclinical investigation. *Nucl. Med. Biol.*, **35**, 365–375.

14
Identifying Biomarkers to Establish Drug Efficacy

J. Suso Platero

14.1
Introduction

14.1.1
Biomarkers

The term *personalized medicine* has been overemphasized for the past several years. The idea behind this notion is that, in the near future, we would be able to tailor specific drug therapies to individual patients. A basic principle of this idea is the ability to find the specific patients that will respond to a drug among all the other patients that have similar medical characteristics. One way to accomplish this is by the use of biomarkers. *Biomarkers* can be loosely defined as any test that will help identify a population of patients. A validated biomarker is defined by the Food and Drug Administration (FDA) "if it is measured in an analytical test system with well-established performance characteristics and there is an established scientific framework or body of evidence that elucidates the physiologic, pharmacologic, toxicologic, or clinical significance of the test results" [1]. Nowadays, there are lots of biomarkers but a few that are validated. Therefore, the promise of personalized medicine has not lived up to the hype that this term has had in the press. There are some examples of these biomarkers in today's world that are validated, like the use of Her2/neu in breast cancer. If the test is positive (Figure 14.1 for an example of a positive staining for the test), breast cancer patients are put in therapy with transtuzumab or herceptin. Another validated biomarker is the presence of the Philadelphia chromosome in chronic myelogenous leukemia (CML) patients. This marker will lead those patients to be treated with Gleevec or Sprycel. Unfortunately, the number of examples is small compared with the number of drugs in the market today, thus keeping us apart from the promise of personalized medicine. The main reason for this discrepancy is that the work of discovering and validating biomarkers is much harder than most people anticipate.

Tumor Angiogenesis – From Molecular Mechanisms to Targeted Therapy. Edited by Francis S. Markland, Stephen Swenson, and Radu Minea
Copyright © 2010 WILEY-VCH Verlag GmbH & Co. KGaA, Weinheim
ISBN: 978-3-527-32091-2

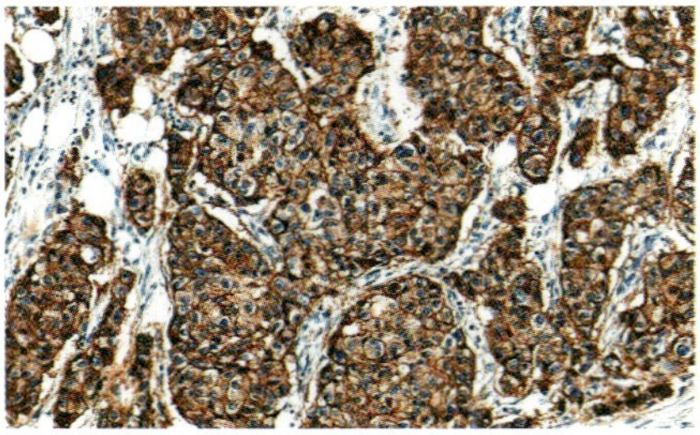

Figure 14.1 Her2/neu immunostaining of a breast cancer sample. Note the brownish colored chicken wire staining around the cancer cells indicating that the tumor is positive for this marker.

14.1.2 Biomarkers in Drug Development

Within the context of drug discovery and development, biomarkers could be classified as follows:

- Predictive biomarkers are those that help differentiate which patients are likely to benefit from a specific therapy. The above examples of Her2/neu and the Philadelphia chromosome are examples of this type of biomarker. They are used to decide which patients to treat with a specific drug.
- Pharmacodynamic biomarkers are those that change upon administration of a drug. One example of this is the change observed in vascular endothelial growth factor (VEGF) upon the administration of Avastin [2]. The higher the dose of the compound, the greater the changed observed with the biomarker, thus establishing the correlation of change in biomarker with increase in the amount of compound.
- A third class could be considered as toxicity biomarkers: those biomarkers that when present increase the side effects of specific treatments. Most of these biomarkers are SNPs (single nucleotide polymorphisms) identified in drug transport genes, like CYP2C9. The presence of this polymorphism will indicate that a change in the amount of drug given to patients is needed to eliminate unwarranted toxic effects. Warfarin is a drug that has this type of biomarkers. Patients who are homozygous for the CYP2C9 SNP need to be treated with lower doses of the drug than those that are heterozygous, due to an increase in the risk of bleeding in those with the mutation.
- Mechanisms of action biomarkers are those that indicate the pathway through which the drug is acting. In the example above regarding the VEGF compound, seeing the increase of the target in plasma could be one of these types of biomarkers.

All the different types of biomarkers could be used in the course of drug development. While not all may be useful all the time, the selection of some of them could help in the development of a compound. Moreover, biomarkers increase the value of drugs since they can help target the right population of patients, rapidly define the drug dose in early clinical trials, and keep the side effects associated with any drug to a minimum; all these will increase the efficacy of the drug. These benefits will lead to a faster, more rational, and efficient way of conducting clinical trials in the near future. Thus, saving time and money to the pharmaceutical companies that are developing these compounds and delivering to the patients' drugs that specifically treat each patient's disease.

So how do we find these biomarkers? In the rest of the chapter, I will discuss several ways that have been employed to identify biomarkers and how, in some occasions, they have been useful in the development of a drug.

14.2
Pathway Analysis as a Tool to Identify Biomarkers

One of the methods that is often forgotten is the old and time-tested method of looking at the pathway in which a specific compound has its action. Interestingly, this approach has led to the identification of both pharmacodynamic and predictive biomarkers. Perhaps, due to the increased number of articles and books published, and the increase in the complexity of pathways, makes this task intimidating. To help in this task, new search engines like PubMed, from National Cancer Institute (NCI) or Ingenuity Pathway Analysis, from Ingenuity Systems, have sprung up to help the researcher navigate the published literature. PubMed is a free service that allows one to search published articles regarding a specific topic or author. The search will point out the papers in which the terms used in the search appear. Then, the work of reading and extracting the information are left for the individual. Ingenuity goes a step further; it will point to you the literature and will build a network using the information provided in such papers. An example of this is shown in Figure 14.2, where the fibroblast growth factor receptor (FGFR) pathway was built using the Ingenuity Pathway Analysis tool.

14.3
Transcriptional Profiling as a Way to Find Biomarkers

The omics revolution that has taken place in the last few years has also given us new ways to search for biomarkers. It allows us now to look at the totality of proteins or mRNAs present in a cell type. Transcriptional profiling is the technique that permits us to compare mRNA profiles in different experimental situations [3]. The interesting part is that the totality of the gene expression in a cell line can be ascertained and compared with other cell types with this experimental technique. Even human samples have been used as the source of mRNA for transcriptional profiling experiments.

This has led to an explosion of the use of this technology in human studies, resulting in viewing human malignancies in a different light. Now, not only we have

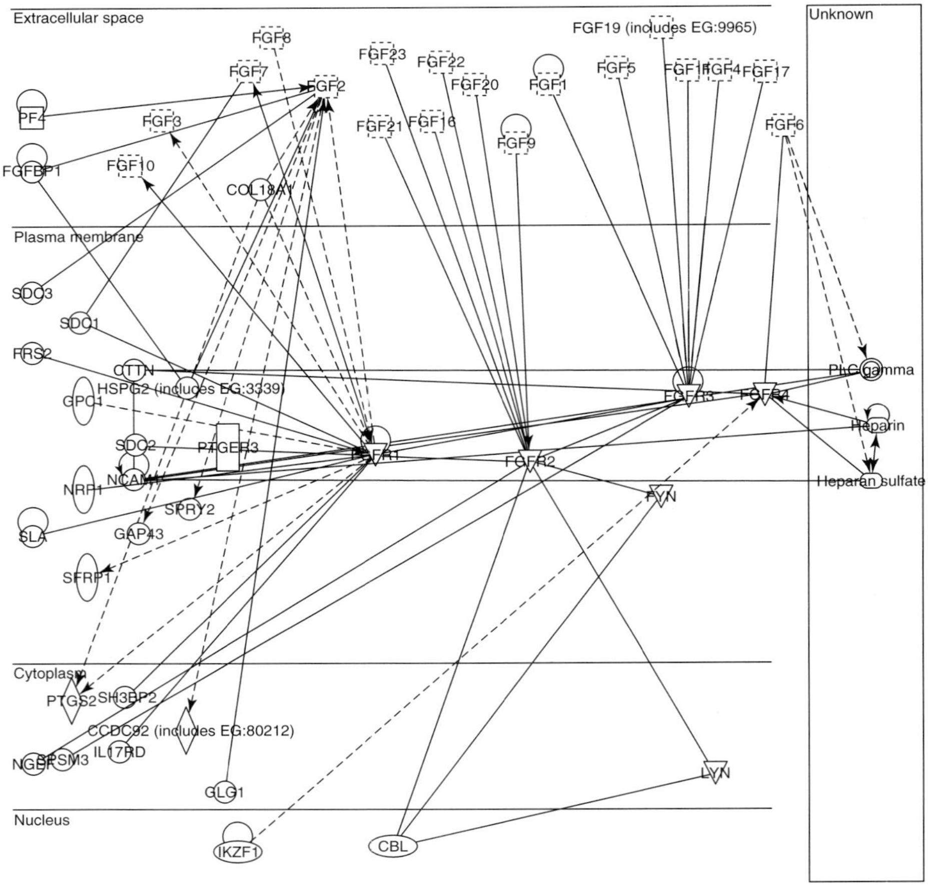

Figure 14.2 FGFR pathway. Use of ingenuity software to depict the FGFR pathway with upstream and downstream proteins involved in the pathway.

access to classical clinical and pathological characterizations of human disease, but also to the whole transcriptome analysis. This has ushered in a new area in human classification of diseases at the molecular level, from classifying the different types of breast cancers [4] to developing a molecular signature that will predict which breast cancer patients are likely to survive [5], to diagnosing lymphoid malignancies [6]. The idea behind classification of the different diseases is that it will lead to a better treatment for each of the malignancies. This idea indeed is bearing fruit as Pusztaia and colleagues have used transcriptional profiling to discover that the overexpression of τ protein in samples from breast cancer patients leads to a greater sensitivity of the cancer to paclitaxel. The result was confirmed by reducing the level of τ via RNA inhibition in cell lines which in itself led to a greater sensitivity of a resistant cell line [7].

14.4 Finding Biomarkers through the Use of Cell Lines

Cell lines are a quick and easy way to find if a specific drug possesses cell killing properties. A cell can be classified as sensitive to a compound if it has the ability to stop cell growth and induce apoptosis. This property can be quantified using the IC50 values, or amount of compound at which 50% of the cells are killed. By finding cells that are sensitive and resistant to a compound, their mRNA can be used as a starting material for transcriptional profiling experiments, as shown below.

14.4.1 Sprycel

Sprycel is a multitargeted kinase inhibitor that is effective against src and bcr-abl family of kinases. The translocation bcr-abl plays a major role in the development of CML. As a consequence of the translocation, a protein fusion is formed having components of bcr and abl proteins. This fusion protein leads to the development of CML [8]. The presence of the Philadelphia chromosome is indicative of the translocation and has been used as a hallmark of this disease. You can think of it as the prognostic biomarker of the disease. Sprycel has been approved by the FDA to treat patients that are positive for the Philadelphia chromosome and have CML. Besides the Philadelphia chromosome, we wonder if there were other biomarkers that predict sensitivity to this compound. In order to identify them, we decided to carry out a transcriptional profiling of cells that were treated with Sprycel. First, we had to differentiate which cell lines were sensitive and which ones were resistant. The cell lines were treated with the compound and the IC50 found for each one of them. Then the IC50 was used to differentiate them into two populations of cells, namely sensitive and resistant (Figure 14.3). Transcriptional profiling was used to find which genes are up or downregulated upon compound treatment either in sensitive or resistant cells. This approach pointed to a signature of genes that was present in the sensitive cells while absent in the resistant ones [9].

Another advantage of using cell lines to generate signatures is the added information obtained to pursue an indication in another type of tumor. There are two drugs on the market that specifically target this fusion protein, Gleevec and Sprycel. Both drugs inhibit the fusion protein and are effective ways of controlling CML. One of the questions that is always asked during drug development is how is your drug different from other drugs that target the same disease? This is an ideal question that can be answered with the use of biomarkers. Going back to our Sprycel gene signature, we interrogated its presence in transcriptional profiles of tissues isolated from other types of cancer. Interestingly a similar signature was found in a subpopulation of breast cancer patients. These patients are triple negative patients, negative for estrogen receptor (ER), negative for progesterone receptor (PR) and negative for Her2/neu, which is a class of patients that medically have little drug options. The inference is that they will also become sensitive to the compound. Therefore, by finding the specific transcriptional signature for Sprycel,

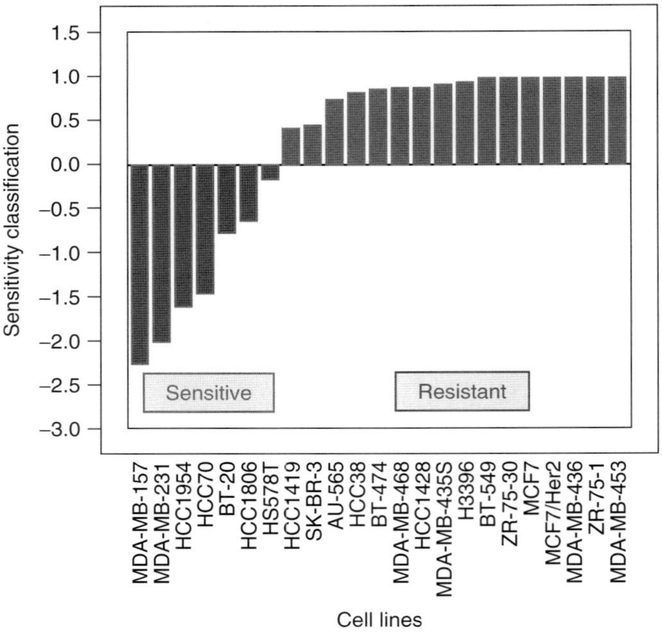

Figure 14.3 Cell lines responding to treatment with Sprycel. Cell lines that are sensitive to the compound are shown in light gray, while those that are resistant are shown in dark gray. (From Huang, F. *et al.* (2007) *Cancer Res.*, **67**(5), 2226–2238).

we were able to find a subpopulation of patients that may benefit from the use of this drug in a different indication, and in the process, help differentiate Sprycel from Gleevec. This hypothesis still needs to be demonstrated in a clinical trial.

14.4.2 **Yondelis**

A similar approach was undertaken by Jimeno and collaborators to find biomarkers for Yondelis, a tetrahydroquinoline alkaloid identified in the Caribbean tunicate *Ecteinascidia turbinate*. This compound binds to guanine in the small groove of DNA, leading to the inhibition of transcription and cell death. Therefore, it possesses antitumor activities [10]. Again, the use of transcription profiling was instrumental in the development of this compound. *In vitro* cell lines were exposed to the compound and transcriptional profiling identified a gene signature that seemed to differentiate sensitive from resistant cells. This pointed the investigators to sarcoma cell lines as being more sensitive. The gene signature was validated by RT-PCR (reverse transcriptase polymerase chain reaction), in tissues from sarcoma patients. This helped the developmental team to direct their efforts in patients with this type of tumor. Furthermore, the gene signature was able to identify a subpopulation of patients within this cancer type that will likely respond better to this compound than to standard therapy. That gene signature is now being evaluated in a clinical trial [10].

14.5 Finding Biomarkers through the Use of Xenografts

More often cell lines are not the appropriate way to find biomarkers, especially if we are targeting the microenvironment around the tumor cells with the compound under investigation. This is the case when we target angiogenesis, where the aim is to reduce the vasculature going into the tumor. A more appropriate model is one that also utilizes the environment around the tumor. This can be achieved with an *in vivo* model, like the use of human xenografts. These are human tumor cell lines that are grown ectopically in immunocompromised mice. These cell lines will grow unimpeded since the mice do not recognize them and cannot destroy them. Drugs are routinely tested in this animal model to look for efficacy in the different human cancer cell lines.

14.5.1 Avastin

One way to look for biomarkers is to observe the process that we try to disrupt; in the case considered here the process is angiogenesis. If it is affected, the expectation will be that the number of vessels will be diminished, that the vessels will lose permeability and neovascularization, the formation of new vessels, indispensable for tumor survival and expansion, will also be disrupted. All those events are seen in xenografts treated with Avastin. Avastin is a human monoclonal antibody that neutralizes the VEGF. Disrupting VEGF signal leads to an inhibition of the growth of tumors and a decrease in the number of vessels [11]. Furthermore, it also leads to a decrease in vascular permeability. Rhodamine labeled bovine serum albumin (BSA) was introduced in xenografts with tumors and the mice treated with Avastin. Measurement of the amount of the labeled protein in the blood confirmed that the xenografts treated with the drug had lower levels than those treated with vehicle control, indicating that the vessels have become leaky. By measuring the microvessel density, using video imaging combined with immune techniques, the thickness of the vessels can be found. A decrease in the treated animals of the thickness of the blood vessels was demonstrated. Also, the vascular or blood volume at the surface of the tumor was significantly smaller at days 1 and 3 in animals treated with compound versus those treated with control [12]. While some of the biomarkers used in this research are indirect, they pointed to a new way of following the efficacy of a compound, by looking at the disruption of angiogenesis. This later on will lead to the development of new imaging techniques in human samples that will be more direct and less invasive. All these measurements are examples of pharmacodynamic biomarkers and also of mechanism of action markers. They measure an effect of the drug being investigated in a specific pathway and they change with increasing drug concentrations.

14.5.2 Sunitinib

Sunitinib is an orally available small molecule tyrosine kinase inhibitor that targets several proteins. It selectively inhibits class III and class IV receptor tyrosine kinases, among them vascular endothelial growth factor receptor (VEGFR)-1, -2, and -3; PDGFR-α and -β; stem cell factor receptor (KIT); and Fms-like tyrosine kinase-3 receptor (FLT3) [13, 14]. It has been approved for the treatment of advanced renal cell carcinoma (RCC) and for gastrointestinal stromal tumor (GIST) after disease progression and resistance or intolerance to Gleevec. Since this small molecule inhibitor targets VEGFR, and because in the past there has been observed a correlation of increasing of pVEGFR in patients treated with anti-VEGF therapy, [15] it seemed likely that a similar effect could have been observed with this compound. In a series of well defined experiments Ebos *et al.* [16] found that increased levels of VEGF, soluble vascular endothelial growth factor receptor (sVEGFR) and placental growth factor (PlGF) correlated with increasing doses of sunitinib, making them good pharmacodynamic markers for this compound. Furthermore, they were able to pinpoint that the increase in sVEGFR was due to an increase in the receptor from both the human cells and from the mouse cells, by engineering the introduced human cell lines with a variant of the protein encoded by the VEGFR gene. Then, they were able to detect both variants in the serum of the animals. This ties together not only the angiogenesis environment as the cause of the increase of the factors but also the human tumor cells secreting those factors.

14.5.3 Brivanib

Xenografts also provide a nice source of material to look for biomarkers of specific compounds. We have been successful in identifying biomarkers to brivanib alaninate, a small molecule inhibitor of VEGFR-2 and FGFR-1 [17, 18]. The way we discovered the biomarkers was as follows: transcriptional profiling was carried out in a cohort of colon cancer patients and genes were identified that coexpressed with VEGFR-2. Xenografts were grown to a certain volume and then treated with the compound or a vehicle with no drug. mRNA was isolated from both and a gene signature was identified that was modified by the compound. We compared it with the signature for the genes that coexpressed with VEGFR-2 and nominated those genes present in both for subsequent experimentation (Figure 14.4 for a flowchart of the approach). The top candidates were further validated by developing immunohistochemistry (IHC) assays and retested in xenografts that were treated with the compound. We found that the decrease associated with two genes, collagen type IV and C1QR1, was also reproducible with the IHC assay (Figure 14.5).

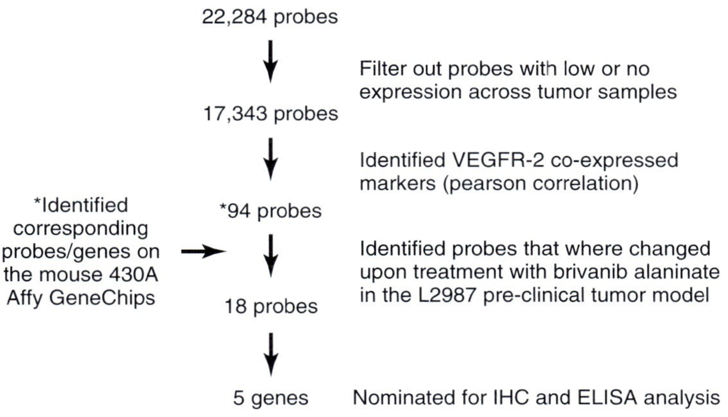

Figure 14.4 Flowchart showing the discovery of pharmacodynamic biomarkers through transcriptional profile in xenografts treated with brivanib. (From Ayers, M. *et al.* (2007) *Cancer Res.*, **67** (14), 6899–6906).

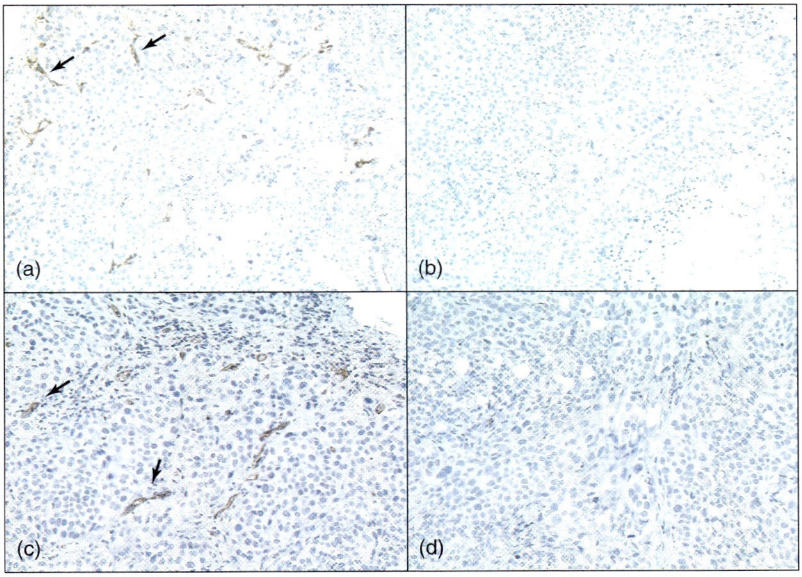

Figure 14.5 Immunohistochemistry staining of L2987 xenografts with collagen type IV (a and b) or C1QR1 (c and d) treated with brivanib (b and d) or control (and c). Arrows point to staining of each marker. (From gb67 (14), 6899–6906).

What was really exciting to us was that both proteins are important components of blood vessels and therefore important players in angiogenesis, thus tying what we knew about the biology of how this compound should work with the signature obtained. We further validated one of those markers by developing an ELISA assay against collagen type IV and found that, in mice, the protein is also downregulated upon compound treatment. This approach identified collagen type IV as a pharmacodynamic marker of brivanib in animal models [19].

A word of caution when interpreting results from transcriptional profiling experiments, it is necessary, almost mandatory, to always double check the results obtained with transcriptional profiling with other methods. This is due to the sensitivity of the assay. Small variations in mRNAs that are expressed at low levels can always trump those of other mRNAs that are expressed at high levels but whose changes are not so great. At the end of the experiment, it is hard to distinguish which one has more value biologically. When we have tried to assert the transcriptional profiling by looking at the protein expression with other methods, like IHC we have found that only about 30% of the mRNAs that change with transcriptional profiling have a detectable change in protein levels.

14.6 Toxicity Biomarkers

Animal models are the standard way of identifying toxicities associated with compounds. Normally, a drug is tested against a battery of animals to detect any unwanted side effects before it is placed in human testing. The testing is carried out with increasing amounts of the compound until side effects are encountered. Then tissues of interest are collected and traditional histopathology is used to assess the phenotypic changes observed. The advantage of using histopathology is the ability of this technique to visualize tissues at the cellular level and thus try to investigate the mechanism by which the injury resulted.

In a typical example, a one month toxicity rat study was carried out in animals treated with a VEGFR inhibitor. Routine histopathology evaluation of H&E slides revealed a drug related renal glomerulopathy. Increased eosinophilia in the glomerular mesangial matrix was found associated with a relative decrease in density and number of cell nuclei. To characterize this finding, IHC was carried out with antibodies against albumin. This biomarker labeled the material thus identifying a problem with reabsorption of excreted albumin. In this case, the biomarker helped in refining and narrowing the hypothesis to be tested later on for the purpose of identifying the mechanism of toxicity [20].

Sometimes this approach is not sufficient to satisfactorily identify the toxicity or the mechanism involved in the side effects, then a transcriptional profiling approach may be undertaken to try to answer some of those questions. Foster and colleagues [21] used a retrospective analysis with transcriptional profiling to investigate the utility of this approach in the drug development process at Bristol-Myers Squibb. They established a workflow in their toxicity studies that included transcriptional profiling. They found that toxicogenomics has utility in the drug safety assessment and that its contributions are frequent with regards to mechanistic classifications. One of their more potentially interesting results is their idea of using the overall transcriptome as a new tool in toxicology, since the overall change in the transcriptome correlated well with changes observed with the typical classical approaches [21].

14.7 Imaging as a Biomarker

14.7.1 Avastin

Imaging has been developed as a biomarker to measure the angiogenic status of tumors in patients undergoing treatment with different compounds. There are several imaging techniques having the advantages that they are noninvasive, have little risk for the patient and directly measure angiogenesis changes. The main techniques that are in used now are: perfusion computed tomography (CT), dynamic magnetic resonance imaging (MRI), and positron electron tomography (PET). Both CT and MRI will provide quantitative and qualitative measurements of tumor vascularity and the choice of one or the other will vary depending on the clinical situation at hand. PET will measure tumor metabolism, oxygenation, as well as blood flow to the tumor. Unfortunately this last technique is limited due to its high cost, the short life of radioisotopes used and the lack of specialized equipment needed for its use [22].

Using CT scans and PET techniques, among a host of other indirect methods, Willet *et al.* [23] showed direct evidence of antivascular effects in patients with rectal cancer that were treated with a single dose of Avastin. The use of CT demonstrated lower tumor perfusion in several patients while 18-fluorodeoxyglucose positron electron tomography (FDG-PET) indicated lower metabolism uptake of some of the tumors treated.

14.7.2 Sorafenib

An interesting observation is that sometimes, pharmacodynamic markers can become predictive markers and even surrogate end points in clinical trials. The use of imaging techniques is one such example [24]. In this case, the investigators use a technique named, dynamic contrast enhanced und Doppler ultrasound (DCE-US), which specifically looks at the size and echostructure of abdominal tumors and the extent of neovascularization. They found a statistically significant correlation between the decreases in contrast uptake exceeding 10% with stability or a decrease in tumor volume and progression-free survival and overall survival in patients with RCC that received sorafenib therapy.

The idea of using imaging as a biomarker was further demonstrated in another study where dynamic contrast enhanced magnetic resonance imaging (DCE-MRI) was used in another cohort of RCC patients undergoing sorafenib treatment. This technique measures tumor perfusion and vascular permeability which can be reproducibly quantitated; changes in those values relate to histologic alterations in tumor microvasculature. Decreased tumor vascular permeability was associated with improved patient outcome, and even more interestingly, baseline tumor vascular permeability which was expected to be a poor prognosis factor, was a predictive marker of favorable response to sorafenib [25].

While the number of patients in these two studies is small, I think they point to an interesting possibility, that imaging could become not only a pharmacodynamic biomarker but a predictive one as well.

14.8 Finding and Validating Biomarkers in Clinical Trials

14.8.1 Pharmacodynamic Markers

14.8.1.1 EGFR Inhibitors

Albanell *et al.* [26] have devised another method for measuring biomarkers in patients undergoing clinical trials. In their case, the skin is used as a surrogate tissue. Gefitinib (Iressa) is a small molecule tyrosine kinase inhibitor of the epidermal growth factor receptor (EGFR). Several proteins present in the EGFR pathway are also found in the skin. IHC can be used as a way to measure them in the skin. The skin provided an ideal organ, since it is easily accessible and multiple measurements can be made with minimum invasive techniques, which greatly improve patient acceptance. Furthermore, the EGFR pathway is active in the skin, so if the compound is acting through the EGFR pathway, one should be able to see its effects. In order to better observe any changes, the authors developed methods to semiquantitatively measure the up regulation or downregulation of proteins in the skin. For downregulation markers, they looked at the decrease in phosphorylated EGFR, the active form of the receptor, a decrease in Ki67, a protein that is present in the S phase of the cell cycle and is associated with proliferation, and of phosphorylated MAPK, a kinase downstream from EGFR. For up regulation, they looked at the expression of p27, a CDK inhibitor that plays a role in the cell cycle, and keratin 1 and phosphorylated STAT 3, markers of keratin differentiation and indicative that the skin cells have exited the cell cycle and started the road to maturation. The decreases and increases of those proteins followed the expected path in the skin. In order to validate their findings in the skin, tumor measurements were also made and found that they change in the same manner as they changed in the skin. Thus, indicating that the mechanism of action of gefitinib in both the skin and the tumor was that of inhibiting the EGFR pathway.

14.8.1.2 Sunitinib

While the previous example used the skin as a surrogate tissue, serum is most commonly used in clinical trials. The ease of access to blood and the availability of different assays to isolate and measure different cells and proteins in those cells make it an ideal system. An example of using this approach is looking at biomarkers in serum of patients treated with sunitinib (Sutent). Serum levels of VEGFR2 and VEGFR3 decreased upon the treatment of RCC patients with sunitinib. This decrease was accompanied by an increase of VEGF, a key ligand of those receptors. Interestingly, the decrease of the soluble receptors and the increase of the ligand correlated with the amount of plasma concentration of sunitinib, indicating that both the receptor and the ligand responded due to the treatment of patients with the drug [27]. Furthermore, an association of response to the treatment with the increase in VEGF and the decrease in VEGFR3 could lead these biomarkers to be treated not only as pharmacodynamic markers but as possible surrogate end points in the future. Obviously, further prospective analysis needs to be conducted before such markers can be called *surrogate*.

14.8.1.3 Brivanib

Another compound for which pharmacodynamic markers have been tested in clinical trials is brivanib. Since the trial was a phase I study, increasing doses of the compound were used to identify the maximum tolerated dose (MTD) of the compound. It was determined that the MTD was 800 mg daily [28]. Following the discovery of collagen type IV in xenografts in mice (as mentioned earlier in the chapter), we developed an assay that was able to measure the change in collagen type IV in human blood. ELISA was performed using blood of patients at days 8 and 26 after the start of therapy with brivanib. At 600 mg and higher doses a decrease of collagen type IV was observed (Figure 14.6). If the patients were grouped into those receiving low doses (180 and 320 mg) and those receiving high doses (600 mg and higher), a statistically significant decrease in the change from baseline was noted for this biomarker in the patients receiving the high dose. At day 8, a decrease of more than 20% with regards to the low dose group ($p = 0.03$) was seen, and this is decreased further at day 26 achieving a reduction of almost 40% ($p = 0.01$, Figure 14.6). This reduction of collagen type IV was also associated with less tumor growth. If we look at the patients that show more than 20% reduction in the levels of collagen type IV compared to their baseline, we observed that the tumors tended not to grow (note that the median is almost 0 in Figure 14.7), while if we look at the patients showing less than 20% reduction, their tumors tended to increase in size (increase of more than 40% this is statistically significant with a $p = 0.008$). Not only did collagen type IV show this decrease, another biomarker that also decreases upon therapy was soluble VEGFR2. We decide to test this as a pharmacodynamic marker because it was the direct target of the compound. Again, ELISAs were developed for this marker and patients' samples collected at days 8 and 26. Separating the patients into the low and high drug groups also showed a decrease of this marker at both days. The decrease observed at day 8 was around 18% ($p = 0.0001$ vs the low dose group) and around 22% for day 26

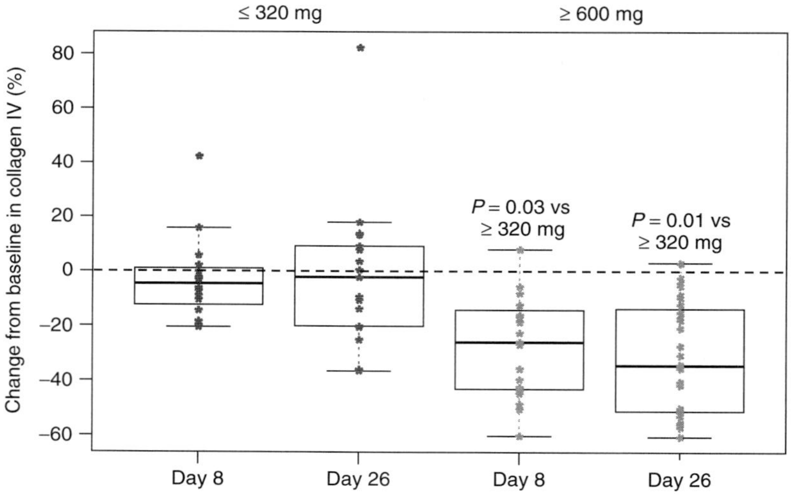

Figure 14.6 Brivanib treatment is associated with a decrease in plasma collagen type IV. The two box plots on the left are for patients receiving low doses of brivanib (320 mg and lower) at days 8 and 26, while the two box plots on the right are for patients receiving high doses of brivanib (600 mg or higher) [29].

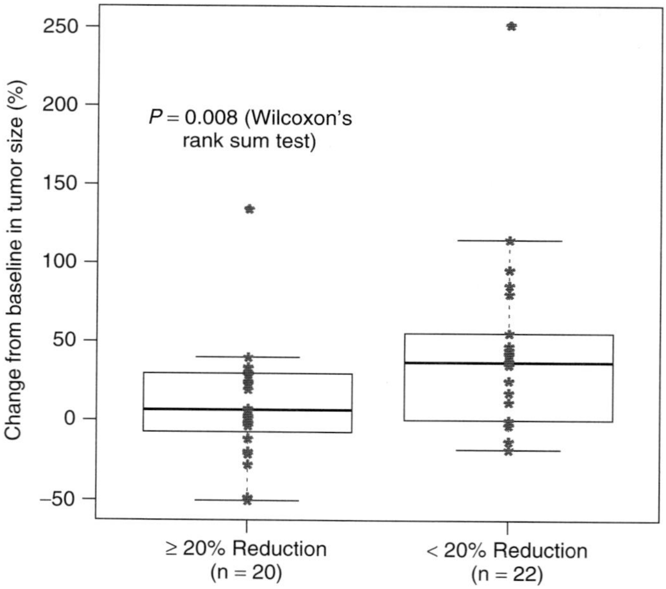

Figure 14.7 Plasma collagen IV reduction $\geq$ 20% on day 26 is associated with less tumor growth [29].

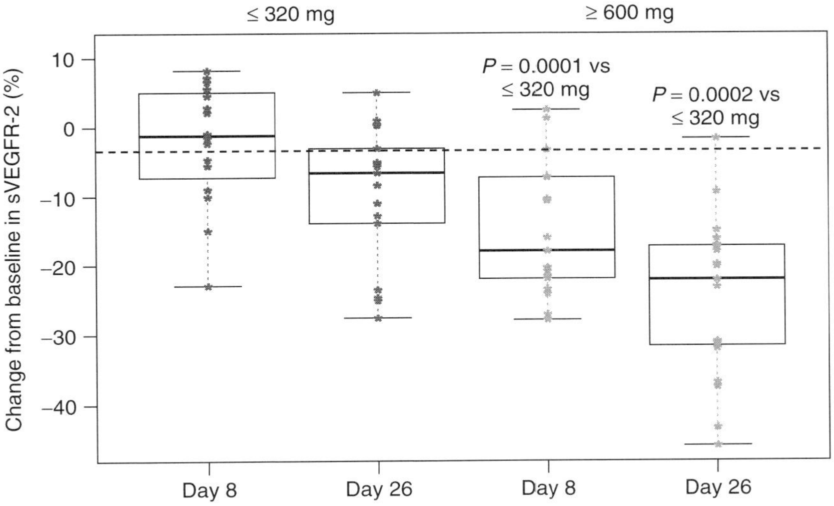

Figure 14.8 Brivanib treatment is associated with decrease in soluble VEGFR-2. Two box plots on the left are from patients receiving low doses of brivanib (320 mg and lower) at days 8 and 26. The two box plots on the right are for patients receiving high doses of brivanib (660 mg or higher) at days 8 and 26 [29].

($p = 0.0002$ vs low group, Figure 14.8) [29]. These results demonstrate that these two biomarkers changed upon treatment of patients with drug in a dose-dependant manner, thus making them to be nice pharmacodynamic markers for brivanib. As an aside, finding that the markers started changing at 600 mg was very helpful since it gave us the information that dose level was the biologically active dose for this compound. This information is extremely useful for a clinical team, since it helps to rapidly find the optimal dose for the compound. The optimal dose is the dose at which the maximum efficacy is obtained with a compound without incurring any toxicity. Again, our use of the known pathway of inhibition of the compound helped us in finding the appropriate pharmacodynamic markers.

14.8.2 Predictive Markers

14.8.2.1 Herceptin

The most commonly cited example of a predictive marker associated with a drug is that of Her2/neu. Breast cancer patients that overexpress this protein in the membrane of their tumor cells tend to respond better to therapy with Herceptin. Herceptin targets the Her2 receptor by inhibiting its activity. Thus, the presence of this protein in a deregulated manner points out which tumors will be more effectively treated. Actually, there are two predictive biomarkers in this case.

Patients in which tumors are scored in an IHC assay with a 3+ and demonstrate a chicken wire type of staining (as seen in Figure 14.1) are considered positive patients. Those that scored as not having staining or light cytoplasmic staining are considered negative. Those that are labeled as 2+ have an additional test to see if the gene is amplified. This is accomplished using fluorescent *in situ* hybridization (FISH). Thus, there are two tests to determine if the tumors of those patients are positive for the overexpression of this receptor, IHC and FISH. Again, this underscores a point we made at the beginning, biomarker discovery and validation is harder than it might appear.

14.8.2.2 **Brivanib**

Going back to our example of brivanib, we investigated which other molecules in the pathway of both tyrosine kinases could provide a view into the inhibition properties of the drug. We decided to look at both receptors, FGFR1 and VEGFR2, and also tested two ligands of the FGFR1 receptor, FGF1 and FGF2. In order to test those proteins, archival tumor biopsy and blood samples were obtained from 43 patients enrolled in a phase 1 clinical trial for brivanib alaninate [28]. Several tumor types were present among the patients treated: 65% had colorectal cancer (CRC), 12% RCC, and the remaining 23% had a range of tumors, including bladder, pancreatic, ovarian, and neuroendocrine. Biopsy samples, obtained prior to treatment, were sectioned and stained for VEGFR2, FGFR1, FGF-1, and FGF-2 expression. While we did not see any correlations with VEGFR2, FGFR1, or FGF-1 a nice correlation was observed with FGF-2 staining (Table 14.1) [29]. If tumors were divided into FGF-2− or FGF-2+ depending on the staining observed (refer Figure 14.9 for representative staining images) a correlation was seen with therapy. Patients were divided into two categories, one receiving low doses of brivanib (defined from 180 to <600 mg) and those receiving high doses of the drug (defined as >600 mg). Interestingly, FGF positivity appeared to predict a patient's response to brivanib. By plotting the change in tumor size from baseline for patients that are either FGF2+ or FGF2− (by IHC), and that have received brivanib treatment, we observed that patients who were FGF2+ have received benefit from this therapy,

Table 14.1 Brivanib treatment is associated with less tumor growth in patients with FGF2+ tumors. Median percent change from baseline in tumor size [29].

FGF2 status	Brivanib (≤ 320 mg)	Brivanib (≥ 600 mg)
FGF2− (n = 21)	45.1	18
FGF2+ (n = 25)	15.2	0

$P = 0.03$ for FGF2+ vs FGF2− (Wilcoxon's rank sum test)

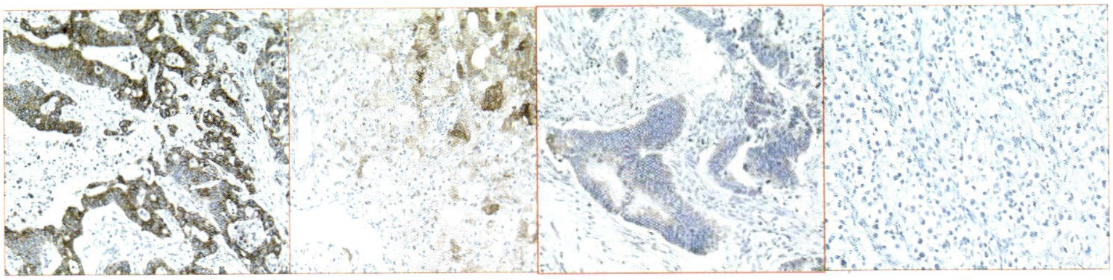

Figure 14.9 Representative images of FGF2 immunohistochemistry staining in samples from patients treated with brivanib. 0 indicates no staining, 1 light, 2 moderate, and 3 strong staining. Tumors staining with 1, 2, or 3 are considered positive [29].

and have a greater decrease in their tumor sizes (Figure 14.10) [29]. Patients were further classified as having progressive disease (PD), stable disease (SD), or a partial response (PR) to brivanib treatment according to standard WHO criteria from CT scans. Significantly more patients with FGF-2+ tumors (15/25) had a PR or SD compared with patients with FGF-2- tumors (5/21; $p = 0.0019$ using Fischer's exact test). In addition, fewer patients with FGF-2+ tumors had PD (10 out of 25) than those with FGF-2− tumors (16 out of 21). This correlation of FGF2 positivity and response to brivanib therapy is translated to a better progression-free survival for those patients whose tumors are positive for FGF2 (Figure 14.11) [29]. If the progression-free survival is broken into those patients receiving low doses of brivanib and we look at the FGF2 status, we noticed that their median day for survival is 71 days. Whereas, when we look at the patients receiving high doses of the compound, the median survival is moved to 106 days. On the other hand, patients in both drug doses that are negative for FGF2 have very similar median survival dates, 54 and 55 for low and high dose respectively. While this progression-free survival does not reach a significant p value ($p = 0.074$ for FGF2+ vs FGF2−), it is a trend that could reach statistical significance if more patients are tested [29].

In conclusion, FGF-2+ tumors have a clear and significant association with a decrease in tumor growth and a survival benefit in brivanib alaninate–treated patients compared with FGF-2− tumors, indicating that FGF2 is a predictive marker for brivanib. To validate this marker, a prospective clinical trial must be run before it can be adopted as such.

14.8.2.3 **EGFR Inhibitors**

Not only can the biomarker be used to predict which patients will benefit from the drug but they can also be used to exclude patients from treatment. Such is the case of the K-ras mutation in patients treated with anti-EGFR therapy. The reason for looking at this gene in this context is that EGFR activates several pathways, the RAS/RAF/MAPk, STAT, and PI3K/AKT [30]. Furthermore, mutations in this

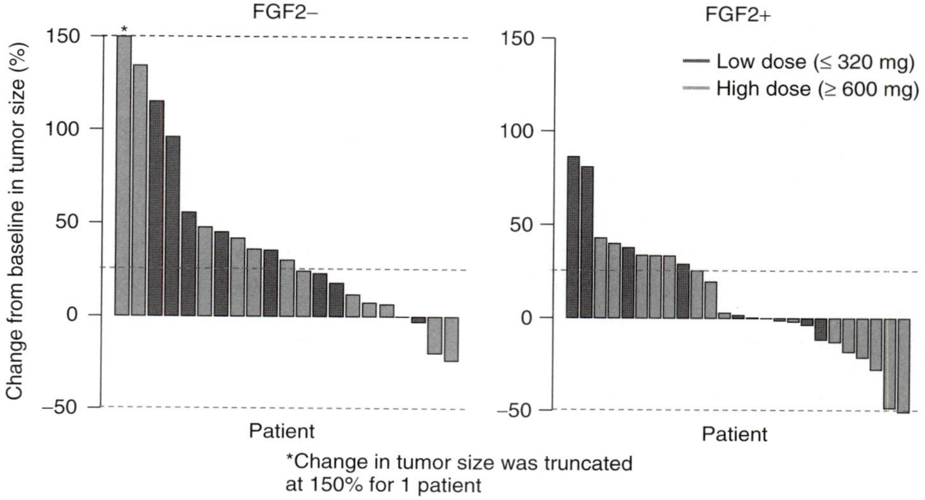

Figure 14.10 Brivanib treatment is associated with less tumor growth in patients with FGF2+ tumors. Graph on the left is for those patients whose tumors were FGF2–, while the one on the right is for those that were FGF2+. Dark gray bars indicate patients receiving low doses of brivanib (180 and 320 mg). Light gray bars indicate patients receiving 600 mg or higher doses [29].

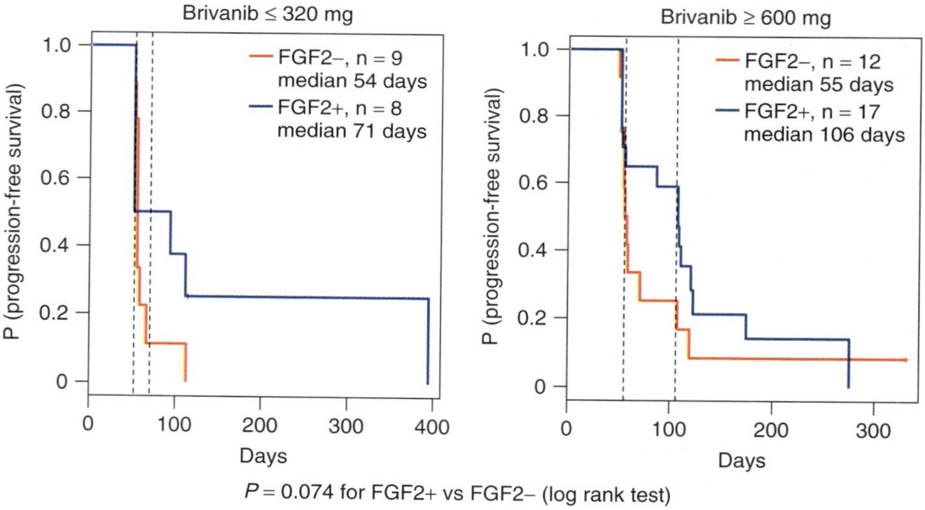

Figure 14.11 Progression-free survival by FGF2 status and brivanib dose. Curves on the left are for those patients receiving low doses (320 mg and lower) while the graph on the right is for those patients in high doses (600 mg or higher). Red lines are patients with tumors negative for FGF2 staining, blue lines are patients with tumors positive for FGF2 [29].

gene have been found in 30–50% of CRC patients and are normally associated with poor prognosis and resistance to EGFR therapy (31, 32). Panitumumab is a fully humanized antibody that targets EGFR. It has been found to be effective in a population of metastatic CRC patients that possess only wild type K-ras. Amado and colleagues [33] tested a cohort of 427 patients having metastatic CRC for the presence of mutations on the K-ras gene. They found that patients carrying mutations on this gene did not benefit from therapy with panitumumab. Of a total of 124 patients that were wild type for K-ras and treated with panitumumab, 17% had a PR, 34% had SD, and 36% had PD. In contrast, those that were mutant for the gene ($n = 84$) and treated with the same compound, none had PR, 12 were SD, and 76% were PD. This also translated in an overall survival benefit of 8.1 months versus 4.9 for those with the wild type K-ras. Not only has this been demonstrated for panitumumab, the same is true for another anti-EGFR therapy, Erbitux [34]. These two examples demonstrate the usefulness of the biomarker to exclude patients taking any anti-EGFR therapy.

14.9 Conclusions

Biomarkers are becoming more common in the process of drug development. While it is not easy to develop and validate them, several examples have shown their utility in the different phases of drug development. From early preclinical use both in cell lines and *in vivo* animals, to their use as pharmacodynamic biomarkers in early clinical trials, to their use as predictive biomarkers in registrational trials, and even later, as their application in everyday use in the clinic, these examples point to the usefulness of the approach and the impact biomarkers are having in the drug development process.

The hope is that biomarkers will make the drug development process faster. Thus, making a reality the goal of finding the right drug for the right patient at the right time, which will result in less costly drug development and less costly drugs for all patients.

References

1. U.S. Department of Health and Human Services Food and Drug Administration Center for Drug Evaluation and Research (CDER) Center for Biologics Evaluation and Research (CBER) Center for Devices and Radiological Health (CDRH) (2005) Guidance for Industry. Pharmacogenomics Data Submissions . Procedural March 2005. *http://www.fda.gov/cder/guidance/6400fnl.pdf.*
2. Margolin, K., Gordon, M.S., Holmgren, E., Gaudreault, J., Novotny, W., Fyfe, G., Adelman, D., Stalter, S., and Breed, J. (2001) Phase Ib trial of intravenous recombinant humanized monoclonal antibody to vascular endothelial growth factor in combination with chemotherapy in patients with advanced cancer: pharmacologic and long-term safety data. *J. Clin. Oncol.*, **19**, 851–856.

3. Eisen, M.B. and Brown, P.O. (1999) DNA arrays for analysis of gene expression. *Methods Enzymol.*, **303**, 179–205.
4. Perou, C.M., Sørlie, T., Eisen, M.B., van de Rijn, M., Jeffrey, S.S., Rees, C.A., Pollack, J.R., Ross, D.T., H, Johnsen., Akslen, L.A., Fluge, O., Pergamenschikov, A., Williams, C., Zhu, S.X., Lønning, P.E., Børresen-Dale, A.L., Brown, P.O., and Botstein, D. (2000) Molecular portraits of human breast tumours. *Nature*, **406** (6797), 747–752.
5. van de Vijver, M., He, Y.D., van't Veer, L.J., Dai, H., Hart, A.A.M., Voskuil, D.W., Schreiber, G.J., Peterse, J.L., Roberts, C., Marton, M.J., Parrish, M., Atsma, D., Witteveen, A., Glas, A., Delahaye, L., van der Velde, T., Bartelink, H., Rodenhuis, S., Rutgers, E.T., Friend, S.H., and Bernards, R. (2002) A gene-expression signature as a predictor of survival in breast cancer. *N. Engl. J. Med.*, **347**, 1999–2009.
6. Wright, G., Tan, B., Rosenwald, A., Hurt, E.H., Wiestner, A., and Staudt, L.M. (2003) A gene expression-based method to diagnose clinically distinct subgroups of diffuse large B cell lymphoma. *Proc. Natl. Acad. Sci. U.S.A.*, **100** (17), 9991–9996.
7. Rouziera, R., Rajan, R., Peter, W., Hessd, K.R., Goldd, D.L., Stece, J., Ayerse, M., Rosse, J.S., Zhangc, P., Buchholzj, T.A., Kuererk, H., Greena, M., Aruna, B., Hortobagyia, G.N., Fraser Symmansc, W., and Pusztaia, L. (2005) Microtubule-associated protein tau: a marker of paclitaxel sensitivity in breast cancer. *Proc. Natl. Acad. Sci.*, **102** (23), 8315–8320.
8. Daley, G.Q., Van Etten, R.A., and Baltimore, D. (1990) Induction of chronic myelogenous leukemia in mice by the P210bcr/abl gene of the Philadelphia chromosome. *Science*, **247** (4944), 824–830.
9. Huang, F., Reeves, K., Han, X., Fairchild, C., Platero, S., Wong, T.W., Lee, F., Shaw, P., and Clark, E. (2007) Identification of candidate molecular markers predicting sensitivity in solid tumors to dasatinib: rationale for patient selection. *Cancer Res.*, **67** (5), 2226–2238.
10. Jimeno, J., Aracil, M., and Tercero, J.C. (2006) Adding pharmacogenomics to the development of new marine-derived anticancer agents. *J. Transl. Med.*, **4** (3) Doi: 10.1186/1479-5876-4-3.
11. Kim, K.J., Li, B., Winer, J., Armanin, M., Gillet, N., Phillips, H.S., and Ferrara, N. (1993) Inhibition of vascular endothelial growth factor-induced angiogenesis suppresses tumour growth in vivo. *Nature*, **362**, 841–844.
12. Yuan, F., Che, Y., Dllian, M., Safabakhsh, N., Ferrara, N., and Jain, R.K. (1996) Time-dependent vascular regression and permeability changes in established human tumor xenografts induced by an anti-vascular endothelial growth factory vascular permeability factor antibody. *Proc. Natl. Acad. Sci. U.S.A.*, **93**, 14765–14770.
13. Mendel, D.B., Laird, A.D., Xin, X., Louie, S.G., Christensen, J.G., Li, G., Schreck, R.E., Abrams, T.J., Ngai, T.J., Lee, L.B., Murray, L.J., Carver, J., Chan, E., Moss, K.G., Haznedar, J.O., Sukbuntherng, J., Blake, R.A., Sun, L., Tang, C., Miller, T., Shirazian, S., McMahon, G., and Cherrington, J.M. (2003) In vivo antitumor activity of SU11248, a novel tyrosine kinase inhibitor targeting vascular endothelial growth factor and platelet-derived growth factor receptors: determination of a pharmacokinetic/pharmacodynamic relationship. *Clin. Cancer Res.*, **9** (1), 327–337.
14. Potapova, O., Laird, A.D., Nannini, A.M., Barone, A., Li, G., Moss, K.G., Cherrington, J.M., and Mendel, D.B. (2006) Contribution of individual targets to the antitumorefficacy of the multitargeted receptor tyrosine kinase inhibitor SU11248. *Mol. Cancer Ther.*, **5** (5), 1286–1289.
15. Wedam, S.B., Low, J.A., Yang, S.X., Chow, C.K., Choyke, P., Danforth, D., Hewitt, S.M., Berman, A., Steinberg, S.M., Liewehr, D.J., Plehn, J., Doshi, A., Thomasson, D., McCarthy, N., Koeppen, H., Sherman, M., Zujewski, J.A., Camphausen, K., Chen, H., and Swain, S.M. (2004) Antiangiogenic and

antitumor effects of bevacizumab in patients with inflammatory and locally advanced breast cancer. *J. Clin. Oncol.*, **24** (5), 769–777.

16. Ebos, J.M., Lee, C.R., Christensen, J.G., Mutsaers, A.J., and Kerbel, R.S. (2007) Multiple circulating proangiogenic factors induced by sunitinib malate are tumor-independent and correlate with antitumor efficacy. *Proc. Natl. Acad. Sci.*, **104** (43), 17069–17074.
17. Bhide, R.S., Cai, Z.-W., Zhang, Y.-Z. *et al.* (2006) Discovery and preclinical studies of (R)-1-(4-(4-fluoro-2-methyl-1Hindol-5-yloxy)-5-methylpyrrolo[2,1-f][1,2,4]triazin-6-yloxy)propan-2-ol (BMS-540215), an in vivo active potent VEGFR-2 inhibitor. *J. Med. Chem.*, **49**, 2143–2146.
18. Cai, Z., Zhang, Y., Borzilleri, R.M., Qian, L., Barbosa, S., Wei, D., Zheng, X., Wu, L., Fan, J., Shi, Z., Wautlet, B.S., Mortillo, S., Jeyaseelan, R.Sr., Kukral, D.W., Kamath, A., Marathe, P., D'Arienzo, C., Derbin, G., Barrish, J.C., Robl, J.A., Hunt, J.T., Lombardo, L.J., Fargnoli, J., and Bhide, R.S. (2008) Discovery of Brivanib Alaninate ((S)-((R)-1-(4-(4-Fluoro-2-methyl-1 Hindol-5-yloxy)-5-methylpyrrolo[2,1-f] [1,2,4]triazin-6-yloxy)propan-2-yl) 2-aminopropanoate), a novel prodrug of dual vascular endothelial growth factor receptor-2 and fibroblast growth factor receptor-1 kinase inhibitor (BMS-540215). *J. Med. Chem.*, **51** (6), 1976–1980.
19. Ayers, M., Fargnoli, J., Lewin, A., Wu, Q., and Platero, J.S. (2007) Discovery and validation of biomarkers that respond to treatment with brivanib alaninate, a small-molecule VEGFR-2/FGFR-1 antagonist. *Cancer Res.*, **67** (14), 6899–6906.
20. Westhouse, R. (2009) in *Molecular Pathology in Drug Discovery and Development* (ed. J.S.Platero), John Wiley & Sons, Inc., pp. 85–111.
21. Foster, W.R., Chen, S.H., He, A., Truong, A., Bhaskaran, V., Nelson, D., Dambach, D., Lehman-McKeeman, L., and Car, B. (2007) A retrospective analysis of toxicogenomics in the safety assessment of drug candidates. *Toxicol. Pathol.*, **35**, 621–635.
22. Jeswani, T. and Padhani, A.R. (2005) Imaging tumour angiogenesis. *Cancer Imaging*, **5**, 131–138.
23. Willet, C., Boucher, Y., di Tomaso, E., Duda, D.G., Munn, L.L., Tong, R.T., Chung, D.C., Sahani, D.V., Kalva, S.P., Kozin, S.V., Mino, M., Cohen, K.S., Scadden, D.T., Hartford, A.C., Fischman, A.J., Clark, J.W., Ryan, D.P., Zhu, A.X., Blaszkowsky, L.S., Chen, H.X., Shellito, P.C., Lauwers, G.Y., and Jain, R.K. (2004) Direct evidence that the VEGF-specific antibody bevacizumab has antivascular effects in human rectal cancer. *Nat. Med.*, **10**, 145–147.
24. Lamuraglia M., Escudier, B., Chami, L., Schwartz, B., Leclere, J., Roche, A., and Lassaua, N. (2006) To predict progression-free survival and overall survival in metastatic renal cancer treated with sorafenib: Pilot study using dynamic contrast-enhanced Doppler ultrasound. *Eur. J. Cancer*, **42**, 2472–2479.
25. Flaherty, K.T., Rosen, M.A., Heitjan, D.F., Gallagher, M.L., Schwartz, B., Schnall, M.D., and Dwyer, P.J.O. (2008) Pilot study of DCE-MRI to predict progression-free survival with sorafenib therapy in renal cell carcinoma. *Cancer Biol. Ther.*, **7** (4), 496–501.
26. Albanell, J., Rojo, F., Averbuch, S., Feyereislova, A., Mascaro, J.M., Herbst, R., LoRusso, P., Rischin, D., Sauleda, S., Gee, J., Nicholson, R.I., and Baselga, J. (2002) Pharmacodynamic studies of the epidermal growth factor receptor inhibitor ZD1839 in skin from cancer patients: histopathologic and molecular consequences of receptor inhibition. *J. Clin. Oncol.*, **20** (1), 110–124.
27. DePrimo, S.E., Bello, C.L., Smeraglia, J., Baum, C.M., Spinella, D., Rini, B.I., Michaelson, M.D., and Motzer, R.J. (2007) Circulating protein biomarkers of pharmacodynamic activity of sunitinib in patients with metastatic renal cell carcinoma: modulation of VEGF and VEGF-related proteins. *J. Transl. Med.*, **5**, 32.
28. Jonker, D.J., Rosen, L.S., Sawyer, M., Wilding, G., Noberasco, C., Jayson, G.,

Rustin, G., McArthur, G., Velasquez, L., and Galbraith, S. (2007) A phase I study of BMS-582664 (brivanib alaninate), an oral dual inhibitor of VEGFR and FGFR tyrosine kinases, in patients (pts) with advanced/metastatic solid tumors: safety, pharmacokinetic (PK), and pharmacodynamic (PD) findings. *J. Clin. Oncol.*, **25**, 3559 (Meeting Abstracts).

29. Platero, J.S., Mokliatchouk, O., Jayson, G.C., Jonker, D.J., Rosen, L.S., Luroe, S., Kelsey, J., Feltquate, D., Velasquez, L., and Galbraith, S. (2008) Correlation of FGF2 tumor expression with tumor response, PFS, and changes in plasma pharmacodynamic (PD) markers following treatment with brivanib alaninate, an oral dual inhibitor of VEGFR and FGFR tyrosine kinases. *J. Clin. Oncol.*, **26** (May 20 Suppl; Abstract 3506).

30. Malumbres, M., and Barbacid, M., (2003) RAS oncogenes: the first 30 years. *Nat. Rev. Cancer*, **3** (6), 459–465.

31. Andreyev, H.J., Norman, A.R., Cunningham, D. *et al.* (2001) Kirsten ras mutations in patients with colorectal cancer: the 'RASCAL II' study. *Br. J. Cancer*, **85**, 692–696.

32. Benvenuti, S., Sartore-Bianchi, A., Di Nicolantonio, F. *et al.* (2007) Oncogenic activation of the RAS/RAF signaling pathway impairs the response of metastatic colorectal cancers to anti-epidermal growth factor receptor antibody therapies. *Cancer Res.*, **67**, 2643–2648.

33. Amado, R., Wolf, M., Peeters, M., Van Cutsem, E., Siena, S., Freeman, D.J., Juan, T., Sikorski, R., Suggs, S., Radinsky, R., Patterson, S.D., and Chang, D.D. (2008) Wild-type KRAS is required for panitumumab efficacy in patients with metastatic colorectal cancer. *J. Clin. Oncol.*, **26**, 1626–1634.

34. Khambata-Ford, S., Garrett, C.R., Meropol, N.J., Basik, M., Harbison, C.T., Wu, S., Wong, T.W., Huang, X., Takimoto, C.H., Godwin, A.K., Tan, B.R., Krishnamurthi, S.S., Burris, H.A.III, Poplin, E.A., Hidalgo, M., Baselga, J., Clark, E.A., and Mauro, D.J. (2007) Expression of epiregulin and amphiregulin and K-ras mutation status predict disease control in metastatic colorectal cancer patients treated with cetuximab. *J. Clin. Oncol.*, **25**, 3230–3237.

Index

Tumor Angiogenesis – From Molecular Mechanisms to Targeted Therapy. Edited by Francis S. Markland, Stephen Swenson, and Radu Minea
Copyright © 2010 WILEY-VCH Verlag GmbH & Co. KGaA, Weinheim
ISBN: 978-3-527-32091-2

u

v